STARCHES

Characterization, Properties, and Applications

STARCHES

Characterization, Properties, and Applications

Edited by
Andréa C. Bertolini

CRC Press
Taylor & Francis Group
Boca Raton   London   New York

CRC Press is an imprint of the
Taylor & Francis Group, an **informa** business

CRC Press
Taylor & Francis Group
6000 Broken Sound Parkway NW, Suite 300
Boca Raton, FL 33487-2742

First issued in paperback 2017

© 2010 by Taylor and Francis Group, LLC
CRC Press is an imprint of Taylor & Francis Group, an Informa business

No claim to original U.S. Government works

ISBN 13: 978-1-138-11341-1 (pbk)
ISBN 13: 978-1-4200-8023-0 (hbk)

Library of Congress Cataloging-in-Publication Data

Starches : characterization, properties, and applications / editor, Andréa C. Bertolini.
   p. ; cm.
  Includes bibliographical references and index.
  ISBN 978-1-4200-8023-0 (hardcover : alk. paper)
  1. Starch. I. Bertolini, Andréa C. (Andréa Curiacos)
  [DNLM: 1. Starch. QU 83 S795 2010]

  TP248.S7S753 2010
  660.6'3--dc22   2009040938

Visit the Taylor & Francis Web site at
http://www.taylorandfrancis.com

and the CRC Press Web site at
http://www.crcpress.com

Contents

Preface

Recent advances have been made in studies of the characterization, uses, and applications of starches. In past years, several modified starches and their derivatives have been used as new food ingredients. However, innovations have proposed the use of starch and new starch-based materials in products others than those manufactured by the food, pharmaceutical, and paper industries. As a consequence, changes in world starch production and diversification in sources of starch have been observed. Also, there has been a considerable impact on research about starch and the techniques proposed for starch characterization, both at the granular and macromolecular levels.

The aim of this book is to contribute a modern overview of trends and advances in the production and applications of starches, with emphasis on some recent techniques used in starch characterization as well as some aspects of production, properties, and biodegradation of starch-based products. This book discusses starch characterization, modified starches, starch-based plastics and nanocomposites, and biodegradation of starch blends. It is a comprehensive reference book for researchers, teachers, and other professionals who are interested in recent advances in starch and starch-based products. The book also may be suitable as a reference or textbook for graduate courses on starches and biopolymers.

An outstanding team of collaborators with complementary expertise from several research areas contributed to this book. This has resulted in an updated overview with an innovative and multidisciplinary approach to trends in production, characterization, properties, and applications of starch. There are ten chapters covering trends in starch production and applications, advances in starch characterization utilizing several techniques, the main new starch-based products, and starch biodegradability. A general discussion is provided about trends in starch production and the application of starch in the food, textile, pharmaceutical, chemical, agricultural, and plastic industries as substitutes for synthetic polymers (Chapter 1). This book also covers recent advances in starch characterization using several techniques such as atomic force microscopy (Chapter 2), high-performance size-exclusion chromatography and sedimentation field flow fractionation (Chapter 3), nuclear magnetic resonance (Chapter 4), and starch thermal transitions (Chapter 5), providing an overview of starch characterization at the granular, macromolecular, and rheological levels. The main industrial concerns about starch applications, such as amylase production employed in starch hydrolysis (Chapter 7) and main modified starches (Chapter 8), are also discussed in this book. Properties and applications of new starch-based products, including starch-based plastics (Chapter 6) and starch-based nanocomposites (Chapter 9), are discussed with emphasis on starch properties, as well as the biodegradability of starch-based blends (Chapter 10).

I would like to acknowledge all of the collaborators for their important contributions and the expertise devoted to this book. Finally, I would also like to thank everyone who contributed, directly or indirectly, to this work.

Andréa C. Bertolini

About the Editor

Andréa Curiacos Bertolini is involved in scientific research on starch and biopolymers. She has a PhD in food science from the Université de Nantes in France, where she worked with a concentration on starch chemistry at the Institute Nationale de Recherche Agronomique (INRA) in partnership with the Centre de Coopération Internationale en Recherche Agronomique pour le Développement (CIRAD). As an invited researcher, she worked at the Fonterra Research Centre in New Zealand and at the University of Idaho in the United States on modified starches and starch rheology.

At present she is a food scientist at the Empresa Brasileira de Pesquisa Agropecuária (EMBRAPA–Brazilian Agricultural Research Corporation, Brazil). Her interests include biopolymers with an emphasis on starch characterization, modified starches, and starch as an ingredient in food and nonfood products.

Contributors

Edith Agama-Acevedo
Centro de Desarrollo de Productos
Bióticos del IPN
Yautepec, Morelos, México

Luis Arturo Bello-Perez
Centro de Desarrollo de Productos
Bióticos del IPN
Yautepec, Morelos, México

James N. BeMiller
Whistler Center for
  Carbohydrate Research
Purdue University
West Lafayette, Indiana, USA

Andréa C. Bertolini
EMBRAPA
Agroindústria de Alimentos
Rio de Janeiro, Brazil

Alain Buleon
Institut National de la
  Recherche Agronomique
Nantes, France

Paul Colonna
Institut National de la
  Recherche Agronomique
Nantes, France

Beatriz Rosana Cordenunsi
Faculdade de Ciências
  Farmacêuticas
Universidade de São Paulo
São Paulo, Brazil

Alain Dufresne
Institut National Polytechnique
Ecole Française de papeterie et des
  Industries Graphiques
Grenoble, France

**Cristina das Graças
  Fassina Guedes**
Universidade São Francisco
Itatiba, Brazil

Kerry C. Huber
School of Food Science
University of Idaho
Moscow, Idaho, USA

Eliton S. Medeiros
EMBRAPA
Instrumentação Agropecuária
São Carlos, Brazil

Marcia Nitschke
Instituto de Química
Universidade de São Paulo
São Carlos, Brazil

William J. Orts
Bioproduct Chemistry and
   Engineering Unit
U.S. Department of Agriculture
Albany, California, USA

Sandra Leticia Rodriguez-Ambriz
Centro de Desarrollo de Productos
   Bióticos del IPN
Yautepec, Morelos, México

Mirna Maria Sanchez-Rivera
Centro de Desarrollo de Productos
   Bióticos del IPN
Yautepec, Morelos, México

Derval dos Santos Rosa
Universidade Federal do ABC
Santo André, Brazil

Renata Antoun Simão
PEMM/COPPE
Universidade Federal do
   Rio de Janeiro
Rio de Janeiro, Brazil

Maria Inês Bruno Tavares
Instituto de Macromoléculas
   Eloisa Mano
Universidade Federal do
   Rio de Janeiro
Rio de Janeiro, Brazil

**Rossana Mara da Silva
   Moreira Thiré**
PEMM/COPPE
Universidade Federal do
   Rio de Janeiro
Rio de Janeiro, Brazil

chapter one

Trends in starch applications

Andréa C. Bertolini
EMBRAPA Agroindústria de Alimentos

Contents

1.1   Introduction: The starch granule

1.1.1   Granular structure

Starch is organized in discrete particles, granules whose size, shape, morphology, composition, and supramolecular structure depend on the botanical source. The diameters of the granules generally range from less than 1 µm to more than 100 µm, and shapes can be regular (e.g., spherical, ovoid, or angular) or quite irregular. Starch granules are partially crystalline particles composed mainly of two homopolymers of glucopyranose with different structures: amylose, which is composed of units of D-glucose linked through α–D–(1–4) linkages and amylopectin,

1

the branching polymer of starch, composed of α–D–(1–4)-linked glucose segments containing glucose units in α–D (1–6) branches. Amylose
is an essentially linear polymer, although evidence has suggested some
branches in its structure. Consequently, 4–9% of the population called
"intermediate material or amylose-like" has been considered part of normal and high amylose starches (Tang, Mitsunaga, and Kawamura, 2006).
Amylose can also present as a hydrophobic helix, allowing the forming of
a complex with free fatty acids, fatty acid components of glycerides, some
alcohols, and iodine (Thomas and Atwell, 2005). Amylopectin is larger
than amylose in most normal starches and their chains are classified as
small chains, with an average degree of polymerization (DP) of about 15,
and large chains, in which the DP is larger than 45. This unique configuration ordered in the packing arrangement contributes to the crystalline
nature of the starch granule. This crystallinity reflects the organization
of amylopectin molecules within the starch granules, whereas amylose
makes up most of the amorphous materials that are randomly distributed among the amylopectin clusters (Blanshard, 1987). Under polarized
light, the starch granule shows a characteristic Maltese cross (Figure 1.1),
reflecting its birefringent structure, which was suggested as a correlation between the molecular and principal optic axes, underscoring that
packing of the amylose and amylopectin is radial (Banks et al., 1972). This
observation can be corroborated by results obtained by x-ray diffraction
(Waigh et al., 1997).

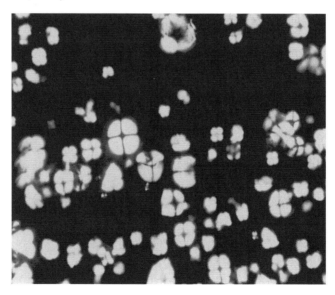

Figure 1.1 Maltese cross in starch granules observed under polarized light.

Consequently, the architecture or structure of a starch granule is influenced in part by the fine structures and ratios of amylose and amylopectin molecules. Finally, the granule architecture in turn determines the accessibility of the starch structure to water and chemical reagents, affecting molecular reaction patterns and properties of modified starches. The chemistry and technology of starch have been reviewed in detail several times (Whistler, BeMiller, and Paschall, 1984; Zobel, 1988; Eliasson, 2004; BeMiller, 2007). Some aspects of granule starch observations are considered in Chapter 2 and details of starch macromolecular structure are discussed in Chapters 3 and 4.

1.1.2   Starch gelatinization

Starch gelatinization involves granule melting in an aqueous medium under heating. In water, granule swelling increases with temperature and it leads to a transfer of water in the suspension to water associated with starch components: amylose and amylopectin. When starch temperature reaches 60–70°C, insoluble granules are disrupted by the energy supplied, resulting in a loss of molecular organization and, consequently, loss of its crystallinity. This process leads to increasing viscosity and starch solubilization, which is a result of irreversible changes such as the disruption of granular and semicrystalline structure, also seen as a loss of birefringence (Douzals et al., 1996). After heating followed by gelatinization, during the cooling phase, starch undergoes retrogradation in which the starch chains tend to reassociate in an ordered structure. It is followed by another rise in viscosity, usually referred to as setback. The pasting profile of previously gelatinized starch does not show a viscosity increase when exposed to changes in temperature, evidence that this starch sample was already gelatinized (Rosa, Guedes, and Pedroso, 2004).

Differences in swelling among native starches have been attributed to interplay of factors such as granule size, crystallinity, amylose–lipid complex content, and interaction among starch chains in the amorphous region. The physicochemical properties of amylose and amylopectin are quite diverse and they contribute in different ways to the pasting properties of starch. In Figure 1.2, the viscoamilographic graph shows the differences of pasting profiles of cassava and corn starches.

Understanding starch thermal transitions and starch rheological behavior is essential for studying starch structure, as well as for proposing new starch applications. The main aspects of rheological and thermal properties of starches are discussed in detail in Chapters 5 and 6.

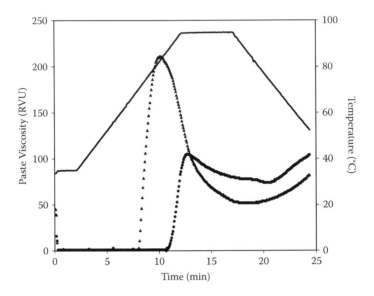

Figure 1.2 Pasting profiles of cassava (▲) and corn (●)starches, under temperature (–) changes.

1.2   Sources and producers of starch

Starches are the major storage polysaccharides in foods of plant origin. Worldwide dry starch production is actually estimated at more than 64 million tons and almost 75 million tons are expected by 2012 (Patil, 2009). Current annual production for primary starch sources is estimated to be 46.1 million tons of corn, 9.1 million tons of cassava, 5.15 million tons of wheat, and 2.45 million tons of potato (Röper and Elvers, 2008).

Starches from cereals come mainly from corn and wheat. European countries are responsible for around 60% and Asia produces 20% of the global production of wheat starch used in dextrin and modified starches. The United States is still the largest producer of corn starch in the world (LMC, 2008). In Western Europe, 46% of produced starch is from corn, 36% from wheat, and 19% from potato starches (Röper and Elvers, 2008), whereas North American production of starch is based almost entirely on corn (LMC, 2008). Actually, the Asian corn starch sector, particularly in China, is growing consistently at over 15% per year, competing with United States and European producers (LMC, 2008).

Because of their lower moisture content, cereals have longer storage times and their starch extraction is easier and faster than in roots and tubers. For example, in cassava, high water consumption is a critical factor in the starch extraction process. In this process, water is used during the grinding, decantation, and washing steps and these large quantities of water are converted to wastewater, which must be treated before being released to the environment. Liquid waste has a high biochemical and chemical oxygen demand; its treatment comprises several steps and requires a long retention time. After extraction and separation, starch moisture content is from 35 to 40%, requiring a great deal of energy in the drying process (Sriroth et al., 2000).

Although cassava production is growing and starch from cassava competes with the corn processors (Patil, 2009), corn still remains the main starch source in the world, followed by cassava, potato, and wheat. More than 70% of starch produced in the world is from corn (Röper and Elvers, 2008) (Figure 1.3). However, starch from roots and tubers shows some particular rheological and physical properties, such as clear gel, high viscosity, and lower retrogradation, which are required in the formulation of specific products. Demand for cassava starch has grown in the past few years and it is actually the most widely traded form of native starch in the world, mainly in Thailand and East Asia. Asia contributes around 90% of cassava starch produced for use in industry (LMC, 2008), with Thailand being the major producing country, followed by China and Indonesia (FAO, 2008). In 2006, around 3.5×10^6 tons of cassava starch were produced in Thailand. Of this amount, 2.3×10^6 tons were exported, with 1.67×10^6 tons as native starch and 638×10^3 tons as modified starches (Röper and Elvers, 2008). In South America, which is responsible for the other 10% of cassava starch production, Brazil is the main producer (FAO, 2008).

Considering starches from roots and tubers, potato starch is the second largest starch source. In Europe, strong support of the grain sector has resulted in decreasing production of potato starch (LMC, 2008). However,

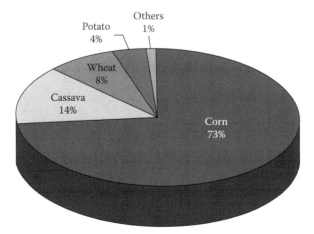

Figure 1.3 Starch production according to botanic sources. Source: Röper and Elvers (2008).

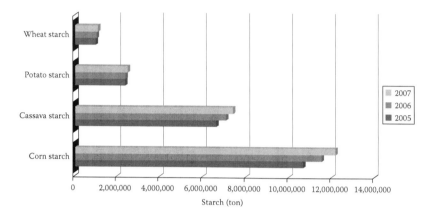

Figure 1.4 Amounts of starches used as food ingredients, dextrins, paper coatings, and adhesives between 2005 and 2007. Source: LCM (2008).

European countries are still responsible for 80% of potato starch production in the world (LMC, 2008), with the Netherlands, France, Belgium, Germany, and Switzerland as the main potato producers (FAO, 2008).

Figure 1.4 illustrates the amounts of starches from several sources used in syrup production or fermented products, used as food ingredients, dextrin, paper coatings, and adhesives for three consecutive years (LMC, 2008). A small amount of rice starch production originates from Europe and Asia (LMC, 2008), while arrowroot, amaranth, banana, mung bean, sweet potato, taro, and yam are considered as minor sources of starch produced in tropical countries.

1.3   Starch applications

Due to its low cost, availability, and ability to impart a broad range of functional properties to food and nonfood products (Jane, 1995; BeMiller, 2007), starch is utilized in several industrial applications (Whistler, BeMiller, and Paschall, 1984). Although the use of starch was always considered essential as a staple food and food ingredient, due to its attractive cost and performance, starch has been one of the most promising candidates for future materials (Rindlav-Westling et al., 1998). Starches are used mainly in the food and paper industries, with 57% of produced starch consumed in the food industries and 43% in the nonfood sector (LMC, 2008). Modified and native starches represent more than 85% of all hydrocolloids used in food systems (Wanous, 2004). In the food industry, the main trend in starch applications remains in syrup production and formulation of ready meals and various sauces.

From total worldwide starch production, excluding starch used to produce syrups and fermented products, more than 30 million tons of natural

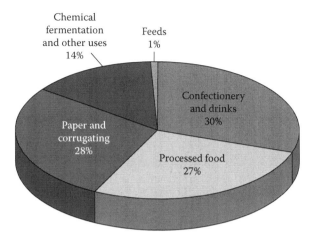

Figure 1.5 Use of starches and their derivatives by European industries. Source: Röper and Elvers (2008).

and modified starches are used in the production of dextrin, food ingredients, paper coatings, and adhesives (LMC, 2008). In Europe, 57% of produced starch is used in sweeteners and hydrolized, 23% as native starch, and 20% as modified starch. In European industries, the use of starch in sweet, drink, and fruit processing is equivalent to 30% of produced starch. Convenience foods, bakery, food ingredients, and dairy products account for 27%; the paper industry consumes 28%; and chemical, fermentation, and other industrial products utilize 14%. Starch usage in feeds is 1% (Röper and Elvers, 2008) (Figure 1.5). However, in the United States most production is not destined for native and modified starch products, but for sweeteners, particularly high fructose corn syrup (LMC, 2008). More than 73% of corn starch produced in North America is used in U.S. production of refinery products such as glycose syrups, high fructose syrups, syrup solids, maltodextrins, and fructose (Röper and Elvers, 2008).

In addition to the demand for starch to produce syrups and dextrin, the demand for modified starches has grown in past years. Modified starches are produced mainly in North America and Europe. Europe is still one of the leading producers of modified starches and its paper and food industries tend to be at the forefront in their use of innovative new products (LMC, 2008). In Table 8.1 of Chapter 8, which provides details about modified starches, the main industrial applications of modified starches are summarized.

Asian markets, among others, demand large amounts of starches used in fermentation. The United States, following the world economy, has focused on starch to biofuel production. However, there is a trend for the demand for starch by nonfood industries to grow faster than demand for starch by food industries.

1.3.1   Food applications

Starch is the main reserve in plants and supplies between 70–80% of the calories consumed by people. In the food industry, starch can be considered as the principal component of several food formulations and it is responsible for important functional properties and nutritional characteristics. However, starches are not only used as a staple food, but also in a wide range of applications in food formulation and the beverage industry. Inasmuch as extensive reviews have been done about starches in the food industry (Eliasson, 2004; BeMiller, 2007), a general overview and main trends of starch applications in food products are discussed here.

1.3.1.1   Texturing, thickening/gelling, retrogradation and freeze–thaw stabilizer, emulsifying agent, gel strength and clarity, and cold-water swelling

In food, starches and their derivatives are used as ingredients, in small or large quantities, to improve the aspect, conservation, or properties. Because of the relation between structure and functionality, starch chemistry and modifying technologies have been studied to obtain structures with specific functionalities. Although starch constitutes an excellent raw material to modify food texture and consistency, starch functionality depends on molecular average weight of amylose and amylopectin, as well as on their molecular organization within the granule. The choice of the right starch within food applications must take into consideration such aspects as food process technology, functional, sensorial, and rheological properties, and co-ingredients.

Native and modified starches are used frequently in foods because of their thickening and gelling properties. Typically, native starch cannot be used at a rate higher than 6% in solids because it imparts such a high viscosity. However, for several applications native starches can show undesirable properties under certain process conditions such as temperature, pH, or pressure. These properties can limit their industrial applications, due to their low resistance to high shear rates, thermal decomposition, high retrogradation, and syneresis. Improvement in these starch properties can be achieved by modifications of starch structure, using chemical, physical, and enzymatic methods.

Physically modified starches, the pregelatinized starches, which do not require cooking, are very popular in the food industry. Swollen but intact starch granules are usually desired in most food starch applications to maintain rheological properties (Miyazaki et al., 2006; Kaur, Singh, and Singh, 2006). They are used to control the cold viscosity and can be cross-linked and stabilized to provide shear and freeze–thaw stability. Granular cold-water soluble starches showed important functional properties for several instant foods, such as a higher viscosity and a soft texture (Bello-Perez et al., 2002). Because they develop viscosity immediately they are typically used in instant dry mixes such as instant puddings, in baby foods, frozen foods, and convenience products. In some applications such as salad dressing, starch can be incorporated into the oil phase to slow down hydration when water is added later, useful in sauce and puree formulation.

Modified starches can present functional derivatives, supporting new product development, improving processing technologies, and influencing market trends. Oxidized starches show reduced swelling power, increased solubility, and lower pasting temperature than native starch (Kuakpetoon and Wang, 2001) and they are used in products demanding softness and transparency, like gum sweets or Arabic gums (Chattopadhyaya, Singhal, and Kulkarni, 1998).

Acetylated starches are used to prevent retrogradation by interrupting the linearity of the amylose or segments of the amylopectin branches, reducing syneresis and loss of texture and consistency during storage of the finished product (Filer, 1998). Acetylating reduces bond strength and increases the swelling power and solubility of the starch granule, improving freeze–thaw stability (Rutenberg and Solarek, 1984). Consequently, these starches have physicochemical characteristics such as low gelatinization temperature, high solubility, and good cooking and storage stabilities, and are widely used in the food industry to overcome undesirable changes in product texture and appearance caused by retrogradation during food processing and storage. They are extensively used in a wide range of foods including baked goods, canned pie fillings, sauces, condensed soups, frozen foods, baby foods, salad dressings, and snack foods.

Modified starches by succinate derivatives are characterized by the tendency to swell in cold water, and to show higher viscosity and lower gelatinization temperature than their base starch. In food areas, succinate derivatives have been recommended as binders and thickening agents in soups, snacks, and canned and refrigerated food products (Wurzburg, 1986), encapsulating flavor oils, emulsion stabilizers, and enhancing oxidation resistance (Trubiano, 1995).

Hydroxypropylated starches improve the shelf life, freeze–thaw stability, cold-storage stability, clarity, and texture properties of starch paste. Because of the slow retrogradation of amylopectin, it is very effective in retarding bread staling and improving bread crumb texture, enhancing breadmaking functionality and bread quality (Miyazaki et al., 2006). Hydroxypropylated starches are also effective thickening and stabilizing agents in low-fat or low-calorie dairy products due to their greater compatibility with milk protein.

Frequently, the structure of starch is reinforced by the application of cross-linking in order to improve the product's texture and its resistance to rupture during processing. Cross-linked starches are more resistant to low pH, rupture, and high temperatures, and show high capacity of holding water (Jarowenko, 1986). However, cross-linking also reduces paste clarity and stability against cold storage. Because cross-linking reduces starch granule rupture, the loss of viscosity, and the formation of a stringy paste during cooking, cross-linked starches are widely used as thickeners in foods, particularly where a high and stable viscosity is needed. Therefore, further modifications such as hydroxypropylation and acetylation are used to change undesirable characteristics of cross-linked starch (Gunaratne and Corke, 2007). Cross-linking of hydroxypropylated starch, for example, imparts viscosity stability and a desired short-textured property of the paste. The main techniques, properties, and trends of modified starches are discussed in Chapter 8.

1.3.1.2   Dextrins, sweeteners, and syrups

Starch hydrolysis has been used in the food industry to convert starches, starch derivatives, and starch saccharification products, which have several applications in food processing. The products from starch hydrolysis are certainly the more versatile ingredients used in the food industry and they are used in a wide range of food products due to both their functional and nutritional properties. They can be obtained through acid catalysis or specific enzyme action. Initially, industrial hydrolysis of starch was done by inorganic acids, however, the use of enzymes has successfully replaced the chemical process. Enzymatic hydrolysis allows greater control and high specificity of reaction, better stability of generated products, lower energy requirements, and does not require neutralization steps. Moreover, the milder enzymatic catalysis conditions reduce unwanted side reactions and the formation of off-flavor and off-color compounds. The starch-processing industry is the second largest segment of the industrial enzyme market, representing approximately 30% of world enzyme production. The major enzymes are amylases, glucoamylases, and glucose isomerases which are used in starch liquefaction, saccharification, and isomerization. Chapter 7 discusses some trends in amylase production used in starch hydrolysis.

Products from starch hydrolysis such as glucose or maltose syrup and maltodextrins and their derivates by isomerization (iso-glucose and fructose) are used in candy, sweet, chocolate, cake, dessert, dairy, and pastry products due to their anticrystallizing, sweetening, and water-holding properties. Sugar syrups and their derivatives are the most popular hydrolized starch used in fermentation processes, beverage formulation, dietetic products, ice creams, soft gum, dairy products, desserts, and candies. Maltodextrins have functional properties and are employed for their thickening properties, prevention of crystallization, control of freezing, and stabilization (Blanshard and Katz, 1995; Loret et al., 2004).

1.3.1.3 Health and functional foods

In the food industry, starch in formulations of health, nutritional, and functional products has been considered one of the most promising markets and it has motivated research in development of new products. This trend supports the development of modern products such as frozen, chilled, low-fat, and gluten-free foods, where starch can be used as the main or co-ingredient.

For example, health nutrition trends for enhanced fiber consumption with emphasis on low glycemic food have pushed the growth of high amylose starch as a source of resistant starch (RS). Resistant starches result from the formation of glycosidic bonds, other than the α–(1–4) or α–(1–6) links by treatments involving heating, cross-linking, hydroxypropylation, or phosphatation, which can reduce the digestibility of the starch (Englyst et al., 1996; Woo and Seib, 2002). Amylose content, water content, storage time and temperature of the starch gels, number of autoclaving/cooling cycles, and the presence of lipids and proteins are some processing factors that can affect the formation of RS (Gidley et al., 1995; Kale et al., 2002). Those starches are resistant to degradation or are not digested/absorbed in the small intestines of healthy individuals, and are fermented in the large intestine (Champ, 1992; Eerlingen and Delcour, 1995), simulating fiber action in the intestinal tract (Englyst et al., 1996). They can replace fibers in food formulation without changing their sensory properties (Lee and Shin, 2006). Studies showed that addition of RS in the diets of rats contributed significantly to the increase of fecal volume (Gee, Johnson, and Lund, 1992; Deckere, Kloots, and Van Amelsvoort, 1992; Cummings et al., 1996) and its *in vitro* fermentation suggests RS induces the cellular proliferation, growth of bacteria, fermentation, and production of short chain fatty acids (Annison and Topping, 1994; Cummings et al., 1996; Englyst et al., 1996; Bird, Brown, and Topping, 2006), indicating that it can behave as a probiotic food. Resistant starches can be classified according to four categories: Type I, represented by the inaccessible starch, which is protected by the wall cell; Type II, the native starch granule found in food that is not cooked, with reduced susceptibility to enzyme action; Type III, the starch fraction not digested, formed by hydrothermal treatments that cause retrogradation; and, Type IV, the chemical modified starch. More discussion about RSs and their structure is provided in Chapter 8.

Another trend in health food is the concern to reduce fat ingestion. A diet rich in fats is associated with an increase in obesity and its effects. Clinical and experimental studies provide evidence of the relationship of the consumption of saturated fat and high blood cholesterol and heart disease (Giese, 1996; Livesey, 2003). However, diets rich in fiber and low in fat have been recommended considering that they would decrease the risk of chronic diseases such as arthritis, cancer, osteoporosis, and heart disease (Lajolo, 2001). Fat, however, is an indispensable nutrient and plays a functional and sensory role in food: it is responsible for the stability, palatability, flavor, texture, structure, brightness, aroma, and sensation of satiety (de Wijk and Prinz, 2006). Consequently, in past years there has been increasing research regarding the improvement of quality and sensory characteristics of low-fat products (Giese, 1996; Livesey, 2003; de Wijk and Prinz, 2006). Several starch derivatives have shown to be promising replacements for fat properties, probably due to their water association systems (Yackel and Cox, 1992; Filer, 1998). To replace fat in food systems, starch modifications must improve their stability in cold water, improve their characteristics as dough or gel texture, or provide emulsifying characteristics through the addition of hydrophobic groups (Alexander, 1995). Starch has been used in several low-fat products to improve some properties such as texture, color, palatability, and stability of foods (Harris and Day, 1993; Esteller, Amaral, and Lannes, 2004). Hydrolyzed starches have been shown to be excellent substitutes for fat, obtaining similar texture to those observed in high-fat products (Harris and Day, 1993; Alexander, 1995; Giese, 1996). Waxy starches and starches from roots and tubers are largely used in low-fat products. Cassava starch derivatives intercrossed with 0.05% $POCl_3$ and acetylated with 8% acetic anhydride has been shown to be an efficient fat replacer, improving the texture of mayonnaise (Teixeira, 2002). Esterification of starches with fatty acids of a linear chain gives the derivative hydrophobic characteristics and emulsifying properties (BeMiller, 1997) and acetylated starches improve stability in low-fat systems (Jarowenko, 1986). However, to replace fat, starches should maintain high moisture content in systems such as mayonnaise, salad dressings, and low-fat spreads. Starches as fat replacers can also be used in bakery products such as cakes, however, they are not recommended for systems with a low moisture content such as cookies and crackers (Giese, 1996; Razavi, Najafi, and Alaee, 2007; Sudha et al., 2007).

Fermented starches are often employed in traditional culinary arts, and they have also been used as ingredients in the food industry. Sour cassava starch is an example of cassava starch modified by fermentation and sun drying, with its characteristic expansion property. For this reason, sour cassava starch is used in bakery products leading to volume expansion without the addition of yeast, biological baking powder, or chemical agents (Bertolini et al., 2001). It is used in breadmaking and biscuit formulation and it can be an alternative for gluten-free products consumed by people who are intolerant to wheat gluten.

1.3.1.4 Edible coatings

Biopolymers such as lipids, proteins, derivatives of cellulose, starch, and other polysaccharides have been tested in the formulation of edible films and coatings. Recently there has been increasing interest in the development of films and edible coatings that can be used to increase the shelf-life of fruits and vegetables (Pagella, Spigno, and de Faveri, 2002). Starch gelatinized with NaOH, glycerol, or sorbitol can be used as plasticizers in edible coatings and films (Larotonda et al., 2004; Cyras, Zenklusen, and Vazquez, 2006), and blends of pectin and high amylose corn starch also have been suggested to produce food coatings and films (Dimantov, Kesselman, and Shimoni, 2004). Some properties of starch-based films and thermoplastic starches are discussed in detail in Chapter 6.

1.3.2 Nonfood applications

In nonfood applications, the most traditional use of starch has been in the paper industry. After food, the paper industry is the largest consumer of starch, estimated at more than 10 million tons per year. Recently, with the increasing cost of cellulose pulp, paper producers have begun to look at lower cost fillers. These fillers often require special modified starches to bind them to the pulp. Consequently, several innovations in this area have increased the share of starch used in those applications (LMC, 2008). Oxidized (Lawala et al., 2005) and enzymatic hydrolized (van der Maarel et al., 2002) starches have wide applications as paper coating, improving the strength and printability of paper. Those starches are also often used in textile and laundry finishing products (Lawala et al., 2005).

Recently, interest in development of biodegradable polymers has grown and, as a consequence, it has resulted in the development of new biodegradable products (Stevens, 1999). Biodegradable polymers are an alternative to overcome the problems related to fossil resources, recycling limitations, and the global environment. Furthermore, biobased material can be recycled through composting, providing support for a sustainable agriculture. This has resulted in a strong trend to replace synthetic materials with biodegradable polymers. However, the high cost of producing biodegradable polymers compared to conventional plastics is still a major problem to be solved but this is expected to decrease with time. Starch is considered as a potential polymer to be used in biodegradable materials because of its low cost, availability, and production from renewable resources. Because it is a thermoplastic material, it has been used to produce blends with nonbiodegradable polymers, biodegradable materials, and injection molding (Dufresne, 2007). Some aspects of thermoplastic starches are covered in Chapter 6.

In recent years, most of the starch-based polymers produced are used for packaging applications, including soluble films for industrial packaging. Loose-fill containers and transparent films are widely used in the packaging industry. Biodegradable bags are used for waste collection and for disposable cutlery, whereas biodegradable films are used for agricultural mulching, pet products, and hygiene, including cotton buds, packaging for toilet paper, overwrap, and backing material for flushable sanitary products. Pencil sharpeners, rulers, cartridges, toys, and plant pots are among the other uses (Alonso et al., 1999; Karim, Norziah, and Seow, 2000). Starch-based plastics are highly hydrophilic and readily disintegrate on contact with water. There is increasing interest in biomaterials in development of biodegradable products. Details about biodegradation of starch-based biomaterials are discussed in Chapter 10.

Starches used as thermoplastic materials have some limitations such as low moisture resistance, high viscosity, and incompatibility with some hydrophobic polymers. In packaging applications, the major drawbacks associated with the use of starch-based materials are their hydrophilic nature, poor mechanic properties, and brittleness, especially if not well plasticized to prevent moisture effects and postprocessing crystallization. Starch-based packages generally absorb moisture and undergo environmental stress cracking, causing significant problems with shelf-life storage. Therefore, any improvement in these properties is of fundamental importance to assure durability of both packages and stored products. Starches modified by chemical treatment have been utilized as an alternative to solve this problem. To enhance viscosity, texture, and stability among many desired functional properties for industrial applications, starches and their derivatives are modified by chemical, physical, and biotechnological means. Consequently, due to the wide range of industrial applications, modified starch derivatives have high added value.

Starch-based and biodegradable polymers have been used in the production of automotive parts (Leão et al., 2007) and nanoparticle-sized corn-derived starch can also be used in a wide range of products (Dufresne, 2007) including as a biofiller to enhance tire performance, reducing the rolling resistance of tires and fuel consumption (Scott, 2002).

In addition to the use of starches in the paper and textile industries and to obtain new biodegradable materials, there is the fuel market. With new trends in the biofuel and bioethanol market, there has been increasing interest in some starchy crops. Biofuel production, particularly in the United States, has put considerable strain on the corn crop and China has suspended further production of ethanol and other starch derivatives from corn to protect domestic food and feed consumption of corn (LMC, 2008). However, in bioethanol production, sugarcane is still more competitive than starch. Consequently, concerning this starch application, it is necessary to improve conversion technology in order to increase the ethanol yield.

1.3   Final remarks

Corn starch remains the most used starch in the world, the United States being its largest producer. However, concerning cassava starch, Thailand is growing as the leading exporter of native starch and starch derivatives. Whereas corn starch is used to produce mainly modified starches and syrups, cassava starch is actually the most widely traded native starch in the world.

Demand for native and modified starches closely follows growth in the food and paper industries. The largest applications of starch in the food industry include syrups, formulation of ready meals, instant food, and sauces, and, in the paper industry, as paper coating. However, starch application is also increasing in health, nutritional, and functional foods, considered as a potential food market. Moreover, due to the increasing demand for raw material by biofuel and bioethanol production sectors, there is decreasing availability of starchy material to be used in the food and paper industries. This tendency will influence policy guidelines on crop production in several countries.

The increased value of modified starch remains very high due to several food and industrial applications. There are several new technologies that can improve the production of modified starches. Due to progress in plant breeding, a wide range of starch sources is available. Nowadays the understanding of starch behavior requires strong knowledge of physical chemistry and rheology because the goal is to provide qualitative information related to starch modification. New varieties of cereal and roots with high resistant starch or high sugar content, potato with high phosphate content, and corn with amylase and other enzymes activated in the grain to improve processing are some examples of these modified starches.

References

Alexander, R.J. 1995. Fat replacers based on starch. *Cereal Foods World.* 40(5):366–368.

Alonso, A.G., Escrig, A.J., Carrón, N.M., Bravo, L., and Calixto, F.S. 1999. Assessment of some parameters involved in the gelatinization and retrodegradation of starch. *Food Chem.* 66:181–187.

Annison, G. and Topping, D.L. 1994. Nutritional role of resistant starch: Chemical structure vs physiological function. *Ann. Rev. Nutrition.* 14:297–320.

Banks, W., Geddes, R., Greenwood, C.T., and Jones, I.G. 1972. Physicochemical studies on starches. *Starch.* 24:245–280.

Bello-Perez, L.A., Contreras-Ramos, S.M., Romero-Manilla, R., Solorza-Feria, J., and Jiménez-Aparicio, A. 2002. Chemical and functional properties of modified starch from banana *Musa paradisiaca L. Agrociencia*. 36(2):169–180.

BeMiller, J.N. 1997. Structure of the starch granule. *J. Appl. Glycosci*. 44:43–49.

BeMiller, J.N. 2007. *Carbohydrate Chemistry for Food Scientists*. St. Paul, MN: AACC International, Inc.

Bertolini, A.C., Mestres, C., Lourdin, D., Della Valle, G., and Colonna, P. 2001. Relationship between thermomechanical properties and baking expansion of sour cassava starch. *J. Sci. Food Agric*. 81(4):429–435.

Bird, A.R., Brown, I.L., and Topping, D.L. 2006. Low and high amylose maize starches acetylated by a commercial or a laboratory process both deliver acetate to the large bowel of rats. *Food Hydrocolloids*. 8:1135–1140.

Blanshard, J.M.V. 1987. Starch granule structure and function: A physicochemical approach. In:. *Starch: Properties and Potential*, T. Galliard (Ed.). Chichester: John Wiley & Sons, 1987, p. 16.

Blanshard, P.H. and Katz, F.R. 1995. Starch hydrolysates. In: *Food Polysaccharides and Their Applications*, A.M. Stephen (Ed.). New York: Marcel Dekker. pp. 99–153.

Champ, M. 1992. Determination of resistant starch in foods and food products: Interlaboratory study. *Euro. J. Clin. Nutrition*. 46(2):51–62.

Chattopadhyaya, S., Singhal, R.S., and Kulkarni, P.R. 1998. Oxidised starch as gum arabic substitute for encapsulation of flavours. *Carbohydrate Polym*. 2:143–144.

Cummings, J.H., Beatty, E.R., Kingman, S.M., Bingham, S.A., and Englyst, H.N. 1996. Digestion and physiological properties of resistant starch in human large bowel. *Brit. J. Nutrition*. 75(5):733–747.

Cyras, V.P., Zenklusen, M.C.T., and Vazquez, A. 2006. Relationship between structure and properties of modified potato starch biodegradable films. *J. Appl. Polym. Sci*. 6:4313–4319.

de Wijk, R.A. and Prinz, J.F. 2006. Mechanisms underlying the role of friction in oral texture. *J. Texture Stud*. 4:413–427.

Deckere, E.A.M., Kloots, W.J., and Van Amelsvoort, J.M.M. 1992. Effects of a diet with resistant starch in the rat. *Euro. J. Clin. Nutrition*. 46(2):121–122.

Dimantov, A., Kesselman, E., and Shimoni, E. 2004. Surface characterization and dissolution properties of high amylose corn starch-pectin coatings. *Food Hydrocolloids*. 1:29–37.

Douzals, J.P., Marechal, P.A., Coquille, J.C., and Gervais, P. 1996. Microscopic study of starch gelatinization under high hydrostatic pressure. *J. Agric. Food Chem*. 44:1403–1408.

Dufresne, A. 2007. Biopolymers on nanocomposites. In: *Biopolymers Technology*, A.C. Bertolini (Ed.). São Paulo: Cultura Academica, pp. 59–84.

Eerlingen, R.C. and Delcour, J.A. 1995. Formation, analysis, structure and properties of type lll enzyme resistant starch. *J. Cereal Sci*. 22:129–138.

Eliasson, A.-C. 2004. *Starch in Food: Structure, Functions and Applications*. Cambridge, UK: Woodhead.

Englyst, H.N., Kingman, S.M., Hudson, G.J., and Cummings, J.H. 1996. Measurement of resistant starch *in vitro* and *in vivo*. *Brit. J. Nutrition*. 75(5):749–755.

Esteller, M.S., Amaral, R.L., and Lannes, S.C.D. 2004. Effect of sugar and fat replacers on the texture of baked goods. *J. Texture Stud*. 4:383–393.

FAO 2008. Food Agriculture Organization. Food Outlook November. ftp://ftp. fao.org/docrep/fao/011/ai474e/ai474e00.pdf.

Filer, L. J. 1998. Modified food starch: An update. *J. Amer. Diet. Assoc.* 88(3).

Gee, J.M., Johnson, J.T., and Lund, E.K. 1992. Physiological properties of resistant starch. *Euro. J. Clin. Nutrition.* 46(2):125.

Gidley, M.J., Cooke, O., Darks, A.H., Hoffmann, R.A., Russel, A.L., and Greenwell, P. 1995. Molecular order and structure in enzyme resistant retrograded starch. *Carbohydrate Polym.* **28**:23–31.

Giese, J. 1996. Fat, oils, and fat replacers. *Food Technol.* 50(4):77–83.

Gunaratne, A. and Corke, H. 2007. Functional properties of hydroxypropylated, cross-linked, and hydroxypropylated cross-linked tuber and root starches. *Cereal Chem.* 84:30–37.

Harris, D. and Day, G. A. 1993. Structure versus functional relationships of a new starch-based fat replacer. *Starch/Stärke.* 45(7):221–226.

Jane, J.-L. 1995. Starch properties, modifications, and applications. *J. Macromol. Sci. Pure Appl. Chem.* A32:751–757.

Jarowenko, W. 1986. Acetylated starch and miscellaneous organic esters. In: *Modified Starches: Properties and Uses*, O.B. Wurzburg (Ed.). Boca Raton, FL: CRC Press, pp. 55–57.

Kale, C.K, Kotecha, P.M., Chavan, J.K., and Kadam, S.S. 2002. Effect of processing conditions of bakery products on formation of resistant starch. *J. Food Sci. Technol.* 5:520–524.

Karim, A.A., Norziah, M.H., and Seow, C.C. 2000. Methods for the study of starch retrogradation. *Food Chem.* 71:9–36.

Kaur, L., Singh, J., and Singh, N. 2006. Effect of cross-linking on some properties of potato (*Solanum tuberosum* L.) starches. *J. Sci. Food Agric.* 86:1945–1954.

Kuakpetoon, D. and Wang, Y.J. 2001. Characterization of different starches oxidized by hypochlorite. *Starch/Stärke.* 53:211–218.

Lajolo, F. M. 2001. *Fibra dietética en iberoamérica: tecnologia y salud (obtención, caracterización, efecto fisiológico y aplicación en alimentos)*. São Paulo: Varela.

Larotonda, F.D.S., Matsui, K.N., Soldi, V., and Laurindo, J.B. 2004. Biodegradable films made from raw and acetylated cassava starch. *Brazil. Arch. Biol. Technol.* 3:477–484.

Lawala, O.S., Adebowaleb, K.O., Ogunsanwoa, B.M., Barbac, L.L., and Ilod, N.S. 2005. Oxidized and acid thinned starch derivatives of hybrid maize: Functional characteristics, wide-angle X-ray diffractometry and thermal properties. *Int. J. Biol. Macromol.* 35:71–79.

Leão, A.L., Ferrão, P.C., Teixeira, R., and Sator, S. 2007. Biopolymers applications on automotive industry. In: *Biopolymers Technology*, A.C. Bertolini (Ed.). São Paulo: Cultura Academica, pp.165–195.

Lee, H. and Shin, M. 2006. Comparison of the properties of wheat flours supplemented with various dietary fibers. *Food Sci. Biotechnol.* 5:746–751.

Livesey, G. 2003. Health potential of polyols as sugar replacers, with emphasis on low glycaemic properties. *Nutrition Res. Rev.* 2:163–191.

LMC. 2008. LMC International's Global Markets for Starch and Fermentation Products Database 2008 report. http://www.lmc.co.uk/Report.aspx?id=6&repID=2983&flag=1537

Loret, C., Meunier, V., Frith, W.J., and Fryer, P.J. 2004. Rheological characterization of the gelation behaviour of maltodextrin aqueous solution. *Carbohydrate Polym.* 57(2):153–163.

Miyazaki, M., Van Hung, P., Maeda, T., and Morita, N. 2006. Recent advances in application of modified starches for breadmaking. *Trends Food Sci. Technol.* 11:591–599.

Pagella, C., Spigno, G., and de Faveri, D.M. 2002. Characterization of starch based edible coatings. *Food Bioprod. Process.* 80(3):193–198.

Patil, S.K. 2009. Global Modified Starch Products Derivatives and Markets A Strategic Review. February 1, 66 pp. - Pub ID: DBAQ2093916http://www.marketresearch.com/product/display.asp?productid=2093916&SID=49302673-446513523-409800162&kw=starch

Razavi, S.M.A., Najafi, M.B.H., and Alaee, Z. 2007. The time independent rheological properties of low fat sesame paste/date syrup blends as a function of fat substitutes and temperature. *Food Hydrocolloids.* 2:198–202.

Rindlav-Westling, Å., Stading, M., Hermansson, A.-M., and Gatenholm, P. 1998. Structure, mechanical and barrier properties of amylose and amylopectin films. *Carbohydrate Polym.* **36**:217–224.

Röper, H. and Elvers, B. 2008. Starch. 3. Economic Aspects. In: *Ullman's Encyclopedia of Industrial Chemistry* (7th ed.). New York: John Wiley & Sons, pp. 21–22.

Rosa, D.S., Guedes, C.G.F., and Pedroso, A.G. 2004. Gelatinized and nongelatinized corn starch/poly(ε-caprolactone) blends: Characterization by rheological, mechanical and morphological properties. *Polímeros: Ciência e Tecnologia.* 14:181–186.

Rutenberg, M.W. and Solarek, D. 1984. Starch derivatives: Production and uses. In *Starch Chemistry and Technology*, R.L. Whistler, J.N. BeMiller, and E.F. Paschall (Eds.). San Diego: Academic Press, pp. 311–388.

Scott, A. 2002. Nanoscale starch challenges silica in fuel-efficient 'green' tires. *Chem. Week.* 164:15–26.

Sriroth, K., Piyachomkwan, K., Wanlapatit, S., and Oates, C. 2000. Cassava starch technology: The Thai experience. *Starch/Stärke.* 52:439–449.

Stevens, M.P. 1999. *Polymer Chemistry: An Introduction.* 3rd ed., New York: Oxford University Press.

Sudha, M.L., Srivastava, A.K., Vetrimani, R., and Leelavathi, K. 2007. Fat replacement in soft dough biscuits: Its implications on dough rheology and biscuit quality. *J. Food Eng.* 3:922–930.

Tang, H., Mitsunaga, T., and Kawamura, Y. 2006. Molecular arrangement in blocklets and starch granule architecture, *Carbohydrate Polym.* 63:555–560.

Teixeira, M.A.V. 2002. *Amidos quimicamente modificados empregados na substituição de gordura em alimentos,* Food Technology doctoral thesis. Universidade Estadual de Campinas, São Paulo, Brazil.

Thomas, D.J. and Atwell, W.A. 2005. *Starches.* St. Paul, MN: Eagan Press.

Trubiano, P.C. 1995. The role of specialty food starches in flavor encapsulation. *Flavor Technol. (ACS Symposium Series)*, pp. 244–253.

Van der Maarel, M.J.E.C., Van der Veen, B., Uitdehaag, J.C.M, Leemhuis, H., and Dijkhuizen, L. 2002. Properties and applications of starch-converting enzymes of the α-amylase family. *J. Biotechnol.* 94:137–155.

Waigh, T.A., Hopkinson, I.M., Donald, A.M., Butler, M.F., Heideçbach, F., and Riekel, C. 1997. Analysis of the native structure of starch granules with X-ray microfocus. Limited diffraction. *Macromolecules.* 30:3813–3820.

Wanous, M.P. 2004. Texturizing and stabilizing, by gum. *Prepared Foods.* 173(1):108–118.

Whistler, L., BeMiller J.M., and Paschall, E.P. (Eds.). 1984. *Starch: Chemistry and Technology* (2nd ed.). New York: Academic Press, pp. 313–314.

Woo, K.S. and Seib, P.A. 2002. Cross-linked resistant starch: Preparation and properties. *Cereal Chem.* 6:819–825.

Wurzburg, O.B. (Ed.) 1986. *Modified Starches: Properties and Uses.* Boca Raton, FL: CRC Press.

Yackel, W.C. and Cox, C. 1992. Application of starch-based fat replacers. *Food Technol.* 46(6):146–148.

Zobel, H.F. 1988. Molecules to granules: A comprehensive starch review. *Starch/ Stärke.* 40:44–50.

Special thanks

We are grateful to Simon Bentley, head of Grains and Starch Research, LMC International, for providing some of the data discussed in this chapter.

chapter two

Characterization of starch granules

An atomic force microscopy approach

Renata Antoun Simão
PEMM/COPPE, Universidade Federal do Rio de Janeiro

Beatriz Rosana Cordenunsi
Faculdade de Ciências Farmacêuticas, Universidade de São Paulo

Contents

2.1   Introduction

It is well established that starch granules are composed of two structurally different polysaccharides of α-D-glucose: amylose and amylopectin. In spite of amylose double helices, the semicrystalline nature of starches is determined by the amylopectin component. The linear amylopectin unit chains (~18–25 units) are relatively short when compared to amylose chains but both can form sixfold left-handed double helices, which in turn associate to form crystalline and semi-crystalline lamellae (Tang, Mitsunaga, and Kawamura, 2006), resulting in a "blocklet model." The blocklet theory, proposed by Nägeli in 1858 (Kossmann and Lloyd, 2000), was rejected in the 1960s and 1970s and reviewed in 1997, after starch granules were imaged by atomic force microscopy (AFM) (Gallant, Bouchet, and Baldwin, 1997). It was observed that repeating units of crystalline and amorphous lamellae are grouped into discrete and elongated structures. It was also observed that, inside on the grains, the polysaccharides are arranged in concentric rings, called growth rings, irradiated from a central hilum. The number and size of these rings are dependent on the botanic origin of the starch and the crop age at harvest time (Tester, Karkalas, and Qi, 2004, Lindeboon, Chang, and Tyler, 2004). A comprehensive review by Tester, Karkalas, and Qi, (2004) integrates aspects of starch composition, interactions, architecture, and functionality. More starch structure details are discussed in Chapter 3.

The internal structure of starch grains has been widely studied by microscopy. Knowledge about shape, size, and growth rings typical of starches from several botanic sources has been acquired by optical and electron microscopy. However, information and inferences about the blocklet and granular structure were obtained in past years by AFM. In spite of the many publications about microscopic observation of the starch granule, there are controversies about the most suitable microscopic or starch granule purification method. Even the better microscopic systems can provide unsatisfactory or insignificant images if the samples are not obtained or treated with enough care. Electronic microscopy techniques require sample insertion into a vacuum environment or coating, which can lead to granule damage due to water loss or surface tension and, consequently, to the observation of artifacts. Techniques for granule setting on the sample holder can lead to artifacts and, consequently, provide results that can be wrongly interpreted. Scanning probe microscopy (SPM) techniques, among which the most employed is atomic force microscopy (AFM), can also lead to artifacts due to the multiple interactions between the sample and the tip used as the analysis probe. However, the AFM has been an important tool in understanding starch structure. In this chapter, we discuss some of the main contributions of microscopic techniques to starch structure elucidation.

2.2   Microscopy as a tool for starch structure elucidation

2.2.1   Optical and electron microscopy

Even if optical microscopy has a characteristic low resolution, important information can be obtained about starch granules regarding their structural integrity, shape, and size. These characteristics are essential for starch characterization according to botanic origin. In spite of its lower resolving power as compared to electronic microscopy, optical microscopy has the advantage of the low cost of the equipment and it is useful due to the ease of sample preparation and micrograph acquisition.

Structures such as growth rings have been observed by light or optical microscopy in large starch granules from potato, banana, and wheat (Gallant, Bouchet, and Pérez, 1992; Cordenunsi, 2004). In contrast to the large granules, it is not possible to observe these rings on smaller granules such as those from rice. With the development of confocal scanning laser microscopy (CLSM) obtaining 3D high-resolution images at different depths, the use of optical microscopy allowed research advances in starch structure and starch behavior during gelatinization (van de Velde, van Riel, and Tromp, 2002). For example, observing potato starch granules in light microscopy, environmental scanning electron microscopy (ESEM), and confocal laser scanning microscopy (CLSM) makes it possible to conclude that growth ring structures in normal or high amylose starches are sharper than those observed in low amylose starches (Blennow et al., 2003).

Scanning electron microscopy (SEM) is a technique with resolving power at least 10 times greater than light microscopy, whereas transmission electron microscopy (TEM) is characterized by a resolving power 1000 times greater. The first elucidative reports on starch granule architecture were TEM and SEM images of starch granules hydrolyzed by acid and enzymes, showing hydrolysis of amorphous and less crystalline regions of the granule (Gallant, Bouchet, and Pérez, 1992).

Preparation of samples is very important in obtaining SEM images. For example, biological samples require a metallic coating, such as gold, to reduce the deposited charge on the sample surface. Techniques for granule setting on the sample holder can lead to artifacts during the cutting or fixing process, resulting in the misinterpretation of images. The use of water-soluble resin, which allows ultrafine cutting without sample deformation, has improved sample preparation conditions. Furthermore, the hydrophilic character of starch granules seems to preserve the granule water content (Helbert and Chanzy, 1996).

Special attention should be paid in order to avoid sample damaging. A high vacuum requirement causes sample dehydration, water loss, and changes in the granular crystallinity. Recently, an environmental scanning electron microscope (ESEM) was developed with a sample chamber kept at constant pressure of 10–20 Pa during sample analysis; high vacuum is no longer required. Surface charges can be compensated by charge exchange with the environmental atmosphere and, in some cases, high-resolution images of insulating samples can be obtained with no surface coating. Sample damage can be reduced by lower accelerating potentials for the primary electron beam, and low-voltage scanning electron microscopes (LVSEM) with optimized conditions such as special electron sources, lenses, vacuum systems, and beam magnetic shielding (Pawley, 1984).

In general, starch structure and its appearance are dramatically changed by TEM or SEM preparation techniques. Nowadays, due to its high resolution power and the possibility of images without complex techniques for fixing granules, AFM is widely used to observe the surface and internal structures of starch granules.

2.2.2 Atomic force microscopy

Atomic force microscopy (AFM) is a surface imaging technique that allows high-resolution images of biological samples without environmental control. This advantage and fast technical developments regarding its microscopic bases, both in hardware and operational condition reliability, allowed this technique to become popular for observation of soft samples. AFM can be operated on starch samples by contact as well as dynamic modes. In contact atomic force microscopy, the tip is kept at a distance from the surface and the net force between the sample and tip is repulsive. Attractive forces are ideal for probing biological samples because forces between atoms within such samples are small and bonds can be broken by shear forces between the sample and tip. This way, most of the different AFM microscopes are also operated in dynamic modes where tip and sample forces change periodically with time. Several modes of operation vibrating the tip close to the surface are called dynamic force microscopy (DFM) in this chapter. Dynamic force microscopy is characterized by vibrating the tip perpendicularly on the sample surface at a frequency close to the resonance frequency of the cantilever. In DFM, shear forces are minimized. Also, in one of the most used DFM modes, the cantilever is vibrated over the sample keeping the oscillation amplitude constant and the signal phase lag can be monitored, providing valuable information about the viscoelastic properties of the sample surface. As consequence, this technique can be widely used in identification of surfaces on polymeric materials. While operating in contact mode, constant distance between probe and surface can also be modulated by adding a small periodic signal to the z-piezo and the image of the force variation can provide information about local surface stiffness.

However, in addition to results about surface topography, AFM also provides information about viscoelasticity and the distribution of known phases of the sample surface. By evaluating phase contrast or force modulation images it is possible to infer the presence and location of proteins in a starch granule, as well as the location, size, and shape of blocklets.

2.2.2.1 AFM sample preparation

Although AFM does not require special sample preparation as already mentioned, some care must be taken concerning the granules' surface.

Starch granules must be left intact and well bound to a fixed surface in order to avoid sample movements while scanning. Both the granule surface and the internal structures of the granules have already been observed by AFM and several techniques for sample fixing had been employed. Spreading starch granules over an adhesive tape fixed onto an AFM sample holder (Szymonska and Krok, 2003; Juszczak, Fortuna, and Krok, 2003a,b) and film of low-viscosity cyanoacrylate polymer (Baldwin et al., 1998) have been employed to fix starch granules. In order to avoid granule movement during scanning, starch samples have also been spread over freshly cleaved mica (Ayoub, Ohtani, and Sugiyama, 2006).

Desiccation of samples can be avoided if they are deposited in substrates with a high affinity to the starch granule. Plasma treatment allows one to change and tune the surface hydrophilicity by attaching specific functional groups to it. Surfaces prepared by this technique lead to the attachment of any granular sample on solid surfaces by van der Waals forces. The main advantage is no adhesive tape is used and the AFM tip can go up and down the granule without risk of tip contamination. Oxygen plasma-modified silicon wafer surfaces were also used as a substrate to fix starch granules from mango seeds previously dispersed in water (Simão et al., 2008).

Preparation of samples by encasing and molding starch granules with resins followed by a section of the sample 1.5 μm thick has been reported (Baker, Miles, and Helbert, 2001; Ridout et al., 2003, 2004, 2006). However, some resins employed in an iodine-colored sample showed a brown color, not blue, indicating that these resins promote partial degradation of amylose. Consequently, the choice of fixing resin must be made carefully in order to avoid its absorption by the starch granule. Encased air-dried starch in rapid-set resins apparently did not promote amylose degradation, preserving the granule structure from the central hilum to growth rings and leading to the possibility of clearly observing starch blocklets (Ridout et al., 2004, 2006).

AFM was also used to observe gelatinized starch films as well as the conformation of isolated amylose and amylopectin. In those experiments, films were obtained by gelatinized starch poured into Petri dishes and dried at 50°C (Rindlav-Westling and Gatenholm, 2003; Thiré, Simão, and Andrade, 2003). AFM images of these films can be obtained simply by fixing them to a rigid substrate with the aid of double-sided tape. Bundles are prepared by filtration of gelatinized starch and their deposition is done on cleaved mica, allowing the use of molecular combing to stretch starch nanounit chains in mica (Zhongdong, Peng, and Kennedy, 2005).

2.3   The starch granule structure: A microscopic view

One of the first published AFM images of starch granules from potato and wheat was published by Baldwin et al. (1998). Comparing the AFM images with those obtained by LVSEM, they observed nodules at the potato granule surface with size distribution ranging from 50 to 200 nm. Similar structures were also observed by SEM and AFM (Gallant, Bouchet, and Baldwin, 1997) and it was suggested that these structures could be related to blocklets constituted by the crystalline amylopectin phase. These blocklets were approximately 25 nm in size in the soft shells and 80–120 nm in the hard shells in wheat starch, with much larger ones in potato starch (200–500 nm) (Gallant, Bouchet, and Baldwin, 1997). As the samples were treated with α-amylase and prepared with colloidal gold coating and vacuum exposure, these results raised conflicting discussions regarding their reliability due to the probable starch degradation. Multiple freezing and thawing of potato starch granules followed by high-resolution noncontact AFM imaging showed internal exposition and lamellar structure of starch granules submitted to high temperatures (Szymonska and Krok, 2003). Based on these observations, two classes of lamella-forming particles were described: one smaller than 20 nm and other 40–50 nm, which they attributed to single or double amylopectin superhelical clusters bundled together into larger blocklets. Finally Ridout et al. (2003, 2004) reinforced the blocklet model by AFM imaging, observing the internal structure of pea starch granules encased in a nonpenetrating matrix of synthetic polymer.

AFM has been also used to study the effect of amylose contents on starch structure. Some researchers suggest that contrast of the images arises due to selective hydration of different granular regions and softer or brighter regions occur the higher the amylose content of the blocklet composition, indicating that hard or dark bands contain no or reduced amounts of amylose (Ridout et al., 2003, 2004). Starch from peas with low amylose content had no change in the crystallinity levels, but it was not possible to visualize the blocklets inside these granules (Ridout el al., 2004, 2006). To obtain reliable AFM images it is necessary to prepare samples with relatively flat surfaces, which can be achieved by the use of sectioned or embedded granules.

AFM was also employed to observe the growth rings in mango starch granules during ripening (Simão et al., 2008). Growth rings can also be clearly observed in banana starch granules as shown both in topographic as well as in phase contrast images presented in Figures 2.1a and b. A clear relationship between both images is shown, but the same plateau can present different phase contrasts, as observed in the topographic image. This is clearly shown in Figure 2.1c, where topography and phase contrast line profiles are compared.

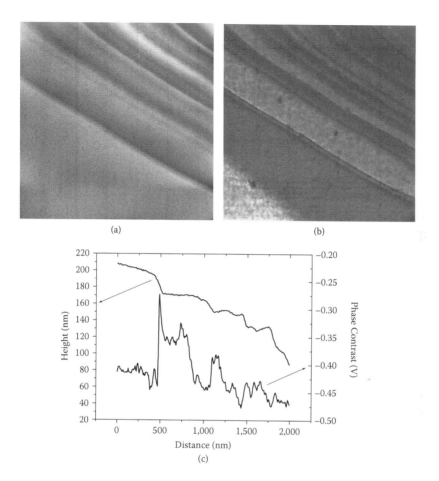

Figure 2.1 Atomic force microscopy (AFM) image of the growth rings in a banana starch granule. The same region is presented as a topographic image (a) and as a phase contrast image (b). Both images are 2.5 × 2.5 μm². Line profile along the diagonal of the image from the lower left corner (c).

Disorder on rice starch granule surfaces promoted by cycles of freezing and thawing was observed by DFM–AFM, showing native granules presenting protrusions of 70 nm in diameter (Ayoub, Ohtani, and Sugiyama, 2006). These results are similar to those previously obtained (Gallant, Bouchet, and Baldwin, 1997; Baldwin et al., 1998) in which protrusions of 10–50 nm were found on the surface of wheat and protrusions 200–500 nm were observed on the surface of potato starch. Depressions, hills, and pores were observed in all granular surfaces of potato, cassava, barley, corn, "waxy" corn, and wheat starch by noncontact AFM (Juszczak, Fortuna, and Krok, 2003a,b).

Observation of the internal structure of the rice starch granule submitted to a plasticizing/lyophilization cycle shows important changes in surface morphology (Ayoub, Ohtani, and Sugiyama, 2006). After two cycles of plasticizing/lyophilization using glycerol, a loosely packed array of 35 nm protrusions was observed, indicating that the process induced the disruption of the crystalline structure of the native starch grains.

Observation by TEM shows that, in early stages of retrogradation, amylose is in a coiled or partly organized conformation. Helices interact forming semicrystalline aggregates separated by amorphous regions. However, amylopectin forms different structures from the ones observed in amylose samples. No significant difference is observed in structure morphology during the early stages of retrogradation, but afterwards the extended chain network forms a fractal-like network joined by an α(1–6) link (Putaux, Buléon, and Chanzy, 2000).

2.4   Starch gelatinization observed by atomic force microscopy

It is well established that starch granules are insoluble in cold water but can undergo significant and irreversible changes upon wet heating. The study of the physical properties and rheological behavior of starch gelatinization is essential for understanding starch structure and supporting its application.

Images from gelatinized starch by DFM–AFM can show film morphology, displaying the differences between partial and complete starch gelatinization (Thiré, Simão, and Andrade, 2003). In this case, a vibrating amplitude higher than 100 nm is employed to ensure that the tip would get into and out of the contamination layer, lowering the shear forces during scanning. Figure 2.2a shows the film morphology from partially gelatinized starch, displaying swollen granules in a continuous matrix. In addition to imaging, AFM allows a detailed description of the surface features with a tridimensional visualization and clear observation of differences between partially and completely gelatinized corn starch, as observed in Figures 2.2a and b, respectively (Thiré, Simão, and Andrade, 2003). The presence of ghost granules is not observed on films prepared under 90 minutes of gelatinization, indicating that the granular structure was completely disrupted (Figures 2.2b and d).

Protrusions of 15 to 35 nm on the superficial layer of starch films were identified as proteins of 60 kDa (Rindlav-Westling and Gatenholm, 2003) and it was suggested that they may be the starch synthase enzyme, which would migrate to the surface resulting in a protein-rich thin layer over the starch film.

Figure 2.2 (a) AFM topographic image of a partially gelatinized corn starch film; (b) AFM topographic image of a fully gelatinized corn starch film; (c) detail of (a) presenting a tridimensional representation of one granule; (d) detail of (b) presenting a tridimensional representation of the disrupted granule. Images in (a) and (b) are 100×100 μm^2 in area.

Because crystalline areas on the film surface and phase separation between amylopectin and amylose were also observed (Rindlav-Westling and Gatenholm, 2003), it is possible to obtain some hints about the wide range of analytical possibilities of AFM on the characterization of starch-based materials. The structure of amylose and amylopectin aggregation on gelatinized starch from potato and rice was also observed by AFM (Dang, Braet, and Copeland, 2006), indicating that, in potato starch, amylose forms aggregates probably during retrodegradation, whereas amylopectin chains show an open extended form. Different aggregation of amylose was observed on rice starch and amylopectin presented some crystalline structures even after partial gelatinization (Dang, Braet, and Copeland, 2006).

With AFM observation it was also possible to conclude that starch granule swelling and rupture during the gelatinization process occur initially due to molecules flowing out of the granule in a densely packed chain (Peng, Zhongdong, and Kennedy, 2007). When the amylose and amylopectin chains are in contact with hot water, swelling takes place and the structure becomes loosely packed in amorphous areas, suggesting that blocklets are made of twists or distortions of nanochains (Peng, Zhongdong, and Kennedy 2007).

2.5 Final remarks

Although starch is one the most abundant and studied natural polymers, there are still many doubts about its arrangement and granule structure. Starch supramolecular structure defines its functional properties, being a characteristic of botanic origin. Microscopic techniques are important tools in starch characterization and they can also provide valuable information about starch degradation during fruit ripening and acid or enzymatic hydrolysis.

The first research observing starch granules by optical microscopic indicated the growth of concentric circles, which are supposed to be directly related to crystalline and semicrystalline rings. Nowadays a comprehensive model of starch granule architecture has gathered hypotheses and data regarding starch structure and behavior. The blocklet theory of densely packed nanometric ordered structures on the starch granule is still under discussion, in addition to evidence provided by SEM and AFM. Several researchers have employed microscopic techniques to observe blocklet location and size, both inside and on the granule surface. Blocklet distribution along the crystalline and semicrystalline rings was successfully modeled based on AFM images. Although microscopy methods have made important contributions to elucidating granule supramolecular structure, the proposed blocklets' internal structure (Gallant, Bouchet and Baldwin, 1997) could not be observed by any of the applied microscopic techniques. On the other hand, further information about granule arrangement and retrogradation of amylose and amylopectin can be obtained by observing starch gelatinization samples. These studies can lead to a better understanding of the chains' release process during gelatinization and also provide support for research focused on developing starch-based materials.

The main challenge faced by the different microscopy techniques in starch research remains undamaged sample preparation. One of the advantages of AFM is that it does not require special sample preparation and it can be accomplished without a strict analysis environment. Furthermore, AFM can provide details about amylose and amylopectin arrangement, contributing to advances in starch granule structure knowledge.

References

Ayoub, A., Ohtani, T., and Sugiyama, S. 2006. Atomic force microscopy investigation of disorder process on rice starch granule surface. *Starch*. 58:475–479.

Baker, A.A., Miles, M.J., and Helbert, W. 2001. Internal structure of the starch granule revealed by AFM. *Carbohydrate Polym.* 330:249–356.

Baldwin, P.M., Adler, J., Davies, M.C., and Melia, C.D. 1998. High resolution imaging of starch granule surfaces by atomic force microscopy. *J. Cereal Sci.* 27:255–265.

Blennow, A., Hansen, M., Schulz, A., Jorgensen, K., Donald, A.M., and Sanderson, J. 2003. The molecular deposition of transgenically modified starch in the starch granule as imaged by functional microscopy. *J. Struct. Biol.* 143:229–241.

Cordenunsi, B.R. 2004. Aspectos químicos, bioquímicos e ultraestruturais do grânulo de amido durante o amadurecimento da banana, thesis, FCF/USP, São Paulo, Brazil.

Dang, J.M.C., Braet, F., and Copeland, L. 2006. Nanostructure analysis of starch components by atomic force microscopy. *J. Microsc.* 224:181–186.

Gallant, D., Bouchet, B., and Baldwin, P.M. 1997. Microscopy of starch: Evidence of a new level of granule organization. *Carbohydrate Polym.* 32:177–191.

Gallant, D., Bouchet, B., and Pérez, S. 1992. Physical characteristics of starch granules and susceptibility to enzymatic degradation. *Euro. J. Clin. Nutr.* 46:S3–S16.

Helbert, W. and Chanzy, H. 1996. The ultrastructure of starch from ultrathin sectioning in melamine resin. *Starch/Stärke.* 48:185–188.

Juszczack, L., Fortuna, T., and Krok, F. 2003a. Non-contact atomic force microscopy of starch granules surface. Part I. Potato and tapioca starches. *Starch/Stärke.* 55:1–7.

Juszczack, L., Fortuna, T., and Krok, F. 2003b. Non-contact atomic force microscopy of starch granules surface. Part II. Selected cereal starches. *Starch/Stärke.* 55:8–12.

Kossmann, J. and Lloyd, J. 2000. Understanding and influencing starch biochemistry. *Crit. Rev. Plant Sci.* 19:171–226.

Lindeboon, N., Chang, P.R., and Tyler, R.T. 2004. Analytical, biochemical and physicochemical aspects of starch granule size, with emphasis on small granule starches: A review. *Starch/Stärke.* 56:89–99.

Pawley, J. 1984. Low voltage scanning electron microscopy. *J. Microscopy* 136:45–68.

Peng, L., Zhongdong, L., and Kennedy, J.F. 2007. The study of starch nano-unit chains in the gelatinization process. *Carbohydrate Polym.* 68:360–366.

Putaux, J.-L., Buléon, A., and Chanzy, H. 2000. Network formation in dilute Amylose and Amylopectin studied by TEM. *Macromolecules.* 33:6416–6422.

Ridout, M.J., Parker, M.L., Hedley, C.L., Bogracheva, T.Y., and Morris, V. 2003. Atomic force microscopy of pea starch granules: Granule architecture of wild-type parent, *r* and *rb* single mutants, and the *rrb* double mutant. *Carbohydrate Res.* 338:2135–2147.

Ridout, M.J., Parker, M.L., Hedley, C.L., Bogracheva, T.Y., and Morris, V. 2004. Atomic force microscopy of pea starch: Origins of image contrast. *Biomacromolecules.* 5:1519–1527.

Ridout, M.J., Parker, M.L., Hedley, C.L., Bogracheva, T.Y., and Morris, V. 2006. Atomic force microscopy of pea starch: Granule architecture of the *rug3-a, rug4-b, rug5-a* and *lam-c* mutants. *Carbohydrate Res.* 65:64–74.

Rindlav-Westling, A. and Gatenholm, P. 2003. Surface composition and morphology of starch, amylose, and amylopectin films. *Biomacromolecules.* 4:166–172.

Simão, R.A., Fioravante, A.P., Peroni, F.H.G., Nascimento, J.R.O., Louro, R.P., Lajolo, F.M., and Cordenunsi, B.R. 2008. Mango starch degradation I: A microscopical view of the granule during ripening. *J. Agric. Food Chem.* 56:7410–7415.

Szymonska, J. and Krok, F. 2003. Potato starch granule nanostructure studied by high resolution non-contact AFM. *Int. J. Biol. Macromol.* 33:1–7.

Tang, H., Mitsunaga, T., and Kawamura, Y. 2006. Molecular arrangement in blocklets and starch granule architecture. *Carbohydrate Polym.* 63:555–560.

Tester, R.F., Karkalas, J., and Qi, X. 2004. Starch: Composition, fine structure and architecture. *J. Cer. Sci.* 39:151–154.

Thiré, R.M.S.M., Simão, R.A., and Andrade, C.T. 2003. High resolution imaging of the microstructure of maize starch films. *Carbohydrate Polym.* 54:149–158.

van de Velde, F., van Riel, J., and Tromp, R.H. 2002. Visualization of starch granule morphologies using confocal scanning laser microscopy (CSLM). *J. Sci. Food Agric.* 82:1528–1536.

Zhongdong, L., Peng, L., and Kennedy, J.F. 2005. The technology of molecular manipulation and modification assisted by microwaves as applied to starch granules. *Carbohydrate Polym.* 61:374–378.

chapter three

Starch macromolecular structure

Luis Arturo Bello-Perez, Sandra Leticia Rodriguez-Ambriz,
Mirna Maria Sanchez-Rivera, and Edith Agama-Acevedo
Centro de Desarrollo de Productos Bióticos del IPN

Contents

3.1 Introduction

3.1.1 Starch structure

Starch granular organization as well as amylose and amylopectin structure depend on the botanical source. Amylose, the linear D-glucose chain, has on average between 500 and 6,000 glucose units that are distributed among 1 to 20 chains. Each chain has shown an average degree of polymerization (DP) of 500. Some spaced branching points were detected in amylose, however, it presents lineal polymer characteristics. The properties of amylose might be explained as its diverse molecular conformations. For example, in neutral solutions the conformation is a random coil.

However, other conformations were reported such as an interrupted helix or deformed helix. Amylopectin, the branching polymer of starch, contains short (DP = 20–25) chains linked to the α–D–(1–6) linkages on the main chain. Studies of the structure of amylopectin showed substantial progress in 1937, with the Haworth hypothesis about the laminated structure of starch (Hood, 1982). In the same year, Staundiger and Husemann suggested the amylopectin structure would be a main chain branched with all branched joined chains. In 1940, based on chemical analysis determining the reduction sugars, Meyer and Bernfeld proposed a model based on random branching and, in 1949, Myrback and Sillen mentioned that the models cited above had different arrangements of the same chains (Hood, 1982). In 1952, Peat proposed the terminology of A-, B-, and C- to designate different chains on the amylopectin structure (Hood, 1982), where the A-chain is joined to the molecule by a reduction end, the B-chain bonds to the A-chain but is also joined to other A- or B-chains at one or more primary hydroxyl groups, and the C-chain is not substituted in the reduction end, being unique to the amylopectin molecule (Greenwood, 1976). This nomenclature was proposed in order to facilitate the comparison of amylopectin models when the chain length is unknown. The A-, B-, and C- terminology led to some mistakes in interpretation of studies about amylopectin structure, therefore using the degree of polymerization (DP) of the chains to describe amylopectin structure was suggested.

Important advances in studies of amylopectin were done by debranching its structure with β-amylase and pullulanase. Robin et al. (1974) proposed the cluster model of amylopectin structure based on the model previously suggested by Nikuni and French (Nikuni, 1978). In this cluster model, the A- and B-chains are presented as linear structures with DP of 15 and 45, respectively. The B-chain is the backbone of amylopectin and extends over two or more clusters, constituted by 2 and 4 A-chains. Those clusters of associated A-chains are responsible for the crystalline regions of starch granules. The intercrystalline areas of the starch granule, also called amorphous areas, are present in intervals of 0.6–0.7 nm and contain the highest amount of α–D–(1–6) linkages, being very susceptible to hydrolysis by acids and enzymes such as pullulanase, isoamylase, and amyloglucosidase. In general, the amylopectin molecule is 1.0–1.5 nm in diameter and 12–40 nm in length.

3.1.2   *Starch components in starch granules*

There is some controversy as to how amylose and amylopectin are packed in starch granules. Biliaderis (1992) proposed the organization of starch components with lipids in the starch granule, whereas amylopectin, and perhaps amylose, are radially oriented toward the granular surface. In this model, the crystallinity of starch is attributed to linear short chains presented in the amylopectin molecules (DP 14–20) and arrangement of these chains is reponsible for the three-dimensional crystalline structure. The three-dimensional structural features of native starch granules can be shown in x-ray diffraction patterns, which are classified as A- or B-type (Imberty et al., 1987; Imberty and Perez, 1988). In the crystalline A starch model, the structure is based on parallel-stranded double helices in the unit cell, and it produces a monoclinic crystal. The model of crystalline B starch is produced by a hexagonal arrangement of the double helices (Imberty and Perez, 1988). Although the geometry of double helices is identical in the A- and B-forms, the two structures differ in the water level and crystalline arrangements (Wu and Sarko, 1978a,b).

It was suggested that the amorphous zone of granular starch is also heterogeneous, consisting of amylose and intercrystalline zones of dense branching in amylopectin. Such a complex morphology plays an important role in thermal characteristics and the plasticization pattern of starch by water, which are dependent on the amorphous phase (Biliaderis, 1992). Studies using colloidal gold-labeled concanavalin A and based on specificity of lectin for the nonreducing endgroups of amylopectin, resulted in positive labeling. However, the amylopectin localization has not yet been proven; a relation was found between the intensity of labeling and abundance of ramified molecules (Gallant et al., 1992).

Some evidence points to *in situ* monoacyl lipids on the starch granule arrangement resulting in complexes with chains (Galliard and Bowler, 1987), which are represented by a V-type x-ray diffraction pattern and small peak at $2\theta = 20°$. However, the amylose–lipid complex formation is not clear. It has been suggested that, when the amylose chain is in the presence of a monoacyl lipid, the amylose helix is formed and the monoacyl lipid is included in this cavity. Another hypothesis is that the double helix exists naturally and then the monoacyl lipid is deposited in the central cavity (Morrison, 1988; Tester, Karkalas, and Qi, 2004).

In the starch structure model proposed by Jane et al. (1992), the most probable location of amylose is in the randomly interspersed radial chains with an increasing level of amylose toward the granule exterior. This model does not consider lipids and proteins, because their exact location and interaction with amylopectin is still not certain. Additionally, it was hypothesized that amylose may be predominantly located in the amorphous zones of the granule and the A-crystalline lattice in the starch structure formed by amylopectin (Gallant, Bouchet, and Baldwin, 1997).

In native starch, amylopectin is considered as predominantly responsible for granule crystallinity, ranging from 15–45% according to botanic origin. Crystallinity is also dependent on the amylopectin chain length and chain ramification (Zobel and Stephen, 1996). The crystalline regions are formed by layers of 120–400 nm thickness, composed of crystalline and semi-crystalline lamellae, which form the granule packing (French, 1984;

Donald et al., 1997). The crystalline lamellae are believed to be ordered in double-helical amylopectin side-chain clusters and are interleaved (alternate) with more amorphous lamellae consisting of the amylopectin branching regions. It is assumed that the crystalline and amorphous lamellae of the amylopectin are organized into larger, more or less spherical structures, which have been termed "blocklets." The blocklet diameters range from 20 to 500 nm, depending on the starch botanical source and location in the granule (Gallant, Bouchet, and Baldwin, 1997) (Figure 3.1).

Understanding starch structure is considered essential to obtain modified starches with target functional properties. The development of new methods of starch characterization based on molar mass, gyration and hydrodynamic radius, branching degree, and chain length distribution can contribute to the knowledge of its structure and, consequently, to its physicochemical and functional properties.

3.2 Starch structure characterization techniques

3.2.1 Solubilization

Solubilization under moderate conditions, always making sure to keep the structure intact, is the first condition for studying starch structure. Starch is soluble in excess water under heating conditions whereas, depending on the temperature and the time of heating, its chains can be depolymerized. For this reason, it is necessary to find a solubilization method that does not damage the starch structure. Methods for the study of starch structure using several solubilization procedures are summarized in Table 3.1.

In the 1990s most starch characterization methods involved a heating step at high temperature, using a boiling water bath, autoclave, steam-jet cooker, extrusion, or microwave oven. Variations were introduced to increase the solubilization level such as DMSO addition, dispersion in KOH or NaOH, a blend of DMSO–LiBr, and ultrasonication. However, in the majority of such research, the amount of soluble starch was not quantified.

Actually, the major part of solubilization procedures involves pretreatment of samples with DMSO and DMSO–ethanol, after the starch dispersion is heated in a microwave oven, autoclave, boiling water bath, or jet cooker, and, in some works, an additional sonication.

Starch recovery after solubilization is also an essential parameter to be considered in sample preparation for starch characterization. Of course, the longer the heating time or heating potency is, the higher the starch solubilization (Fishman and Hoagland, 1994), but the risk of sample depolymerization increases with high temperatures. A starch solubilization range between 48.9 and 71.2% can be obtained using microwave heating for 90 seconds (Fishman and Hoagland, 1994), whereas a starch solubilization range between 59 and 81% can be obtained using microwave heating

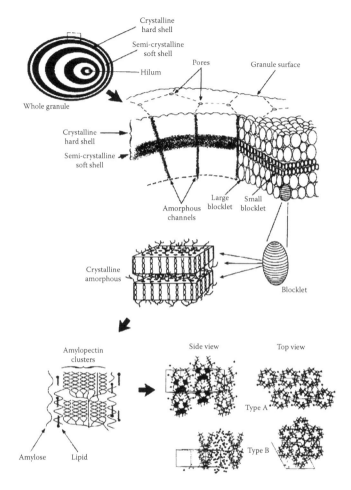

Figure 3.1 Overview of starch granule structure. At the lowest level of granule organization (upper left), the alternating crystalline (hard) and semi-crystalline (soft) shells are shown (dark and light, respectively). The shells are thinner toward the granule exterior (due to increasing surface area to be added by constant growth rate) and the hilum is shown off-center. At a higher level of structure the blocklet structure is shown, in association with amorphous radial channels. Blocklet size is smaller in the semi-crystalline shells than in the crystalline shells. At the next highest level of the structure one blocklet is shown containing several amorphous crystalline lamellae. In the next diagram the starch amylopectin polymer in the lamellae is shown. The next image (from Blanshard, 1987) reminds us that amylose–lipid (and protein) feature in the organization of the amylopectin chains. At the highest level of order, the crystal structures of the starch polymers are shown (modified by Gallant et al., 1992; from Imberty et al., 1987 and Imberty and Perez, 1988).

Table 3.1 Macromolecular Characteristics of Starches

Sample	Solubility method	Solubility (%)	Mw (g/mol)	R_G (nm)	Technique	Reference
Maize, pea, potato (starch)	Autoclaving		$16.7\times88.0 \times 10^6$	165–235	SEC/MALLS	Aberle et al. (1994)
Maize, pea, potato (amylose)			53.8–112×10^6	170–241		
Maize, pea, potato (amylopectin)			5.45–9.6×10^6	19.4–60.1		
Cassava, wheat, maize, potato and pea (amylose)	KOH		340–1050×10^3	44.5–89.9	DLS	Roger et al. (1996)
Potato and waxy corn (starch)	Autoclaving different times.		37.5–360×10^6	121–320	sdFFF/MALLS	Hanselmann et al. (1996)
Corn, 5 and 7 eurolyn (amylopectin)	Microwave 35 s		7.9×10^6–2.7×10^8	110–259	SLS	Bello-Perez et al. (1998a)
Maize, wheat, potato (starch)	Microwave different times	87–100	0.45×10^{-7}–2.2×10^{-7}	96–251	HPSEC-MALLS	Bello-Perez et al. (1998b)
Amaranth (starch)	DMSO-microwave different times	83.7–100	0.76×10^{-7} 69×10^7	60–334	HPSEC-MALLS Light scattering	Bello-Perez et al. (1998c)
Soft and hard wheat cultivars (amylopectin)	NaOH 70°C, neutralized with HCl		29–349×10^6	95–340	HPSEC-MALLS-RI	You et al. (1999)
Amylose fractions synthesized	KOH-water		1.1×10^2–1.2×10^6	9.0–63.2	SEC-MALLS	Roger et al. (2000)

Sample	Solvent/treatment		Molecular weight	DP	Method/system	Reference
Wheat (amylopectin)	DMSO		$6.1\text{–}6.97 \times 10^{10}$	300–324	GPC (Sepharose CL-2B gel) HPSEC-MALLS-RIHPAE-ENZ-PAD	Franco et al. (2002)
Bean, rice, canna, barley, wheat, banana, potato (amylopectin)	DMSO		$1.3\text{–}56.8 \times 10^{8}$	201–782	HPSEC-MALLS system	Yoo and Jane (2002a)
Yam tubers, maize, wheat, potato, cassava (starch)	DMSO-microwave 38 s	72.0–98.0	$0.75\text{–}3.41 \times 10^{8}$	195–396	HPSEC-MALLS-RI system	Rolland-Sabate et al. (2003)
Waxy, normal, high amylose corn (amyloses) Waxy, normal, high amylose corn (amylopectin)	DMSO, boiling 1 h, and stirring 8 h	78.5–100	$1.4\text{–}3.13 \times 10^{6}$ $197\text{–}254 \times 10^{6}$	83–164 214–247	SEC-MALL-RI system	Han and Lim (2004a)
Corn (amylose) Corn (amylopectin)	NaOH	78.5–96.7	$2.0\text{–}2.9 \times 10^{6}$ $46\text{–}126 \times 10^{6}$	81.6–150 211	SEC-MALL-RI system	Han and Lim (2004b)
Corn (amylose) Corn (amylopectin)	Water bath 1 h		1.5–3.1 197–254	83–164 214–247	SEC-MALL-RI system	Han et al. (2005)
Apple (amylopectin)			$4.63\text{–}11.10 \times 10^{8}$	406–435	SEC-MALL-RI system	Stevenson et al. (2006)
Jicama from Texas and Mexico (amylopectin)			$2.44\text{–}6.23 \times 10^{8}$	295–420	HPSEC-MALLS-RI	Stevenson et al. (2007)

Notes: SEC: Size exclusion chromatography; MALLS: multiangle laser light-scattering; RI: refractive index; DLS: dynamic light scattering; SLS: static light scattering; GPC: gel permeation chromatography; HPSEC: high-performance size exclusion chromatography; SdFFF: sedimentation field flow fractionation.

for 80 seconds (Fishman et al., 1996). Microwave heating for 35 seconds at 900 W promoted a starch solubilization range between 87 and 96%, without apparent structure degradation (Bello-Perez et al., 1998a). With slight modifications, this method has been used in research on starch structure from several botanical sources (Tetchi et al., 2007; Rolland-Sabate et al., 2007), where the solubilization recovery values ranged between 96.7 and 100%.

Some works suggested solubilization increasing with a rise in amylose content (Fishman and Hoagland, 1994), probably due to the difficulty of amylopectin solubilization, which often results in gel remaining on the bottom of the centrifuge tube. If the solubilization is affected by the amylose contents, it seems clear that the solubilization rate changes according to the botanical source of the starch. For waxy and normal corn starches a solubilization rate between 97.5 and 100% was reported (Han, Lim, and Lim, 2005), and for rice starches the solubility was between 49.1 and 87.4% (Zhong et al., 2006), according to the solvent. However, most research on starch solubilization does not report the percentage of solubilization obtained from the original sample weighed in the experiment. It is important to note that the solubilization rate is the first step to verify whether the starch samples were completely dissolved and the success of the starch characterization depends on it.

3.2.2 Gel filtration chromatography

Gel filtration or low pressure chromatography was the first technique used for the characterization of starch structure. This technique is based on the size exclusion principle, where the sample is introduced in a long column (1 m or longer; diameter between 1.25–2.54 cm), and its retention time is as short as the particles are large. After elution, sample fractions are collected and total carbohydrates and reducing sugar content are quantified.

Gel filtration of hydrolyzed samples by debranching enzymes was one of the main tools in studies of starch structure. One of the first works with this approach was done by Biliaderis (1982), using elution in a Sepharose CL-2B column of debranching amylopectin from pea and waxy maize starches. It allowed the authors to identify two populations corresponding at the A- and B-chains of the cluster model. Using a similar technique (with Sepharose CL-2B and Bio-gel P–6 columns), it was possible to calculate the molecular weight of amylose from potato and normal and high-amylose corn starches, and amylopectin branch chain length from high-amylose corn, waxy corn, and normal rice starches, and also correlate its influence on paste properties of starch (Jane and Chen, 1992). Improvement of peak resolution by a gel filtration system with a long column (2 m and 1.25 cm diameter) can be obtained on evaluation of chain length distribution and degree of polymerization of debranching amylopectin (Paredes-Lopez, Bello-Perez, and Lopez, 1994).

Reports of chain length distribution of amylopectin are shown in Table 3.1, obtained from different gels, elution rates, solvents, and temperatures. Unfortunately, gel filtration chromatography needs a long time to run the samples (approximately 12–15 hours per sample), careful gel preparation in the column to avoid trouble with fissures and/or bubbles which cause low resolution, and a large number of samples to allow total carbohydrate quantification.

3.2.3   Traditional light scattering (LS)

Light-scattering equipment has been used for the past 30 years in structural and molecular characterization of polymers. This technique was established after Debye (1947) introduced the theoretical concepts, contributing significantly to the elucidation of the macromolecule structure and allowing direct measurement of the molecular weight and size. Information about polymer structure can be obtained from the full curve of the particle scattering factor (Hanselmann et al., 1996; Burchard, 1997; Bello-Perez, Roger, and Paredes-Lopez, 1998) and special attention has been given to the thermodynamic properties of solutions on molar mass and mean square radius of gyration. Light-scattering measurements can be done by static or dynamic light scattering. The information obtained by the light-scattering technique is useful in understanding the polymer structural changes in specific systems and to explain their behavioral functionality, such as in semi-dilute solutions (Galinsky and Burchard, 1996) and their angular dependence on static (Galinsky and Burchard, 1997a) and dynamic (Galinsky and Burchard, 1997b) light scattering. With static light scattering, molar mass and gyration radius (Rg) are calculated by root mean square distance of the scattering elements (as monomeric units or chain segments) from the center of mass. On dynamic light scattering, the hydrodynamic radius (Rh) is obtained from the translational diffusion coefficient (D) using the Stokes–Einstein relationship (Burchard, 1993). Also, by dynamic light scattering measurement, virial coefficient A_2 (which characterizes the interactions between polymers and solvent) and the intrinsic viscosity (η) are based on the ratio of volume to molecular weight (Mw) (Galinsky and Burchard, 1995, 1997b; Han, Lim, and Lim, 2005).

Information about internal structure and flexibility can also be obtained by light scattering (Burchard 1997; Galinsky and Burchard, 1997a,b). An intrinsic feature of the polymer–solvent system, which results from the conformation of polysaccharide in solution, can be determined by the structural ρ-factor (which is the ratio of the two radii; Rg/Rh = ρ) or by the molecular weight ratio of the two mentioned radii (Roger et al., 1996). Several ρ values for different molecular architectures of polymers were reported (Burchard, 1992), with the high ratio ($\rho > 2.0$) the consequence of a rigid rod architecture polymeric, and the low ratio ($\rho = 0.778$) the homogeneous sphere architecture.

The light-scattering technique has been also used to evaluate structural characteristics of depolymerized starches. It allows the comparison of conformational structures considering ABC monomers of polycondensation models, which denote different functional groups on polymer structure (Burchard, 1972). As previously noted, the sample preparation heating treatment has a strong influence on starch structure results. For example, using different heating times in sample preparation, Bello-Perez et al. (1998c) obtained amylopectin from amaranth starch with Mw of 69×10^7 g/mol when heated in a microwave oven for 35 seconds and Mw of 7.6×10^6 g/mol when heated for 90 seconds, with Rg and Rh values between 334 nm (35 seconds) and 114 nm (90 seconds), and between 273 nm (35 seconds) and 130 nm (90 seconds), respectively. Data from Kratky plots, fractal dimensions, and relaxation rate dependence on **q** obtained by static and dynamic light scattering showed significant differences in internal structure and amylose content, according to the heating treatment (Roger, Bello-Perez, and Colonna, 1999).

Comparing dilute aqueous solutions of starches from maize and pea analyzed by static light scattering, it is possible to conclude that starches with high amylose content (amylomaize) present different angular dependences from those with high amylopectin content (waxy). In this case, the molar mass for amylopectin ranged between 60.9 and 112.2×10^6 g/mol with Rg between 170.8 and 241.5 nm and for amylose the molar mass ranged between 2.09 and 19.6×10^6 g/mol) with low Rg (between 19.4 and 60.1 nm), giving evidence of a pronounced aggregation of amylose chains (Aberle et al., 1994).

The Zimm plot does not allow a sufficiently accurate extrapolation to a zero scattering angle ($q^2 = 0$), therefore it has also been proposed to obtain the molar mass and gyration radius of total starch, amylose, and amylopectin using a Berry plot. The differences between the diverse starch structures can be particularly perceptible in the Kratky plots, where the particle-scattering factor (Pq) is plotted as $u^2 Pq$ versus u. In this kind of diagram, the asymptotic region at high u values is accomplished with increasing the molecular weight. Therefore, even a small change in the particle-scattering factor in the u domain, which is related to the internal structure, can be detected (Aberle et al., 1994).

In sample preparation, both static and dynamic light scattering are used to study the stability in water of starch dispersion after solubilization in a microwave oven. In general, the Mw, Rg, and Rh values decrease when storage time increases, probably because of the increase of starch depolymerization during storage. However, samples with a high amylose level (50 and 70% of amylose) stored over 72 hours decrease; this pattern can be due to retrogradation (Bello-Perez et al., 1998a). The fractal dimension and branching level of amylopectin can be estimated by treatment of Rg–Mw data. Kratky plots (where the internal structure might be predicted) can also show depolymerization during storage time (Bello-Perez et al., 1998b). Amylopectin content is the major parameter governing the behavior of starches in solution; the contribution of amylose becomes noticeable when the weight fraction exceeds 0.5 (50% of amylose in the starch) (Roger, Bello-Perez, and Colonna, 1999).

Using static light scattering with the Zimm plot fit, Mw and Rg of amyloses from potato starch ranged between 1.08 and 9.42×10^5, and 24 and 71 nm, respectively (Radosta, Haberer, and Vorwerg, 2001). In this case, the second virial coefficient (A_2) was positive and relatively large, indicating that DMSO is a good solvent for amylose. However, synthetic amyloses and transgenic potato starches with different amylose evaluated by Berry plot showed Mw and Rg ranged between 1.08 and 9.42×10^5 g/mol, and between 24 and 71 nm, respectively. However, no significant differences in the Mw or Rg of the transgenic and traditional potato starches were observed (Radosta, Haberer, and Vorwerg, 2001).

Using the same principle of light scattering, a multiangle laser light scattering (MALLS) detector was designed to couple with a chromatographic system (e.g., high-performance liquid chromatography) and measure the intensity of the scattered light emitted by the molecules present in the sample at different scattering angles. Consequently, it is possible to monitor the scattering continuously by means of several detectors mounted at different angles (Thielking and Kulicke, 1996). Each particle-scattering function corresponds to a fraction of the population distribution, which contains a polymer fraction of low polydispersity. In this detection, the distribution of starch fractions can be quantitatively evaluated by scattered light and angular dependence, defined by the particle-scattering factor $P(q)$, and it can also provide details about the fractions' molar mass (Hanselmann et al., 1996). The light-scattering detector can be used in microbatch mode wherein the constituents in solution are measured together. The molecular characterization of amylopectin from wheat by microbatch mode of the MALLS, showed Mw ranged from 29×10^6 to 349×10^6 g/mol and the Rg between 122 and 340 nm, obtained by Berry plot with a third-order polynomial fit (You, Fiedorowicz, and Lim, 1999).

As expected, solubilization of samples must be done under suitable conditions, considering time and temperature. With nonconventional starches (banana, mango, and okenia) solubilized at different times in a microwave oven using the microbatch mode, longer heating time resulted in a decrease of Mw and Rg due to depolymerization of amylopectin (Millan-Testa et al., 2005).

3.2.4   *High-performance size-exclusion chromatography (HPSEC)*

Separation of polymers by HPSEC is done on the basis of their effective diameter and molecular weight (Jackson et al., 1988). First, research using HPSEC on polymers described a refractive index detector (RI), two Zorbax columns (Kobayashi, Schwartz, and Lineback, 1986), and four Shodex columns (Jackson et al., 1988). Later, Bondagel columns in series coupled to RI (Yuan, Thompson, and Boyer, 1993) and intrinsic viscosity detectors (Fishman and Hoagland, 1994) were employed. The advantage of HPSEC is the shorter time (30–60 minutes/sample) than gel filtration (12–16 hours) and better resolution.

In HPSEC analysis, standards of molar mass such as dextran and pullulan are required as parameters to evaluate the molecular structure and molar mass of samples. In systems coupled with a multiangle (MALLS) detector carrying out a scan, the dynamic behavior of the molecule in solution gives information about the molar mass, gyration radii, and second virial coefficient. A mathematical fit adjusted according to Berry (Figure 3.2), Zimm, or Debye is applied to determine the above-mentioned parameters. The HPLC-MALLS-RI system is widely employed to relate the structural characteristics of starches to new varieties (Yoo and Jane 2002b; Chiou et al., 2005), native or modified starches (Yoo and Jane, 2002a; Millan-Testa et al., 2005; Stevenson, Domoto, and Jane, 2006; Stevenson, Jane, and Inglett, 2007) (Figures 3.3 and 3.4), and their functional properties (Table 3.1).

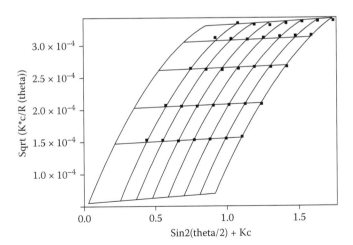

Figure 3.2 Berry plot of barley starch dissolved 50 s in microwave analyzed with the microbatch mode of the MALLS detector.

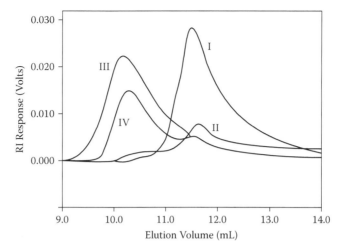

Figure 3.3 HPSEC profiles of starch from amylose of maize treated with DMSO (I), amylopectin maize (II), normal maize (III), and Eurylon 7 starch (IV) (Bello-Perez et al., 1998a).

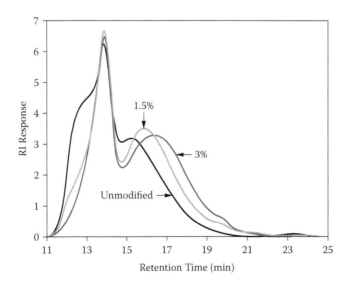

Figure 3.4 Normalized high-performance size-exclusion chromatographs of unmodified and oxidized starches (1.5 and 3% active chlorine).

3.2.5 Sedimentation field flow fractionation with MALLS-RI

Traditionally, the molar mass and its distribution have been determined by high-performance size-exclusion chromatography in combination with MALLS and RI. With the advances in sedimentation flow field-flow fractionation (SdFFF) it is possible to fractionate molecules according to diffusion coefficients, which can be empirically related to their molar mass (Thielking and Kulicke, 1996). In the SdFFF method, particle separation is done according to hydrodynamic volume and density, induced by a gravitational field oriented perpendicular to the flow direction. A combination of online light-scattering detection with this fractionation method allows the determination of the molar mass and the radius of gyration in each eluted fraction (Hanselmann et al., 1996), providing information about the structural properties of amylopectin and its relative molecular weight (Roger, Bello-Perez, and Colonna, 1999).

This technique has been also used to evaluate the performance of sample preparation conditions. For example, starch from waxy corn treated at 20 minutes of heating showed the Mw of 3.6×10^6 Da, with the Rg of 320 nm (Hanselmann et al., 1996). The SdFFF-MALLS-RI system has been also used in research to evaluate the amylose/amylopectin ratio in starches; molar mass of 4.5×10^8 g/mol and 3.2×10^7 g/mol, with Rg value of 334 and 112 nm, respectively, were obtained for waxy and normal starches (Roger, Baud, and Colonna, 2001). In analysis using SdFFF separation, Mw and Rg showed dependence with the cross-flow in normal and waxy barley starches. For normal barley starch Mw ranged between 133 and 1.9×10^6 g/mol and Rg between 183 and 123 nm, and for waxy starch Mw was between 187 and 216×10^6 g/mol, and Rg between 197 and 270 nm. Amylose from normal starch presented Mw of 2.3×10^6 g/mol and Rg of 107 nm, whereas amylopectin has Mw 280×10^6 and Rg 260 nm and amylopectin from waxy starch has Mw 360×10^6 g/mol and Rg of 267 nm (You et al., 2002). Using this technique, it was also possible to observe in jet-cooked amylopectin from cationic potato the decrease of Mw from 34 to 2.6×10^7 g/mol and Rg values from 380 to 90 nm when jet-cooking temperature was raised from 110 to 140°C (Moding, Nilson, and Wahlund, 2006).

The SdFFF-MALLS-RI system also yields information about starch structure and its conformation. Using Kratky plots, Moding, Nilson, and Wahlund (2006) showed that, at low ionic strength, starch behaves as a flexible chain with a high degree of branching and, that with increasing ionic strength, it shows a more compact internal structure. Differences in internal and global amylopectin structure according to botanical origin can also be observed by the SdFFF-MALLS-RI system; waxy wheat starch appeared to show the most homogeneous structure, whereas normal maize and cassava amylopectin showed the most heterogeneous one (Rolland-Sabate et al., 2007).

Whatever method is used, chromatography is the most often employed technique to characterize starch structure. However, the choice of method is a function of the sample's nature, parameters to be evaluated, and experimental conditions. Some of the main advantages and disadvantages of chromatography methods used in starch characterization are reported in Table 3.2.

3.3   Analyzing starch structure

3.3.1   Amylose and amylopectin fractionating

Fractionation of amylose and amylopectin is one of the most important steps in the characterization of starch structure. Often, Schoch's (1964) traditional method of starch fractionation is used with slight modification according to the amylose/amylopectin ratio.

In high amylose starch fractionation, a 90% DMSO solution is used to solubilize the samples, followed by precipitation in methyl alcohol, redissolution in distilled water, and filtration to remove insoluble material. In this method, the pH solution is adjusted with phosphate buffer and starch samples are autoclaved (121°C for 3 hours) and stirred in a water bath (96°C) for 2 hours to disperse starch molecules. *n*-Butyl alcohol is used to form an amylose–butyl–alcohol complex, which is recovered by centrifugation followed by amylose purification (Wang and White, 1994). The amylopectin remaining in the supernatant is concentrated by evaporation and then treated twice with *n*-butyl alcohol and methyl alcohol to remove amylose residues (Jane and Chen, 1992).

The method proposed by Banks and Greenwood (1967) using thymol and butanol amylose complexation is often used on starches from cassava, amaranth, potato, wheat, normal maize, and smooth-seeded peas. Amylopectin from amaranth isolated with the method proposed by Banks and Greewood (1967) and evaluated by HPSEC-MALLS-RI was considered of high purity with molar mass of 10^8 g/mol (Paredes-Lopez, Bello-Perez, and Lopez, 1994; Bello-Perez et al., 1996a, 1998c). However, Roger and Colonna (1993) found the presence of large aggregates of amylopectin in purified amylose solution, suggesting additional ultracentrifugation steps to remove them (Roger et al., 1996). On cereal starches (wheat and maize), modification of the thymol and butanol method with further purification steps with ultracentrifuge at 10,000 g for 1 hour and prewarming at 40°C was also suggested by Takeda, Hizukuri, and Juliano (1986).

Other techniques of amylose and amylopectin fractionation are carried out by gel filtration chromatography. However, after fractionation

Table 3.2 Advantages and Disadvantages of Different Separation Methods

Methods	Advantage/disadvantage
Gel filtration chromatography	• Separation involves tedious fraction collection and chemical analysis that requires at least one day per sample to complete. The time required limits the number and type of samples that can be analyzed. • Fractionation can be obtained between 8–10 h. • Each collected fraction is analyzed for total carbohydrate and blue value. Measured absorbance spectrophotometric. • Carbohydrate assays can be laborious, time-consuming, and susceptible to errors. Some chemicals used are hazardous. • Risk of starch degradation and retrogradation. Temperature variation.
HPSEC-RI	• Separation in much shorter times than conventional chromatography. • Separation based on effective diameter and molecular weight. • Molecular degradation could occur due to the pressure, shear, and friction prevailing in the column. • Starch Mw standards are not available. • Pullulan and dextran calibration standards are used to simulate starch, but this determination is relative; the molecular structure of the standards used must be close to the polymer studied with molecular weight in same range. • Calibration method for GPC (log Mw vs. retention volume) has serious limitations in application to samples that are different from the polymers employed as standards. • Reliable narrow distribution standards are available for only very few polymer materials. • Standards of one polymer do not behave in solution in the same manner as another polymer.
Traditional light scattering	• Large apparatus occupies considerable space. • Separate SLS and DLS detectors required to analyze the samples. • Many disposable materials required for each sample to avoid signal interference. • Manual change to angle possession for measurements. • Very laborious to process the data. • The time required is approximately 2 hours to run each sample.

Table 3.2 Advantages and Disadvantages of Different Separation Methods (Continued)

Methods	Advantage/disadvantage
HPSEC-MALLS-RI	• Rapid method to separate and monitor starch components and debranched fragments. • Molecular parameters absolute and obtained directly without calibration standards. • Small samples (10 µg/mg) can be analyzed in a short time with less day-to-day variation than traditional techniques. • Measurement of many samples in a short time. • The most descriptive method because it relies on the motion or dynamics of the scatters.
SdFFF-MALLS-RI	• Molecular parameters obtained directly without calibration standards. • Separate molecules according to their diffusion coefficients, which can be empirically related to their molar mass.

Sources: Ong et al. (1994); Kobayashi et al. (1986); Roger and Colonna (1993); Batey and Curtin (1996); Timpa (1991); Jackson et al. (1988); Hanselmann et al. (1996); Han et al. (2005); Thielking and Kulicke (1996).

with Schoch (1942) and Banks and Greenwood (1967) methods, gel filtration chromatography can be used to evaluate the purity of both macromolecules (Jane and Chen, 1992). Then, the separated fractions containing the amylopectin and amylose can be pooled, concentrated, and freeze-dried.

3.3.2 Amylose structure

Research on amylase structure has been carried out using HPSEC coupled with two detectors: multiangle laser light scattering and refractive index. Amylose powders isolated from starches with diverse botanical origins (cassava, potato, wheat, normal maize, and smooth-seeded peas) by the thymol and *n*-butanol complexation method, dispersed in 1 M KOH and diluted in 0.1 M KOH, have been characterized by combined systems of size-exclusion chromatography and MALLS (Roger and Colonna, 1993). In this method, pullulan and dextran commercial standards are used to assess the validity of the HPSEC-MALLS experiments and a semi-logarithmic plot of Mw versus elution volume at the maximum of the RI peak as the reference. For the same elution volume (i.e., hydrodynamic size), the branched polymer exhibits a higher Mw than the linear one due to the higher density compactness, leading to some conclusions about the sample structure. By this method, it was possible to conclude that amylose from cassava and wheat presented a higher average molar mass (1.2 and 1.1×10^6 g/mol, respectively) than commercial amylose (4×10^5 g/mol) (Roger and Colonna, 1993), whereas Takeda, Shitaozono, and Hizukuri (1988) reported Mw 4×10^5 g/mol for amylose from corn starch. However, decrease of Mw in amylose from cassava starch from 1.2×10^6 g/mol to 1.05×10^6 g/mol after ultracentrifugation allowed Roger and Colonna (1993) to conclude that large aggregates eluted at low elution volume often cause contamination of the amylose solution in the amylopectin fraction, which can be reduced by ultracentrifugation (Roger et al., 1996).

SLS and HPSEC-MALLS-RI systems are also employed in starch structure characterization. Both were used to studied amyloses synthesized by phosphorylase from malto oligomers as the primer and glucose-1-phosphate as the substrate, where the Mw ranged between 1.3×10^5 and 1.0×10^6 g/mol and Rg was between 24 and 71 nm. Synthetic amyloses show a broader unimodal molar mass distribution, with a polydispersity similar to that obtained from native starch (Radosta, Haberer, and Vorwerg, 2001).

However, the parameter most commonly used to evaluate amylose structure is the molecular weight and dynamic behavior in solution. Using SLS and HPSEC-MALLS-RI systems, Mw values of amylose from three cultivars of rice starch ranged from 3.12 and 3.44×106 g/mol with Rg between 86 and 103.3 nm (Zhong et al. 2006). These systems have also been employed to study the impact of temperature on the amylose structure during its fractionating. Pea starch, which has a high amylose content, showed that Mw (15.7 and 18.5×10^6 g/mol) and recovery percentage of amylose increase as isolation temperature rises (Vorwerg, Radosta, and Leibnitz, 2002).

3.3.3 Amylopectin structure

Amylopectin is a large molecule (10^7 and 10^8 Da) with a high branching degree (Hanselmann et al., 1996). Several authors agree that the chain length distribution and the molar or weight ratio of amylopectin play an important role in starch functional properties. It has been suggested that, in some products elaborated with wheat starch, amylopectin fraction would be the main factor responsible for functional characteristics of the final product (Kobayashi, Schwartz, and Lineback, 1986).

For example, amylopectin from three corn starches (amylose content of 36, 41, and 59%) with different proportions of short (DP 17–22) and long (DP 99–110) chains (Takeda, Takeda, and Hizukuri, 1993) present different gelatinization behavior. In amylopectin, the higher the proportion of short chains, the lower the gelatinization temperature is (Jane et al., 1999). Conversely, the amylopectin with a high level of long branching chains presents a high gelatinization temperature and a strong tendency to retrogradation (Jane et al., 1999; Chang and Lin, 2007).

As with amylose, amylopectin structure is usually evaluated by chromatographic methods. The first structural studies of the chain length distribution of amylopectin were carried out using gel filtration chromatography (Bio-Gel P-10, 100–200 mesh, Sepharose CL-2B) by debranching the chains with isoamylase or pullulanase (Biliaderis, Grant, and Vose, 1981; Biliaderis, 1982; Hizukuri, 1986; Paredes-Lopez, Bello-Perez, and Lopez, 1994). The debranched components of starch were determined by the total sugar and reducing endgroup content (Dubois et al., 1956; Ong et al., 1994). The average degree of polymerization of the eluted fraction can be determined by dividing the total carbohydrate concentration (quantified by phenol-sulfuric acid) by its reducing capacity to obtain the molar ratios of chain populations (Biliaderis, 1982; Paredes-Lopez, Bello-Perez, and Lopez, 1994).

Debranched amylopectin from different species (potato, cassava, and waxy rice) evaluated by HPLC-RI shows polymodal distribution, indicating that B-chains are separated in different groups (B1, B2, B3, and B4) (Hizukuri, 1986). The proposed distribution is consistent with the cluster model, which best explains the amylopectin structure: chains B1 and A make a single cluster; chains in fractions B2 and B3 extend into two and three clusters, respectively; and chains in fraction B4 stretch across more than four clusters.

Amylopectin molecular weights from normal maize, smooth pea (Aberle et al., 1994), amaranth starch (Bello-Perez et al., 1998c), and waxy corn starch (Millard et al., 1997) were evaluated by the HPSEC-MALLS system. However, Rg is difficult to evaluate in amylopectin by the HPSEC-MALLS system because heterogeneities in chromatograms may be related to branching density of amylopectin. To better evaluate amylopectin structure by chromatographic methods, some studies have been focused on sample preparation, mainly on the solubilization step. Amylopectin isolated from normal maize solubilized by microwave heating for 35 seconds and analyzed by the HPSEC-MALLS-RI system, showed a Mw 20×10^{-7} g/mol and Rg of 227 nm (Bello-Perez et al., 1998b). However, samples show different chromatography profiles and molecular weight if solubilized with (Mw 2.2×10^{8} g/mol) or without DMSO (Mw 9.4×10^{7} g/mol) (Bello-Perez et al., 1998a). Concerning the typical botanic structure, an HPSEC profile of debranched cassava starches shows trimodal distribution and structural differences in these populations (Charles et al., 2005).

Ion exchange chromatography, specifically anion exchange chromatography (HPAEC), coupled with pulsed amperometric detector (PAD) has allowed us to learn the chain length distribution of the individual components of debranching amylopectin. Amylopectin from different botanical sources debranched with isoamylase shows chains between 10 and 13 degrees of polymerization (Koizumi, Fukuda, and Hizukuri, 1991; Suzuki et al., 1992). Comparisons of the debranching pattern of amylopectin from diverse botanical sources with isoamylase and pullulanase using the HPAEC-PAD equipment have been an effective technique for evaluating chain length distribution (Bello-Perez et al., 1996b). In four groups of chains in several cultivars of rice starch debranched with isoamylase, one was constituted of A-chains and the other three groups corresponded to the B-chains of the cluster model (Patindol and Wang, 2002). Figure 3.5 shows a modified HPAEC-PAD chromatographic profile obtained using a postcolumn amyloglucosidase reactor; this system was named HPAEC-ENZ-PAD (Wong and Jane, 1997). The chain length distribution of debranching amylopectin from soft wheat starches obtained by this technique was DP 25.6 and 26.9 and some cultivars had amylopectin with more long chains (DP ≥ 37) and longer average chain length (DP 26.2–26.9). The structural characteristics were related to the thermal and pasting pattern; the gelatinization temperature, paste peak viscosity, and shear thinning increased with increasing branch chain length of amylopectin (Franco et al., 2002).

Amylopectin isolated from banana and analyzed by the HPAEC-ENZ-PAD system showed a higher proportion (24.0%) of longer branch chains (DP ≥ 37) than amylopectin from other botanical sources (normal and waxy maize and rice, barley, and waxy amaranth) (Jane et al., 1999). Using this

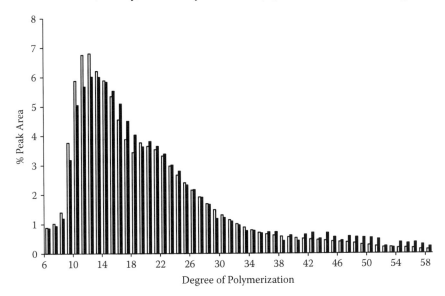

Figure 3.5 Amylopectin chain length distribution of normal maize (□) and barley (■) starch, measured by high-performance anion exchange chromatography (HPAEC) with pulsed amperometric detection (PAD).

technique it was also possible to correlate functional properties with morphological and structural characteristics on A- and B-type starch granules from cereal starches, indicating that large granules (A) have a higher proportion of longer chains than small granules (B) (Ao and Jane, 2007).

3.4   Final remarks

Starch structure research allows the elucidation mechanism and the identification of specific functional properties. This research is important in understanding the physicochemical and functional starch properties. This knowledge is essential to support breeding programs focused on starch biosynthesis and new varieties with specific starch functionality. It allows us to obtain new starches with improved functionality which can be used to develop new products. The proteomic approach is an important tool for studying starch biosynthesis, relating it to the starch structure.

References

Aberle, T., Burchard, W., Vorwerg, W., and Radosta, T.S. 1994. Conformational contributions of amylose and amylopectin to the structural properties of starches from various sources. *Starch/Stärke*. 46:329–335.

Ao, Z. and Jane, J. 2007. Characterization and modeling of the A- and B-granule starches of wheat, triticale and barley. *Carbohydrate Polym*. 67:46–55.

Banks, W. and Greewood, C.T. 1967. The fractionation of laboratory-isolated cereal starches using dimethyl sulphoxide. *Starch/Stärke*. 19:394–398.

Batey I.L. and Curtin B.M. 1996. Measurement of amylose/amylopectin ratio by high-performance liquid chromatography. *Starch/Starke*. 48:338–344.

Bello-Perez, L.A., Colonna, P., Roger P., and Paredes-Lopez, O. 1998c. Macromolecular features of amaranth starch. *Cereal Chem*. 75(4):395–402.

Bello-Perez, L.A., Paredes-Lopez, O., Roger, P., and Colonna, P. 1996a. Molecular characterization of some amylopectins. *Cereal Chem*. 73:12–17.

Bello-Perez, L.A., Paredes-Lopez, O., Roger, P., and Colonna, P. 1996b. Amylopectin: properties and fine structure. *Food Chem*. 56(2):171–176.

Bello-Perez, L.A., Roger P., and Paredes-Lopez, O. 1998. Structural properties of starches dissolved by microwave heating. *Starch/Stärke*. 50:137–141.

Bello-Perez, L.A., Roger, P., Baud, B., and Colonna, P. 1998a. Macromolecular features of starches determined by aqueous high-performance size exclusion chromatography. *J. Cereal Sci*. 27:267–278.

Bello-Perez, L.A., Roger, P., Colonna, P., and Paredes-Lopez, O. 1998b. Laser light scattering of high amylose and high amylopectin materials, stability in water alter microwave dispersion. *Carbohydrate Polym*. 37:383–394.

Biliaderis, C.G., 1982. Physical characteristics, enzymatic, digestibility, and structure of chemically modified smooth pea and waxy maize starches. *J. Agric. Food Chem*. 30:925–930.

Biliaderis, C.G. 1992. Structure and phase transitions of starch in food systems. *Food Technol*. 46(6):98–109.

Biliaderis, C.G., Grant, D.R., and Vose, J.R. 1981. Structural characterization of legume starches. I. Studies on amylose, amylopectin, and beta-limit dextrins. *Cereal Chem.* 58:496–502.

Blanshard, J.M.V. 1987 Starch granule structure and function: A physicochemical approach. In: *Starch: Properties and Potentials, Critical Reports on Applied Chemistry*, Vol. 13, T. Galliard (Ed.). Chichester, UK: John Wiley & Sons, pp. 16–54.

Burchard, W. 1972. Angular distribution of Rayleigh scattering from branched poly-condensates. Amylopectin and glycogen types. *Macromolecules.* 5:604–610.

Burchard, W. 1992. Biochemistry. In: *Static Dynamic Light Scattering Aproaches to Structure Determination of Biopolymers*, S.E. Harding, D.E. Satelle, and V.A. Bloom Ficia (Eds.). UK: Redwood Press, pp. 8–22.

Burchard, W. 1993. Plant polymeric carbohydrates. In: *Solution Properties of Plant Polysaccharides as a Function of Their Chemical Structure*, F. Meyer, D.J. Manners, and W. Seibel (Eds.). Cambridge, UK: The Royal Society of Chemistry, pp. 215–232.

Burchard, W. 1997. Particle scattering factors of some branched polymers. *Macromolecules.* 10(5):920–927.

Chang, Y.H. and Lin, J.H. 2007. Effects of molecular size and structure of amylo-pectin on the retrogradation thermal properties of waxy rice and waxy corn-starches. *Food Hydrocolloids.* 21:645–653

Charles, A.L., Yung, H.G., Chang, W.C.K., Klanaroth, S., and Tzou, C.H. 2005. Influence of amylopectin structure and amylose content on the gelling prop-erties of five cultivars of cassava starches. *J. Agric. Food Chem.* 53:2717–2725.

Chiou, H., Fellows, C. M., Gilbert, R.G., and Fitzgerald, M.A. 2005. Study of rice-starch structure by dynamic light scattering in aqueous solution. *Carbohydrate Polym.* 61:61–71.

Debye, P. 1947. Molecular-weight determination by light scattering. *J. Phys. Colloidal Chem.* 51:18–32.

Donald, A.M., Waigh, T.A., Jenkins, P.J., Gidley, M.J., Debet, M., and Smith, A. 1997. Internal structure of starch granules revealed by scattering studies. In: *Starch: Structure and Functionality*, P.J. Frazier, A.M. Donald, and P. Richmond (Eds.). Cambridge, UK: The Royal Society of Chemistry, pp.172–179.

Dubois, M., Gilles, K.A., Hamilton, J.K., Rebers, P.A., and Smith, F. 1956. Colorimetric method for determination of sugars and related substances. *Anal. Chem.* 28:350–356.

Fishman, M.L. and Hoagland, P.D. 1994. Characterization of starches dissolved in water by microwave heating in a high pressure vessel. *Carbohydrate Polym.* 23:175–183.

Fishman, M.L., Rodriguez, L., and Chau, H.K. 1996. Molar masses and sizes of starches by high-performance size exclusion chromatography with on-line multiangle laser light scattering detection. *J. Agric. Food Chem.* 44:3182–3188.

Franco, C.M.L., Kit-Sum, W., Sang-ho, Y., and Jay-lin, J. 2002. Structural and functional characteristics of selected soft wheat starches. *Cereal Chem.* 79(2):243–248.

French, D. 1984. Organization of starch granules. In: *Starch: Chemistry and Technology*, R.L. Whistler, J.N. BeMiller, E.F. Paschall (Eds.). Orlando, FL: Academic Press, pp. 183–247.

Galinsky, G. and Burchard W. 1995. Starch fractions as examples for nonrandomly branched macromolecules. 1. Dimensional properties. *Macromolecules.* 28:2363–2370.

Galinsky, G. and Burchard, W. 1996. Starch fractions as examples for nonrandomly branched macromolecules. 2. Behavior in the semidilute region. *Macromolecules.* 29:1498–1506.

Galinsky, G. and Burchard W. 1997a. Starch fractions as examples for nonrandomly branched macromolecules. 3. Angular dependence in static light scattering. *Macromolecules.* 30:4445–4453.

Galinsky, G. and Burchard W. 1997b. Starch fractions as examples for nonrandomly branched macromolecules. 4. Angular dependence in dynamic light scattering. *Macromolecules.* 30:6966–6973.

Gallant, D.J., Bouchet, B., and Baldwin, P.M. 1997. Microscopy of starch: Evidence of a new level of granule organization. *Carbohydrate Polym.* 32:177–191.

Gallant, D.J., Bouchet, B., Buleón, A., and Perez, S. 1992. Physical characteristics of starch granules and susceptibility to enzymatic degradation. *Euro. J. Clin. Nutrition* 46(2):3–16.

Galliard, T. and Bowler, P. 1987. Morphology and composition of starch. In: *Starch Properties and Potential*, T. Galliard (Ed.). Chichester: John Wiley & Sons, pp. 55–78.

Greenwood, C.T. 1976. Starch. In: *Advances in Cereal Science and Technology*, Y. Pomeraz (Ed.). St. Paul, MN: American Association of Cereal Chemists, pp. 119–157.

Han, J.A. and Lim, S.T. 2004a. Structural changes of corn starches by heating and stirring in DMSO measured by SEC-MALLS-RI system. *Carbohydrate Polym.* 55:265–272.

Han, J.A. and Lim, S.T. 2004b. Structural changes of corn starches during alkaline dissolution by vortexing. *Carbohydrate Polym.* 55:193–199.

Han, J.A., Lim, H., and Lim, S.T. 2005. Comparison between size exclusion chromatography and micro-batch analyses of corn starches in DMSO using light scattering detector. *Starch/Stärke.* 57:262–267.

Hanselmann, R., Burchard, W., Ehrat, M., and Widmer, H.M. 1996. Structural properties of fractioned starch polymers and their dependence on the dissolution process. *Macromolecules.* 29:3277–3282.

Hizukuri, S. 1986. Polymodal distribution of the chain lengths of amylopectins and its significance. *Carbohydrate Res.* 147:342–3477.

Hood, L.F. 1982. Current concepts of starch structure. In: *Food Carbohydrates*, D.R. Lineback and G.E. Ingleit (Eds.).Westport, CT: AVI, pp. 217–269.

Imberty, A. and Perez, S. 1988. A revisit to three dimensional structure of B-type starch. *Biopolymers.* 27:1205–1221.

Imberty, A., Chanzy H., Perez S., Buleon A., and Tran, V. 1987. New three-dimensional structure for A-type starch. *Macromolecules.* 20:2634–2636.

Imberty, A., Chanzy, H., Perez, S., Buleon, A., and Tran, V. 1988. The double-helical nature of the crystalline part of A-starch. *J. Molec. Biol.* 201:365–378.

Jackson, D.S., Choto-Owen, C., Waniska R.D., and Rooney L.W. 1988. Characterization of starch cooked in alkali by aqueous high-performance size-exclusion chromatography. *Cereal Chem.* 65:493–496.

Jane, J. and Chen, J. 1992. Effect of amylose molecular size and amylopectin branch chain length on paste properties of starch. *Cereal Chem.* 69:60–65.

Jane, J., Chen, Y.Y., Lee, L.F., McPherson, A.E., Wong, K.S., Radosavljevic, M., and Kasemsuwan, T. 1999. Effects of amylopectin branch chain length and amylose content on the gelatinization and pasting properties of starch. *Cereal Chem.* 76(5):629–637.

Jane, J.L., Xu, A., Radosavljevic, M., and Seib, P.A. 1992. Location of amylose in normal starch granules. I. Susceptibility of amylose and amylopectin to cross-linking reagents. *Cereal Chem.* 69:405–409.

Kobayashi, S., Schwartz, S.J., and Lineback, D.R. 1986. Comparison of the structures of amylopectins from different wheat varieties. *Cereal Chem.* 63(2):71–74.

Koizumi, K., Fukuda, M., and Hizukuri, S. 1991. Estimation of the distribution of chain length of amylopectins by high-performance liquid chromatography with pulsed amperometric detection. *J. Chromatog.* 585:233–238.

Millan-Testa, C.E., Mendez-Montealvo, M.G., Ottenhof, M.A., Farhat, I.A., and Bello-Perez, L.A. 2005. Determination of the molecular and structural characteristics of okenia, mango and banana starches. *J. Agric. Food Chem.* 53:495–501.

Millard, M.M., Dintzis, F.R., Willett, J.L., and Klavons, J.A. 1997. Light-scattering molecular weights and intrinsic viscosities of processed waxy maize starches in 90% dimethyl sulfoxide and H_2O. *Cereal Chem.* 74:687–691.

Moding, G., Nilson, P.-O., and Wahlund, K.-G. 2006. Influence of jet-cooking temperature and ionic strength on size and structure of cationic potato amylopectin starch as measured by asymetrical flow field-flow fractionation multi-angle light scattering. *Starch/Stärke.* 58:55–65.

Morrison, W. R. 1988. Lipids in cereal starches: A review. *J. Cereal Sci.* 8:1–15.

Nikuni Z. 1978. Studies on starch granules. *Starch/Stärke.* 30:105–111.

Ong, M.H., Jumel, K., Tokarczuk, P.F., Blanshard, J.M.V., and Harding, S.E. 1994. Simultaneous determinations of the molecular weight distributions of amyloses and the fine structures of amylopectins of native starches. *Carbohydrate Res.* 260:99–117.

Paredes-Lopez, O., Bello-Perez, L.A., and Lopez, M.G. 1994. Amylopectin: Structural, gelatinization and retrogradation studies. *Food Chem.* 50:411–417.

Patindol, J. and Wang, Y. 2002. Fine structures of starches from long-grain rice cultivars with different functionality. *Cereal Chem.* 79:465–469.

Radosta, S., Haberer, M., and Vorwerg, W. 2001. Molecular characteristics of amylose and starch in dimethyl sulfoxide. *Biomacromolecules.* 2:970–978.

Robin, J.P., Mercier, C., Charbonniere, R., and Guilbot, A. 1974. Lintnerized starches. Gel filtration and enzymatic studies of insoluble residues from prolonged acid treatment of potato starch. *Cereal Chem.* 51:389–406.

Roger, P. and Colonna, P. 1993. Evidence of the presence of large aggregates contaminating amylose solutions. *Carbohydrate Polym.* 21:83–89.

Roger, P., Axelos, M.A.V., and Colonna, P. 2000. SEC-MALLS and SANS studies applied to solution behavior of linear R-glucans. *Macromolecules.* 33:2446–2455.

Roger, P., Baud, B., and Colonna, P. 2001. Characterization of starch polysaccharides by field flow fractionation-multi angle laser light scattering-differential refractometer index. *J. Chromatog.* 917:179–185.

Roger, P., Bello-Perez, L.A., and Colonna, P. 1999. Contribution of amylose and amylopectin to the light scattering behaviour of starches in aqueous solution. *Polymers.* 40:6897–6909.

Roger, P., Tran, V., Lesect, J., and Colonna, P. 1996. Isolation and characterization of single chain amylose. *J. Cereal Sci.* 24:247–262.

Rolland-Sabate, A., Amani, G.G., Dufour, D., Guilois, S., and Colonna, P. 2003. Macromolecular characteristics of ten yam (Dioscorea spp) starches. *J. Sci. Food Agric.* 83:927–936.

Rolland-Sabate, A., Colonna, P., Mendez-Montealvo, G., and Planchot, V. 2007. Branching features of amylopectins and glycogen determined by asymmetrical flow field flow fractionation coupled with multiangle laser light scattering. *Biomacromolecules.* 8:2520–2532.

Schoch, T.J. 1942. Fractionation of starch by selective precipitation with butanol. *J. Amer. Chem. Soc.* 64:2954–2956.

Schoch, T.J. 1964. Iodometric determination of amylose. Potentiometric titration: Standard method. In: *Methods in Carbohydrate Chemistry.* Vol. IV, R.L. Whistler (Ed.). Orlando, FL: Academic Press, pp. 157–160.

Stevenson, D.G., Jane, J., and Inglett, G.E. 2007. Characterization of jicama (Mexican potato) (*Pachyrhzus aerosus* L. Urban) starch from taproots grown in USA and México. *Starch/Stärke.* 59:132–140.

Stevenson, G.D., Domoto, A.P., and Jane, J. 2006. Structures and functional properties of apple (*Malus domestica* Borkh). *Carbohydrate Polym.* 63:432–441.

Suzuki, A., Kaneyama, M., Shibanuma, K., Takeda, Y., Abe, J., and Hizukuri, S. 1992. Characterization of lotus starch. *Cereal Chem.* 69:309–315.

Takeda, Y., Hizukuri, S., and Juliano, B. 1986. Purification and structure of amylose from rice starch. *Carbohydrate Res.* 148(2):299–308.

Takeda, Y., Shitaozono, T., and Hizukuri, S. 1988. Molecular structure of corn starch. *Starch/Stärke.* 40:51–54.

Takeda, C., Takeda, Y., and Hizukuri S. 1993. Structure of the amylopectin fraction of amylomaize. *Carbohydrate Res.* 246:273–281.

Tester, R.F., Karkalas, J., and Qi, X. 2004. Starch: Composition, fine structure and architecture. *J. Cereal Sci.* 39:151–165.

Tetchi, F.A., Rolland-Sabate, A., Amani, G.N., and Colonna, P. 2007. Molecular and physicochemical characterisation of starches from yam, cocoyam, cassava, sweet potato and ginger produced in the Ivory Coast. *J. Sci. Food Agric.* 87:1906–1916.

Thielking, H. and Kulicke, W.M. 1996. On-line coupling of flow field-flow fractionation and multiangle laser light scattering for the characterization of macromolecules in aqueous solution as illustrated by sulfonated polystyrene samples. *Anal. Chem.* 68:1169–1173.

Timpa, J.D. 1991. Application of universal calibration in gel permeation chromatography for molecular weight determinations on plant cell wall polymers: Cotton fiber. *J. Agric. Food Chem.* 39:270–275.

Vorwerg, W., Radosta, S., and Leibnitz, E. 2002. Study of a preparative-scale process for the production of amylose. *Carbohydrate Polym.* 47:181–189.

Wang, L.Z. and White, P.J. 1994. Structure and properties of amylose, amylopectin, and intermediate materials of oat starches. *Cereal Chem.* 71:263–268.

Wong, K.S. and Jane, J. 1997. Quantitative analysis of debranched amylopectin by HPAEC-PAD with a post-column enzyme reactor. *J. Liquid Chromatog.* 20:297–310.

Wu, H.C.H. and Sarko, A. 1978a. The crystal structures of A- B- and C-type polymorphs of amylose and starch. *Starch/Stäerke.* 30:73-78.

Wu, H.C.H. and Sarko, A. 1978b. The double-helical molecular structure of crystalline A-amylose. *Carbohydrate Res.* 61:27–40.

Yoo, S.H. and Jane, J. 2002a. Molecular weights and gyration radii of amylopectins determined by high-performance size-exclusion chromatography equipped with multi-angle laser-light scattering and refractive index detectors. *Carbohydrate Polym.* 49:307–314.

Yoo, S.H. and Jane, J.B. 2002b. Structural and physical characteristics of waxy and other wheat starches. *Carbohydrate Polym.* 49:297–305.

You, S., Fiedorowicz, M., and Lim, S. 1999. Molecular characterization of wheat amylopectins by multiangle laser light scattering analysis. *Cereal Chem.* 76:116–121.

You, S., Stevenson S.G., Izydorczyk M.S., and Preston K.R. 2002. Separation and characterization of barley starch polymers by a flow field-flow fractionation technique in combination with multiangle light scattering and differential refractive index detection. *Cereal Chem.* 79:624–630.

Yuan, R.C., Thompson, D.B., and Boyer, C.D. 1993. The fine structure of amylopectin in relation to gelatinization and retrogradation behavior of maize starches from three *wx*-containing genotypes in two inbred lines. *Cereal Chem.* 70:81–89.

Zhong, F., Yokoyama, W., Wang, Q., and Shoemaker, C.F. 2006. Rice starch amylopectin and amylose: molecular weight and solubility in dimethyl sulfoxide-based solvents. *J. Agric. Food Chem.* 54:2320–2326.

Zobel, H.F. and Stephen, A.M. 1996. Starch: Structure, analysis and application. In: *Food Polysaccharides and Their Application*, A.M. Stephen (Ed.). New York: Marcel Dekker, pp.19–66.

chapter four

Solid-state NMR applied to starch evaluation

Maria Inês Bruno Tavares

Instituto de Macromoléculas, Universidade Federal do Rio de Janeiro

Contents

4.1   Introduction

The nuclear magnetic resonance spectroscopy (NMR) technique, in both the solution and solid states, is a useful tool in the chemical characterization of polymers. NMR spectrometers with strong magnetic fields employing pulse radiofrequency (RF) allow us to perform characterization on several biopolymers, mainly on amorphous solid samples (Komoroski, 1986; McBrierty and Parker, 1993; Harris, 2000), and special NMR tools have been developed to improve solid resolution. Consequently, important and precise information about linking among structure, microstructure, and properties can be obtained by NMR (Sanders and Hunter, 1993; Bovey and Mirau, 1996; Harris, 1996; Schmidt-Rohr and Spiess, 1999).

Because NMR solid state can provide details about crystalline structure and differences on domain composition (Harris, 1996; Silva and Tavares, 1996), it becomes particularly important in starch characterization. Because of the NMR solid-state accuracy in evaluating amorphous and heterogeneous materials, polymorphism, and phase transition (Harris, 2000), several researchers have employed it as a tool for understanding starch and polysaccharide structure, crystallinity, molecular organization/orientation, and molecular dynamics (Gidley and Bociek, 1985; Cheethan and Tao, 1998; Tang, Godward, and Hills, 2000; Conte, Spaccini, and Piccolo, 2004). NMR relaxation times have been successfully used to provide details about starch structure and heterogeneity (Bovey and Mirau, 1996; Silva and Tavares, 1996; Harris, 1996; Tavares, 2000, 2003). The relaxation rate depends on sample composition, its moisture content, and distribution of crystalline and amorphous phases along the domains. Proton spin-lattice relaxation time can also provide details about domain distinctions and structural features (Costa et al., 2007a,b; Souza and Tavares, 2002).

4.2   Principles of NMR

The NMR technique is based on the principle that, if a nucleus with angular moment (P) and magnetic moment (μ) is placed in a probe in a static magnetic field (B_0), the angular moment takes up an orientation along the direction of the magnetic field (z), depending on the energy levels. The nucleus starts processing in the various energy levels, such as α (less energy) and β (high energy). Thus, the excess nucleus population, defined as vector magnetization (Mo), in the α energy level can be irradiated to the β energy level, because the same frequency precession, called Larmor frequency (ν), is applied to this nucleus by the B_1 field, which transfers the magnetization for the xy plane. When the radio frequency emission is switched off, the nuclear spins are deflected from their position through an angle θ; this magnetization will return to the fundamental state and the relaxation process can begin. In this process two relaxation times occur simultaneously: one called "transversal relaxation" or "spin–spin," which occurs in the xy plane at time constant T_2, and another one denominated "longitudinal relaxation" or "spin–lattice," which occurs at the z axis at time constant T_1. The NMR signal is detected after RF has been switched off by a free induction decay process (FID), which results in a free precession of nuclear spins that results in its return to the equilibrium.

The T_1 and T_2 relaxation processes are governed by fluctuating magnetic fields associated with the molecular motions; both relaxations are assumed to be first-order processes. T_1 relaxation is an enthalpic process, because this process involves energy absorption. A temporal evolution of a transversal relaxation (T_2) process differs from the longitudinal foundation. It is equivalent to the loss of the coherence phase between individual magnetic moments on precession, and results in an increase in entropy (Gidley and Bociek, 1985; Cheng, 1991; Stejskal and Memory, 1994; Bovey and Mirau, 1996; Silva et al., 2002; Souza and Tavares, 2002; Tang and Hills, 2003; Tavares, 2003; Conte, Spaccini, and Piccolo, 2004; Nascimento, Tavares, and Nascimento, 2007; Costa et al., 2007a,b). These relaxation times are measured in distinct ways: T_1 is determined through an inversion-recovery pulse sequence, occurring along the z axis and depending on a strong magnetic field, the MHz frequency observation scale, whereas T_2 is determined by pulse sequence, occurring in the xy plane and is also dependent on the external magnetic field.

4.2.1 NMR solid state

NMR solid-state parameters, such as pulse sequences, can provide details about behavior and molecular mobility of samples, allowing data gathering in different time scales, such as MHz and kHz. It allows development of specific NMR methods according to the sample nature, as done with starch samples (Harris, 2000).

The solid-state analyses are different from liquid analyses for two reasons. The first one is related to the signal width. In NMR solid state the signals are wider than those of the liquids, mainly in polymer samples due to the high molecular weight, hydrogen bonds, and monomer links. The second reason is associated with the response of the measurements; normally in the solid state it is possible to have more information than in the liquid state because the whole sample is analyzed without solvent influence. As a consequence more details can be obtained from the chemical structure and molecular dynamic (Edbon, 1980; Cheng, 1991; Costa, Oliveira, and Tavares, 1998; Costa et al., 2007a; Tang, Godward, and Hills, 2000; Tavares, 2000; Conte, Spaccini, and Piccolo, 2004; Souza and Tavares, 2002). Furthermore, when the sample is insoluble or has low solubilization, it needs to be analyzed by solid state. NMR solid state also has the advantage of providing information about the relationship between structure and properties, such as sample homogeneity, phase dispersion, and interaction of components. Moreover, it is a nondestructive and noninvasive technique.

NMR spectra obtained from liquid samples often provide narrow signals, with better resolution than the solid ones due to chemical shift isotropy. Because samples are in solution, all interactions such as shielding, dipolar coupling, and indirect coupling depending on nuclear orientation in the magnetic field B_0 are compensated by isotropic movements. Therefore, the sample nature, its solvent interaction, and external magnetic field strength can influence the results (Conte, Spaccini, and Piccolo, 2004; Tang, Godward, and Hills, 2000; Costa et al., 2007a; Tavares 2000, 2003).

In solid samples, molecular mobility is smaller than in liquid samples. However, the majority of samples (excluding the crystals) have no molecular orientation of the NMR signal, causing a wideness signal due to the random distribution of the signal orientation. This factor comes from the chemical shift anisotropy as well as from the strong dipolar interaction between hydrogen and ^{13}C nuclei. Sample structure and the observed nuclei are also fundamental factors that can contribute to spectral resolution in the solid state.

The spectrum obtained from the NMR solid state results from the molecular dynamic behavior of the sample, which reflects its heterogeneity, the interactions of components, and relationship of the molecular dynamic characteristics. Sometimes the wideness of NMR signals in the solid state does not reveal information about the molecular structure. The development of NMR solid-state methods produced solid-state spectra with narrow signals, enabling its use in studies of polymeric systems, including starches.

4.2.2   High resolution in NMR solid state

The influence of factors on solid-state signals is defined as a sum of several Hamiltonian contributions, according to the following expression (4.1).

$$\mathbf{H_{RMN} = H_Z + H_{RF} + H_{CSA} + H_D + H_J + H_Q} \qquad (4.1)$$

where $\mathbf{H_Z}$ = Zeeman interaction, $\mathbf{H_{RF}}$ = radiofrequency, $\mathbf{H_{CSA}}$ = chemical shift anisotropy, $\mathbf{H_D}$ = dipolar interaction, $\mathbf{H_J}$ = coupling constant, and $\mathbf{H_Q}$ = quadrupole moment.

When a spin ½ nuclei is observed, for example, ^{13}C, the main contribution in signal width comes from $\mathbf{H_{CSA}}$ and $\mathbf{H_D}$. Because this contribution can be fractionated as proposed in expression (4.1), signal resolution of spectra from the NMR solid state can be improved, reducing the wideness arising from mathematical methods (Komoroski, 1986; Stejskal and Memory, 1994; Bovey and Mirau, 1996; Conte, Spaccini, and Piccolo, 2004).

4.2.2.1   Magic angle spinning (MAS)

Combining factors such as strong dipolar interaction between H and ^{13}C, the restricting solids' molecular mobility, and chemical shift anisotropy result in a wide signal in the NMR solid sample, with line width around 20 kHz. Supression of dipole–dipole interaction results in narrowing the signal to 5 kHz and elimination of chemical shift anisotropy changes the signal width to 100 Hz, allowing an NMR solid-state spectrum with narrow signals. Both dipolar interaction and chemical shift anisotropy are dependent on the term $3\cos^2\theta - 1$. The suppression or reduction of both effects is done by sample spinning at high frequency at an angle of

54.74°, with magnetic field direction able to eliminate the 3 $\cos^2\theta - 1$, allowing a strong dipolar decoupling that results in significant narrowing of the NMR signal (Costa, Oliveira, and Tavares, 1998; Conte, Spaccini, and Piccolo, 2004; Stejskal and Memory, 1994).

The pulse sequence employed is single (basic):

$$\text{Hydrogen decoupling}$$

$$\text{observed} = [90°x \rightarrow \text{FID} - t]_n$$

where t is the recycle delay between 90° pulses and n is the number of scans.

The recycle delay (t) is directly associated with the spin-lattice relaxation times of the different nuclei types. Thus, variations of this parameter can provide information about sample molecular mobility.

All nuclei in resonance phenomenon can be analyzed by MAS technique, which employs the single pulse sequence. However, to observe the nuclei with a quadrupole moment, it is necessary to use specific techniques or probes and stronger magnetic fields, because the NMR signal widths of those nuclei are wider than the nuclei with dipolar moments. Observing nuclei with dipolar moments by MAS technique allows obtaining narrow signals and high spectral resolution. Because polymers have high molecular weight and consequently wide NMR signals, their NMR solid-state spectra provide information about the chemical structure of the molecular mobility chains. However, fine structures such as those of polymers can only be seen in NMR solution analysis (Gidley and Bociek, 1985; McBrierty and Parker, 1993).

MAS is basically employed on solid-state samples; thus it can be used for both qualitative and quantitative results. Therefore, due to the high relaxation time values, it necessarily requires a long time to perform quantitative analyses. In solids the spin–spin relaxation time is short, resulting in fast FID decay. Consequently, in this case, the spin–lattice relaxation is very long, mainly for nuclei with a rare spin. Thus, variation in the spectral parameters such as pulse sequence, allows obtaining information about material homogeneity, compatibility, and molecular mobilities of starches and other polymers (Gidley and Bociek, 1985; Tang and Hills, 2003). On amorphous polymers such as starch, this technique allows us to observe the molecular mobility and the presence of minor compounds with low molecular weights, such as lipids. These compounds with low molecular weights and water are responsible for the changes in the molecular mobility of starches (Stejskal and Memory, 1994).

Figure 4.1 shows the ^{13}C MAS NMR spectrum of flour obtained from the mango seed with a short recycle delay (0.3 ms). The width signals on the spectrum are due to the heterogeneous and amorphous sample nature

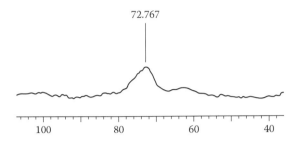

Figure 4.1 ^{13}C MAS NMR spectrum of seed flour from bourbon mango.

and the main signal detected refers to the 2, 3, and 5 carbons from the hexose structure, identified in the high molecular mobility domain, which is probably located in the amorphous region of the starch granule.

4.2.2.2　*Cross-polarization magic angle spinning (CPMAS)*

The cross-polarization technique is employed on NMR spectra of the nucleus considered as rare spins, which are less abundant in nature, such as isotope ^{13}C (Sanders and Hunter, 1993; Stejskal and Memory, 1994; Conte, Spaccini, and Piccolo, 2004). Its objective is to make the analysis time shorter than required in basic MAS analysis. It is based on a polarization transfer of abundant spin (^{1}H) to rare spin (^{13}C); because ^{13}C and ^{1}H nuclei are in thermal contact for the time established (contact time) for the cross-polarization transfer to occur. During this time the nuclei are kept in contact at an identical frequency precession in the Hartman–Hahn condition (expression 4.2).

$$\omega_H B_H = \omega_C B_C \tag{4.2}$$

The cross-polarization technique combined with magic angle spinning (CPMAS) and high power hydrogen decouple allows us to obtain NMR solid-state high-resolution spectra of rare spin nuclei. Because the hydrogen nucleus is in the relaxation process, the intensity of signals increases and the time required for obtaining the signal is shorter than in MAS. Figure 4.2 exhibits the ^{13}C CPMAS–NMR spectrum of flour from the mango bourbon seed, where three signals are detected: one located at 73 ppm, due to CH–OH groups, another one located at 62 ppm related to CH$_2$–OH groups, and a third one located at 101 ppm, which refers to anomeric carbon CH–O–CH from the glycoside link. The applied pulse sequence to obtain NMR spectra via CPMAS is the same as that applied in the MAS technique; however, the introduction of Hartman–Hahn conditions provides a cross-polarization transfer between the proton and ^{13}C, promoting an increase in the ^{13}C intensity signal, whereas the proton intensity signal decreases. Thus, CPMAS and high-power hydrogen

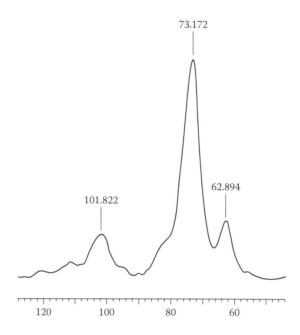

Figure 4.2 ^{13}C CPMAS NMR spectrum of seed flour from mango bourbon.

decoupling can provide NMR solid-state spectra with details about the sample's molecular structure and homogeneity. With this technique, changes in the signal width and in the chemical shift values are related to the changes in molecular mobility. For example, comparing ^{13}C MAS (Figure 4.1) and ^{13}C CPMAS (Figure 4.2) NMR spectra, two domains with different molecular mobility in the sample can be observed, as well as a third domain, which is probably a result of the interaction of both these domains.

4.2.2.3   *Variable contact time (VCT) experiment*

This technique promotes a scanning of contact times, which generates a series of ^{13}C NMR spectra with different contact times. These results provide details about sample heterogeneity, molecular mobility, intermolecular interactions, and the value of proton spin–lattice in the rotating frame ($T_1\rho H$) (Costa, Oliveira, and Tavares, 1988; Tavares, 2003), which is obtained from the ^{13}C decays during the variable contact time and hydrogen nuclei relaxation process. Figure 4.3 shows spectra obtained from VCT analysis of flour from seed mango, where it is possible to identify a series of ^{13}C spectra with contact times varying from short (200 ms, A) to long values (8,000 ms, G). Flours from cereals or seeds present one rigid domain that controls the relaxation process; the decay is more abrupt at the end of the experiment, indicating that the rigid domain is interacting with

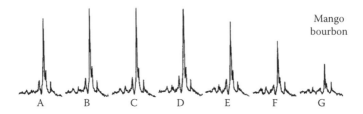

Mango
bourbon

A   B   C   D   E   F   G

Figure 4.3 ^{13}C VCT NMR spectra of seed flour from mango bourbon: (A = 200, B = 400, C = 800, D = 1000, E = 2000, F = 4000, G = 8000 ms).

Table 4.1 $T_1\rho H$ Values for Assigned Carbons

Carbon type	δ (ppm)	$T_1\rho H$ (ms)
C–O–C	101.8	3
CH–O	73.1	5
CH₂OH	62.8	7

the mobile domain. Successive decays controlled by the protons' spatial proximity on the ^{13}C spectra are observed, where it is possible to obtain the proton spin–lattice relaxation time in the rotating frame. It shows the molecular mobility and the interactions between domains, with size at 4 nm. The decay intensity of the ^{13}C signals during this experiment can be quick, slow, or intermediate; it depends on the molecular mobility and spatial interactions on the samples.

Table 4.1 shows the values of this parameter for the assigned carbons of the flour from seed mango, which are 101.8 ppm (anomeric carbon), 73.1 ppm (2, 3, 4, and 5 CH–O carbons, from the glycosidic ring), and 62.8 ppm (carbon 6, CH₂–OH). According to these relaxation data, the increase of the proton spin–lattice relaxation time in the rotating frame value is promoted by an increase of carbon molecular mobility, resulting in the high level of freedom rotation and translation of the observed nuclei.

4.2.3 Low magnetic field

In low-field NMR spectrometers such as 23 MHz, due to low intensity, the obtained spectra do not provide information on chemical shift at solid state. Thus, these spectrometers are not used in studies of chemical structure, but they are useful to determine the relaxation time, mainly in food samples. In NMR analysis, the thermal equilibrium of the spin systems is disturbed when radiofrequency resonance is applied and, when it is switched off, a relaxation mechanism occurs.

Two relaxation parameters are often used to determine molecular mobility, T_1H and T_2H. In low-field NMR proton relaxation, parameters T_1 and T_2 can be directly measured employing their proper pulse sequences, providing information about interactions in the sample (Cheng, 1991; Stejskal and Memory, 1994; Bovey and Mirau, 1996; Harris, 1996; Silva et al., 2002). Proton T_1 relaxation is related to the return of Mo (initial magnetization aligned to Bo) magnetization to equilibrium after radiofrequency has been switched off. Proton T_2 is related to the inverse of signal width, and occurs due to the loss of phase precession coherence of the nucleus after it has been excited in the direction of the applied magnetic field (Bovey and Mirau, 1996; Harris, 1996). However, relaxation times determined by NMR allow evaluating the plasticization effect. It is also possible to compare the macromolecular structure in each domain. T_1H and T_2H relaxation times can be measured in a wide range of temperatures and they allow the observation of plasticizing properties and macromolecular mobility on samples including starch, flours, and polysaccharides (Gidley and Bociek, 1985; Harris, 2000; Tang, Godward, and Hills 2000; Souza and Tavares, 2002).

4.2.3.1 T_1H determination

The pulse sequence inversion recovery is the most accurate NMR method to measure the proton spin–lattice relaxation time (T_1H). This relaxation process is dependent on the magnetic field and occurs on the z axis. In this process the energy excess (thermal energy) is transferred directly to the lattice as dipolar interaction. During this relaxation, an enthalpy decrease occurs in the spin nuclear system (Schmidt-Rohr and Spiess, 1999). The equilibrium undergoes magnetization, resulting in a first-order process at constant rate (R_1) and constant time (T_1). From the values determined for this parameter it is possible to evaluate the molecular dynamics of samples and also their homogeneity, which is very useful in studies of samples such as starch granules (Stejskal and Memory, 1994; Harris, 2000; Tavares, 2003; Silva et al., 2002).

The applied pulse sequence is described as follows.

$$\text{Observed nuclei} \quad [180°_X - \tau - 90°_X]_n$$

where τ is a time interval between pulses.

Table 4.2 shows the values of proton spin–lattice relaxation time on starches from cereals and fruit seeds, obtained by an inversion recovery pulse sequence in an NMR low-field spectrometer. The values suggest that the starches from cereals present similar behavior, such as observed on corn and wheat starches, even if starch from wheat is more rigid than the starch from corn due to its macromolecular arrangement and also to the quantity, form, and water distribution in the starch granules. Therefore,

Table 4.2 Proton Spin–Lattice Relaxation
Times for Starches

Starch	T_1H (ms)
Corn	4
	84
	124
Wheat	7
	97
	137
Cumbaru	1
	38
	172

when compared with starches from cereal, starches from seed fruit present some particularities such as higher free-water mobility, higher intermediate molecular mobility (probably due to the higher movement freedom of carbon 6, which can be in different arrangements with different amorphous region distributions), and a region with higher rigidity composed of a crystalline and constricted amorphous region rich in fibers.

4.2.3.2 T_2 determination

In solid samples, T_2H relaxation time can be obtained by the signal width. However, the heterogeneity of the magnetic field causes different nuclei processes' magnetizations at different rates when the radiofrequency is switched off at a determined time (τ). This time is chosen as a function of the sample mobility. Therefore, this lack of homogeneity is refocused after a 180° pulse is applied, turning one magnetization (Komoroski, 1986; Stejskal and Memory, 1994; Souza and Tavares, 2002). The pulse sequence employed to determine the T_2 relaxation parameter is Carr–Purcell–Meiboom–Gill (CPMG), which is described below:

$$(90°_x - \tau - 180°_x) \text{ decay}$$

This relaxation time, as well as spin–lattice relaxation time, provides information about sample mobility at the molecular level, allowing the evaluation of the molecular dynamic. Sometimes, the T_2H parameter corroborated with the T_1H can provide details about sample behavior. Table 4.3 shows the spin–spin relaxation times for starches from cereals and fruit seed, using an inversion recovery pulse sequence in a NMR low field. Two values of T_2H are observed in cereal starches, whereas three values are detected in starches from fruit seed, suggesting that these starches show different structural behavior according to

Table 4.3 Spin–Spin Relaxation Times for
Starches

Starch	T_2H (ms)
Corn	6
	98
Wheat	6
	87
Cumbaru	3
	68
	134

their molecular organization and arrangements, inter- and intramolecular interactions, and water distribution on the starch granule. The
T_2H behavior was more expected for starch derived from fruit seed, in
relation to the distinction of molecular mobility of the material regions,
compared to cereal starches. Therefore, the same behavior of T_1H was
in evidence for T_2H.

4.3   Final remarks

NMR solid state can be useful in the characterization of starches, allowing us to obtain information about chemical structure, molecular dynamics behavior, and macromolecular interactions. It can be an alternative in
food analysis, because results are accurate, rapid, and reproducible, providing information about the complex matrix. In starch characterization,
NMR can give complementary information to x-diffractograms and thermal analysis, contributing to knowledge about starch structure, morphology, and dynamic behavior.

References

Bovey, F.A. and Mirau, P.A. 1996. *NMR of Polymers*. New York: Academic Press.
Cheethan, N.W.H. and Tao, L. 1998. Solid state NMR studies on the structural and
    conformation properties of natural maize starches. *Carbohydrate Polymers*. 36:
    285–292.
Cheng, H.N. 1991. NMR characterization of polymers. In: *Modern Methods of
    Polymer Characterization*, H.G. Barth and J.W. Mays (Eds.).113: 409–493. Series
    Chemical Analysis. Winefordner.
Conte, P., Spaccini, P., and Piccolo, A. 2004. State of the art of CPMAS ^{13}C-NMR
    spectroscopy applied to natural organic matter. *Prog. Nuclear Magnet. Reson.
    Spectros.* 44: 215–223.
Costa, D.A., Oliveira, C.M.F., and Tavares, M.I.B. 1998. Solid-state nuclear mag-
    netic resonance of the PA6/PC, PA6/PPO and PA6/PC/PPO blends. *J. Appl.
    Polym. Sci.*, 69:129–136.

Costa, P.M., Tavares, M.I.B., Bathista, A.L.B.S., Silva, E.O., and Nogueira, J.S. 2007a. High resolution NMR study of tropical fruit seed starches. *J. Appl. Polym. Sci.,* 105: 973–977.

Costa P.M., Tavares, M.I.B., Silva, E.O., Bathista, A.L.B.S., Nogueira, J.S., Filho, N.P., Ferreira, A.G., Barison, A., Daolio, C., and Vizzotto, L. 2007b. NMR and X-ray studies of derived from tropical fruit seed gelatinization process. *Int. J. Polym. Mater.* 56:1135–1143.

Edbon, J.R. 1980. *Developments in Polymer Characterisation*, J.V. Dawking (Ed). London: Applied Science.

Gidley, M.J. and Bociek, S.M.J. 1985. Molecular organization in starches: A ^{13}C CPMAS NMR study. *J. Am. Chem. Soc.* 107:7040–7044.

Harris, R.K. 1996. NMR studies of solid polymer. In: *Polymer Spectroscopy*, A.H. Fawcell (Ed.). UK: John Wiley & Sons.

Harris, R.K. 2000. Recent advances in solid state NMR. In: *Proceedings of the Fifth International Conference on Applications of Magnetic Resonance in Food Science*, Aveiro, Portugal, Vol. I, 1–11, University of Aveiro, Portugal.

Komoroski, R.A. 1986. In: *High Resolution NMR Spectroscopy of Synthetic Polymers in Bulk*, R.A. Komoroski (Ed.). Deerfield Beach, FL: VCH.

McBrierty, V. and Parker, K. J. 1993. *Nuclear Magnetic Resonance in Solid Polymers.* Cambridge, UK: Cambridge University Press.

Nascimento, A.M.R., Tavares, M.I.B., and Nascimento, R.S. 2007. Solid state NMR study of *Couma utilis* seeds. *Int. J. Polym. Mater.* 56:365–386.

Sanders J.K.M. and Hunter, B.K. 1993. *Modern NMR Spectroscopy: A Guide for Chemists*, 2nd ed. Oxford: Oxford University Press. ,

Schmidt-Rohr, K. and Spiess, H. W. 1999. *Multidimensional Solid-State NMR and Polymers*, 3rd ed., New York: Academic Press.

Silva, E.O., Tavares, M.I.B., Bathista, A.L.B.S., Filho, N.P., and Nogueira, J.S. 2002. High resolution ^{13}C NMR study of natural resins. *J. Appl. Polym. Sci.* 86:1848–1854.

Silva, N.M. and Tavares, M.I.B. 1996. Characterization of EPDM/atactic polypropylene blends by high resolution solid-state NMR. *J. Appl. Polym. Sci.* 60:663–667.

Souza, C.M.G. and Tavares, M.I.B. 2002. High resolution solid state NMR and SEM study of the interaction behavior of poly(ethylene-co-vinyl acetate)/poly(vinyl acetate) blends. *J. Appl. Polym. Sci.* 86:116–124.

Stejskal, E.O. and Memory, J.D. 1994. *High Resolution NMR in the Solid State.* New York: Oxford University Press.

Tang, H. and Hills, B.P. 2003. Use of ^{13}C MAS NMR to study domain structure and dynamics of polysaccharides in the native starch granules. *Biomacromolecules.* 4:1269–1276.

Tang, H., Godward, J., and Hills, B. 2000. The distribution of water in native starch granule: A multinuclear NMR study. *Carbohydrate Polym.* 43:375–387.

Tavares, M.I.B. 2000. NMR Molecular dynamic study of high crystalline polymers. *Polym. Test.* 19:899–904.

Tavares, M.I.B. 2003. High-resolution ^{13}C-nuclear magnetic resonance study of heterogeneous amorphous polymers. *J. Appl. Polym. Sci.* 87:473–476.

chapter five

Thermal transitions of starches

Paul Colonna and Alain Buleon
Institut National de la Recherche Agronomique

Contents

5.1   Introduction

Starch is used in a wide range of industrial applications. A first family of starch uses corresponds to foods and feeds. Native starch is by far the most consumed polysaccharide in the human diet; traditional staple foods such as cereals, roots, and tubers are the main sources of starch. Uses of starch after disruption of the native starch granules, such as gelatinized, hydrolized, or modified starches, led to ingredients and foods, either homemade or manufactured. These modifications also affect sensorial properties (texture, viscosity, visual aspect, flavor) as well as nutritional ones.

Starch has always been considered as an essential carbohydrate food energy source for the human population. A major change in starch perception occurred at the end of the last century, with the evidence of resistant starch, contributing to the fiber fraction. The rate of glucose release by the starchy structures, the level of resistant starch, and the functional properties are determined by the state of organization of amylose and amylopectin in the foods. Many public health authorities and food organizations such as the World Health Organization, the USDA Nutrition Advisory Program (2005), the French National Program for Nutrition and Human Health (2006–2008), the Japanese National Institute of Health and Nutrition, and the British Nutrition Foundation all recognize starch as a beneficial carbohydrate: starch, with a recommended nutrient intake of 150g/day, should be a predominant nutrient.

As most starches are processed before eating, thermal transitions of starch represent a key point in the control of starch functions. In industry, starch used in the manufacturing of adhesives, paper, textiles, and new materials either in blends or in composites depends upon its organization and interactions with other components. Each final application of starch has to be interpreted as an optimization of a given process–product goal. Consequently, thermal transitions take a central place in obtaining a matrix exhibiting properties in a large range of conditions (e.g., temperature, moisture, time length, presence of biotic agents) from a granular material.

Whatever the final uses, the mechanisms involved in these thermal transitions are the same. However, the resulting structures strongly depend upon the starch hydrothermal history and kinetics. Moreover, associated properties are related to phase transition, glass transition, physical aging, and to the plasticizing effect of water.

This chapter summarizes current knowledge of the thermal transitions of starches with a view to covering the latest improvements in physicochemical methods and the point at which they can be useful for all applications. It focuses on the influence of water, heat, and shearing on starch transformation, including melting/gelatinization and obtaining of starchy glasses or gels.

5.2 Disorganization of starch granules

It should be considered that the native initial state is more or less a theoretical state for starches coming from plant sources.

During milling of cereal grains, some starch granules undergo mechanical and thermal damage, governed by process intensity and hardness of the grain. As a consequence, when processed at the same conditions, hard wheat is more susceptible to mechanical damage than soft wheat. This starch damage is defined as any change in granule structure that can result in reduced resistance of amylases, acid, or thermal actions. It leads to quicker hydration and a higher amylase susceptibility, factors that play an essential role during starch hydrolysis, gelatinization temperature, dough formation, and fermentation. According to the technology, there is an optimum level of starch damage that promotes dough stickiness and gas production during fermentation, improving bread-making conditions.

Because extraction conditions and bread-making technology are used on starches from tubers and roots, granule damage is less relevant for these starches than those from cereal.

5.2.1 Phenomenological changes of gelatinization, pasting, and melting

5.2.1.1 In excess moisture (water:starch ratio > 1.5)

Under polarizing microscope observation, native starch granules show a dark birefringence cross ("Maltese cross"), which is characteristic of a spherulitic macromolecule organization inside each granule. The loss of the birefringence during gelatinization reflects the disappearance of the radial organization of amylopectin and amylose chains.

Gelatinization corresponds to irreversible disruption of molecular order within the starch granules observed when aqueous dilute suspensions of starch granules are heated above 60°C (Blanshard, 1987). A loss of birefringence occurs in a temperature range characteristic of each type of starch according to its botanic origin: potato starch from 56 to 66°C, wheat starch from 52 to 63°C, maize starch from 62 to 72°C, and rice starch from 66 to 77°C.

During gelatinization, several events take place simultaneously: diffusion of water inside the starch granule with a limited swelling, disappearance of birefringence, loss of crystallinity of the granule, endothermal phase transitions, predominant swelling of the granule after the loss of birefringence, and a decrease in the relaxation times of the water molecules.

Gelatinization exhibits two-stage behavior. The first is a limited swelling and a low level of solubilization, located around the gelatinization temperature of 60–75°C. Tester and Morrison (1990) stated that swelling is primarily the property of amylopectin and that amylose and lipids inhibit swelling of starches. The second step occurs above 90°C: granules swell and disrupt, leading to a more or less complete solubilization (30–60%).

In swelling and gelatinization, the starch granules may undergo changes into various shapes. A tangential swelling has been observed for normal cereal starches, supposedly due to the hydration and lateral expansion of lateral crystallites. For Marchant and Blanshard (1978), gelatinization is a semi-cooperative process, due to the relationship between the hydration-facilitated melting of the crystalline regions and the swelling owing to further hydration of disordered polymer chains in the amorphous zones. The undispersed fraction of the granules swells by absorbing 10 to 30 times its weight in water as a function of temperature, according to the type of starch. All laboratory procedures are far from industrial conditions of gelatinization, where shearing and high heating rate play an important function.

During gelatinization, changes on a microscopic scale occur simultaneously with nanoscale changes. The loss of crystallinity can be detected as an endothermic event by differential scanning calorimetry (DSC). Amylose-rich starches yield very broad endotherms, with higher melting temperatures (between 80 and 130°C) (Colonna et al., 1982; Inouchi et al., 1991). In addition, a second reversible endothermic transition is observed near 100°C for lipid-containing cereal starches. This transition is usually assigned to the melting of the amylose–lipid complex.

The use of synchrotron radiation allows the diffraction diagrams of starch to be picked up at very short intervals upon heating, and thus the kinetics of starch melting and amylose complexation to be followed. Garcia et al. (1996) and Jenkins and Donald (1998) have already used synchrotron radiation to study the gelatinization/melting behavior of wheat and cassava starches, respectively. Each type of starch gives an endotherm characterized by its enthalpic change (10–20 J g⁻¹) with characteristic temperatures (beginning, midpoint, end).

The enthalpy of gelatinization is a net thermodynamic quantity corresponding to the granule swelling, crystallite melting, hydration, and even recrystallization (exothermic) when complexable lipids are present. For most cereal starches which contain complexable lipids, an additional reversible endothermic transition is observed at 95–105°C, which represents a disordering transition of amylose–lipid complexes. These complexes are formed when amylose chains are released during gelatinization.

Changes occurring by gelatinization also can be observed by x-ray diffractometry, where A and B patterns disappear and a V pattern, which is attributed to the amylose–lipid complex, is observed, except for starches that are either waxy or do not contain lipids. X-ray diffractometry gives quantitative information about the residual crystallinity (Le Bail et al., 1999).

After gelatinization with infrared spectroscopy, the intensity of the band at 1047 cm^{-1} decreases mainly due to the line-broadening process of the bands at 1047 and 1022 cm^{-1}. The reverse process involving reappearance at 1047cm^{-1} is usally observed during gelation. Therefore the ratio 1047/1022 cm^{-1} is often interpreted as related to the amount of short-range order (double-helix content) relative to the amorphous content.

Initially, the breakage of the hydrogen bonds from crystalline zones gives a huge water absorption band, followed by a leaching of macromolecules of lower molecular weight, mainly amylose. This event can be mainly explained by the difference in diffusional coefficients: DT$_z$ values are 1.2 × 10^{-7}, 8–15 × 10^{-12}, and 1.7 × 10^{-12} m^2.s^{-1} for water, amylose, and amylopectin, respectively. Pasting is the rheological description of this event. Starch pastes correspond to suspension of swollen and empty starch granules, also called "ghosts," and solubilized amylose and amylopectin macromolecules outside the ghosts.

5.2.1.2   In reduced moisture (water:starch ratio < 1)

The designation "starch melting" includes the losses of crystallinity and native granular structures. Melting occurs at low moisture (as low as 11% w/w) and high temperature (as high as 180°C), where shear and heat enable the formation of a viscoelastic melt (Vergnes, Della Valle, and Colonna, 2003).

By submitting low-moisture starch to high shear stresses and temperatures, a macroscopic homogeneous molten phase is obtained, due to starch melting. Such phenomena are achieved on extruders and their extent may vary according to the parameters resulting in the versatility of this process: screw and die geometries and arrangement, barrel temperature, screw speed, and water addition. The viscous behavior of starches can change according to amylose content and it can be adjusted to an empirical model (Vergnes et al., 1987) with two parameters: moisture content and mechanical energy. An increasing sensitivity to water content and mechanical treatment is observed as amylopectin content increases. This trend is in agreement with the most significant plasticizing action on highly branched macromolecules noticed by Lourdin, Della Valle, and Colonna (1995). However, amylose leads to higher values of viscosity and more pronounced shear thinning behavior, which may be related to the larger sensitivity of amylopectin to macromolecular degradation due to its higher molecular weight.

5.2.2   Mechanisms of gelatinization and melting

Starch gelatinization is generally considered to occur in an excess of water (more than 100% added water on dry basis), whereas melting corresponds to the disappearance of native starch crystallinity at low hydration.

Starches present a melting process that is irreversible, kinetically controlled, and mediated by water plasticization. During the melting and gelatinization, successive endothermic transitions are observed according to water content and type of starch (Donovan, 1979; Biliaderis et al., 1986a; Russel, 1987). The first peak, identified as the gelatinization peak, observed above 66% (w/w), disappears progressively whereas further endotherms are progressively shifted toward higher temperatures as the water content decreases.

In many starch-based foods, water content determines the type of preparation and food; in baked goods the water content is 50–60% (w/w), and in French cookies it is <30%. Therefore, incomplete melting explains the residual starch granules observed in French cookies and bread crusts, in contrast to bread crumbs, where starch is completely gelatinized.

Melting temperature (Tm) is often mistaken for the gelatinization temperature. DSC studies have shown that there is no discontinuity differentiating gelatinization and melting. Experimental melting temperatures are known to vary as a function of water content according to the Flory (1953) equation for volume fractions of water between 0.1 and 0.7, where the inverse of melting temperature is a positive linear relation to volume fraction of water.

As in melting temperature, melting enthalpy depends on water content and starch origin. At low water content, multiple endothermic events are observed, which are more relevant to melting transitions, such as in extrusion cooking. The theoretical melting temperature of perfect crystallites with no water is between 160°C and 210°C, with the corresponding enthalpy between 160 and 875 J mg^{-1}, depending on the botanical origin of the starch. Using spherulites of DP 15 over a large range of volume fraction of water (0.4 to 0.95), Whittam, Noel, and Ring (1990) observed a higher melting temperature for A-type crystallites than B-type ones, suggesting that A-type spherolites present higher crystallinity than B-type. In short chain amylose, for A- and B-type crystallites, the melting behavior follows a typical polymer–diluent system behavior (Prasad and Mandelkern, 1989): on reducing water content, the melting temperature of both allomorphs increases.

All these interpretations are based upon a direct interaction between starch crystallites and water. Although the Flory equation fits well with the experimental results, true equilibrium between crystallites and solvent is not present during gelatinization, making the application of this equilibrium thermodynamics questionable. Furthermore Donovan (1979) has proposed that the coupling between the crystallites and the amorphous zones would be involved in multiple peaks observed in low-moisture conditions. Crochet et al. (2005) concluded that amorphous zones of starch granules have little influence on the gelatinization process, which is considered a thermodynamically controlled dissolution process. In contrast, at higher water content the cooperative nature of granule gelatinization is found at low water content, with the conversion of a crystalline double-helical to coiled conformation.

In studies of starch behavior at low water content, another approach was developed by Slade and Levine (1988) who introduced the glass transition concept as a key event determining the change in polymer mobility. According to these authors, because temperature increases, molecular motion is initiated, enabling molecules to slide past one another. At this point, the polymer becomes rubbery and flexible and presents a viscosity around $10^{12}Pa.s^{-1}$; this physical event reflects the increase of short segments' (3–20 monomers) mobility on the polymer backbone and it is called the glass transition temperature (T_g). Above T_g the material becomes rubber-like (mobile); structural transformations are allowed in the amorphous zones. The melting of interconnected microcrystallites depends on a glass-to-rubber transition of the amorphous zones on which water exerts a plasticizing effect. Below T_g, segmental mobility of polymer chains is frozen in a random conformation, rendering starch phases solid and glassy.

More than one melting endotherm at limited moisture content may be attributed to heterogeneous moisture distribution on the granule or to microcrystalline domains with different thermal stabilities. However, melting of isolated crystallites remains explained by the Flory law. Slade and Levine (1988) suggested that a range of T_g should be observed in the amorphous zones as a function of the constituents constrained by the crystallites. It does not take into account that gelatinization is a broad event for a population of starch granules, but a short event (1–1.5°C) when considering one granule.

Despite the use of modulated temperature on differential scanning calorimetry, unfortunately Tan et al. (2004) were unable to observe a plasticization of starch granules prior to gelatinization.

Glass transition temperature and melting temperature can also define the boundaries within crystallization of amylose–lipid complexes; annealing or melting of metastable crystallites occurs without loss of granular integrity. Annealing produces an ordering of the crystallites by propagation.

The plasticizers depress the T_g of polymer according to the Couchman and Karasz (1987) equation (Lourdin et al., 1997). The magnitude of this depression depends on the amount and the nature of these molecules. Zeleznak and Hoseney (1987) published T_g data on native and pregelatinized wheat starches, where increasing water content in dried starch after addition of 10% water resulted in depressed T_g from 151°C to 137°C, respectively. Above 30% water T_g remains stable (63 to 65°C), which corresponds to the minimum water requirement for the plasticization. It must be recalled that T_g for water is −134°C. However, some gap in temperature between experimental results and theory was observed by Whittam, Noel, and Ring (1991).

Recently, a side-chain liquid crystal model was proposed to approach the melting/gelatinization process (Waigh et al., 2000a,b). In this model, amylopectin is suggested as a structural analogue to a synthetic side chain liquid crystalline polymer with three distinct components: rigid units (mesogens) corresponding to double helices, flexible spacers, and a flexible backbone (Waigh et al., 1998, 2000a). In the nematic state, helices are not aligned into lamellae while in the smectic state; as in the granules at native conditions, the mesogens are aligned which leads to a 9 nm repeat between the lamellar lengths. At low water content (<5% w/w), the amylopectin helices are in a glassy nematic state and the rigid crystalline parts (mesogens) are somewhat disordered and the single peak corresponds to a helix–coil transition. At intermediate water content, the first DSC endotherm is attributed to dislocations between double helices leading to a smectic–nematic transition. The second is the helix–coil transition with irreversible disentanglement of double helices. Finally, in excess water, lamellar breakup and disentanglement of double helices occur simultaneously. The gelatinization appears as the result of four processes: (i) cleavage of existing starch–starch–OH bonds (endothermic), (ii) formation of starch–solvent–OH bonds (exothermic), (iii) the unwinding helix–coil transition of amylopectin helices (endothermic), and (iv) the formation of amylose–lipid complexes. Therefore the magnitude of the ΔH reflects the level of helices more than the crystallinity level.

Solvent and temperature are main parameters controlling long-term changes of starchy products. In parallel, the contribution of processes can be approached by using different solvents (Tan et al., 2004). One output of this new approach is to focus attention on solvent mobility, which will determine the conditions and possibilities of molecular rearrangements on a no-equilibrium basis. However, the main physical properties of T_m–T_g are governed particularly by the Williams–Landel–Ferry theory. Consequently, rates of all physical changes increase exponentially as a function of $T_{storage}$–T_g (i.e., temperature of storage minus temperature of glass transition) (Slade and Levine, 1987).

5.2.3 Functional properties

Some functional properties of starches comprise thickening behavior and enzyme susceptibility of gels or pastes. These properties are essential to the choice of processing technology and starch-based product applications and therefore have received much attention.

Due to their complex composition, starch pastes are difficult to model. However, important differences in rheological properties are observed according to the botanical origin of the starch. The different gelatinization procedures are believed to result in variable proportions of amylose and amylopectin solutions and swollen particles. Consequently, rheological behavior of starch pastes will depend upon the relative proportions of these two different systems, which are determined by concentration, heating rate, shearing, and starch damage. In the following some aspects of starch pastes and their rheological behaviors are discussed.

5.2.3.1 *Starch pastes in excess moisture (water: starch ratio >1.5)*

At low concentrations of starch (<10% w/w) pastes from starch exhibit shear thinning with thixiotropic behavior. Shear stress (τ versus shear rate γ) curves are linear over the range of 10–1000 s^{-1} and the apparent viscosity (η) can be represented as a function of flow behavior index <1. At shear rates below 10 s^{-1}, a yield stress is observed, which is typical of suspensions (Doublier, 1987).

For highly concentrated starch pastes (>10% w/w), flow behaviors depend upon cooking conditions, culminating in either shear thickening or shear thinning. These starch pastes can be considered as concentrated suspensions of closely packed deformable gel particles.

5.2.3.2 *Starch pastes in reduced moisture (water: starch ratio <1)*

Upon extrusion, the melt expands and cools rapidly due to vaporization of moisture, eventually settling into an expanded solid foam (Vergnes, Della Valle, and Colonna, 2003). Starch with $T_m/T_g \ll 1$ has a high glass-forming tendency, which results in a large free volume requirement and, thus, a large temperature increase required for mobility.

Expansion, sometimes called puffing, is the phenomenon by which foods, mainly cereals, acquire a porous structure, like a solid foam, due to transient heat and vapor transfers. Pore size and distribution, thickness, and mechanical properties of the wall's material will define, in turn, the texture of the product. This texture is strongly influenced by how the pores are generated. In the last ten years, there has been much attention paid to the basic phenomena of expansion: nucleation, bubble growth, water evaporation, coalescence, and shrinkage (Della Valle et al., 1997).

This solid foam is sensitive to moisture plasticization (moisture induces a lowering of T_g) especially in the 0–10% water content range; crispness is lost when water activity shifts above a water activity range between 0.35 and 0.5. Consequently, the plasticization effect on extrudate products is essential to define the final product characteristics. For example, it is known that for corn flour extrudates, an increase in T_g is related to an increase in the sensory–textural attribute of crispness (Katelunc and Breslauer, 1993).

5.3 Gelation

Pastes and molten materials are metastable nonequilibrium states and undergo structural transformations when cooling and during storage.

This process is called gelation, where starch gels and pastes show changes in their structure and rheological properties.

5.3.1 Structural changes

After the gelatinization–melting phenomena and upon cooling below T_m, amylose and amylopectin always produce gels with a B-type x-ray diffraction pattern.

During storage, amylose gel then constitutes an interconnected three-dimensional network, formed with an interchain association over the length of ~13 nm. These crystallites composed of long constitutive chains (DP 35–40) are acid resistant and exhibit a high fusion temperature (140–150°C) with an enthalpy change of 9–24 mJ.mg^{-1}. These chain segments inside the crystallites are obliquely disposed to the microfiber axis (Leloup, Colonna, and Ring, 1992). This model allows us to relate the microfibrous structure observed by electronic microscopy to macromolecular characteristics of the associated regions determined by acid hydrolysis. The length of the crystallites ranges from 8 to 18 nm, which is in agreement with the filament width observed by electron microscopy (20 ± 10 nm).

A network strand would therefore consist of contiguous associated blocks, aligned along the length axis of the microfiber. Double helices would then be linked to each other by loops of amorphous amylose segments, dangling in the gel pores. This fraction would be responsible for the hydrodynamic behavior of the amylose gels. This model suggests the occurrence of both parallel and antiparallel packings, which correspond to two local energetic minima of amylose double helices (Imberty et al., 1988). This packing of crystalline blocks is related to the coil conformation of amylose chains in solution (Ring, I'Anson, and Morris, 1985).

After gel formation and during storage, amylopectin forms thin crystallites which melt at 45–55°C (enthalpy transition 3–15 mJ.mg^{-1}). These crystallites are composed mainly of DP 15 linear chains. They consist of single- or multiple-branched molecules that originate from the S chain clusters of the molecule (Ring et al., 1987), resulting in a gel which is more easily degraded by enzymic action than amylose and presents poor mechanical properties. The key structural feature is the existence of short S chains consisting of both outer (A chain length CL 12–16) or inner (B1 CL 20–24) (Hizukuri, 1985, 1986; Hanashiro, Abe, and Hizukuri, 1996). These S chains can be considered as the limiting factor defining crystallite thickness. Size distribution of these chains also influences the melting temperatures of amylopectin gels.

If the chain distribution changes the melting temperature, it is expected that it also inteferes in water acessiblity. This is an important detail, because water entrapped in these three-dimensional networks can diffuse macromolecular probes as enzymes. Accessibility has been studied by diffusion until equilibrium: probe size and macromolecule concentration are the two main parameters. Accessibility is a reasonably continuous function of the hydrodynamic radius. For probes where the hydrodynamic radius is <10Å, the volume of solvent trapped within the network is completely accessible to the probe. The diffusion coefficient of bovine serum albumin in amylose and amylopectin gelled networks also decreases with increasing polysaccharide concentration; the diffusion coefficient decreased from 5.5×10^{-11} m^2.s^{-1} to 1.5×10^{-11} m^2.s^{-1} over the concentration range 5–15% w/w of gelling polymer (Leloup, Colonna, and Ring, 1990).

Modeling of the gel network has been carried out via a collection of immobilized, randomly spaced, rigid rods. The particle diffuses through the network by taking directionally random steps: if collision with the network occurs, the random step is not completed and the diffusion is retarded. More recently, hydrodynamic screening has been thought to be a more appropriate description of the physical process responsible for the retardation of diffusion (Leloup, Colonna, and Ring, 1990). The relationship between the average mesh size of the semi-dilute polymer solution and the hydrodynamic radius of the probe particle is important. When the hydrodynamic radius is much lower than the average mesh size, a continuum matrix is observed. The Stokes–Einstein relationship may be used to describe the diffusion process with the viscosity polymer solution.

5.3.2 *Mechanisms of gel formation*

Opaque gels are obtained upon cooling aqueous dispersions of concentrated starch as a result of a ternary phase separation occurring in the water–amylose–amylopectin system, which is followed by a nucleation leading to the development of a "thin" three-dimensional amylose network following this phase separation (Doublier and Choplin, 1989; Gidley and Bulpin, 1989). At this stage, elementary junction zones are established locally between the macromolecules. These should locally adopt a left-handed, parallel-stranded, double-helical conformation. On aging, starch gels develop a B-type crystallinity resulting from an aggregation process in a parallel register of the elementary junction zones.

A rapid quench of a starch paste or a melt leads to the formation of a vitreous glass with a brittle behavior. A similar state is obtained by rapid removal of water as in extrusion cooking at the expansion stage, or during the processing of pregelatinized starch, water being a plasticizer of starch. When cooled starch melts at high kinetic rates below T_g, gelation can be prevented and replaced by the formation of a glass. The behavior of a semi-crystalline solid is governed by its melting (T_m) and glass transition (T_g) temperatures.

Starch retrogradation and gelation occur within the temperature range of $T_g < T < T_m$. In that case, the rate of crystallization increases exponentially with $\Delta T = T - T_g$ (Slade and Levine, 1991b). When the storage temperature is cycled between low temperature (6°C), facilitating the nucleation, and a higher temperature (40°C), facilitating the growth of crystallites, the quality of the crystallites is improved and they melt at higher temperatures (Silverio et al., 2000). Therefore gelation by cooling occurs in two steps: (i) a phase separation that produces polymer-rich and polymer-deficient regions, and (ii) a crystallization within this polymer-rich region.

Miles, Morris, and Ring, (1985), Miles et al. (1985), and Doublier and Choplin (1989) studying amylose gelation, first observed an increase in turbidity ascribed to a phase separation process. In a subsequent phase, an increase in elasticity was recorded, followed on a longer time scale by a crystallization process. In this gelation model, the two latter phases correspond to the establishment of the gel junction zones. Conversely, Gidley and Bulpin (1989) and Clark et al. (1989) showed that monodisperse amylose (DP~2500) exhibits an immediate increase in elasticity before the development of turbidity. They suggested that the formation of junction zones leads to a phase separation phenomenon.

If the concentration is sufficiently high, the polymer-rich regions form an interconnected gel network. Otherwise, precipitation occurs. For amylose, gelation occurs only in semi-dilute regions, which correspond to a concentration of 1.5%. For amylopectin, chain entanglements begin at 0.9%; however, the liquid–solid transition is gradual with three defined stages. The polymer concentration range of 6–10% (w/w) is a critical zone above which solid behavior is evident (Ortega-Ojedo, Larsson, and Eliasson, 2004).

Incompatibility of gelling polymers is a fairly common phenomenon and usually leads to mixed gels formed of two phases. The polymer compositions of the continuous and dispersed phases are dependent on the ratio concentration of these polymers. For a particular value of this ratio, called the "phase inversion point," the continuous matrix becomes the discontinuous filler and vice versa. Starch gels obtained from a total dispersion of starch macromolecules can be considered as a particular case of two phase-mixed gels. Amylose and amylopectin are incompatible in solution (Kalichevsky and Ring, 1987) at 80°C. For gels with starch concentrations higher than 3% w/w, phase separation occurs below 90°C, giving two phases, each composed essentially of 70% w/w of either amylose or amylopectin. Therefore mixed gels are composed of a continuous network of one polymer (amylose or amylopectin) entrapping droplets of the other polymer.

Based upon rheological and thermal properties and hydrolytic resistance of amylose-rich mixed gels, Leloup, Colonna, and Buleon (1991) and

Ortega-Ojedo, Larsson, and Eliasson (2004) concluded that the transition in the gels' behavior observed for an amylose–amylopectin mixture with a ratio of 30:70 (weight basis) corresponds to the inversion in the polymeric composition of the continuous and discontinuous phases. Mixed amylose–amylopectin gels have a continuous amylopectin matrix below this threshold but a continuous amylose matrix above. This behavior is emphasized for gelatinized starch granules, where amylose leaching reinforces this exclusion. This threshold can be affected by the preparation conditions of the gel and the botanical origin of the starch.

There are also some indirect ways to evaluate starch gelation in gels. For example, starchy products prepared by casting, extrusion, or freeze drying of starch, amylose, and amylopectin, evaluated by solid-state C^{13} CPMAS/NMR through the decomposition of the C1 signal (Paris et al., 2001), reflect the influence of the intrinsic primary structure (linear or branched) and of the preparation procedure in conformational changes interpreted in terms of distribution of average glycosidic linkages, dihedral angles, and local order. Resonance peaks at 103.3 and 94.4 ppm are associated, respectively, with an angular conformation similar to the single helices present in V_a type structures (A line) and to more constrained conformations favored by drastic methods of preparation such as freeze drying. At 102.9 ppm, resonance peaks are associated with the helical conformation present in V_h and isopropanol structures and ranges from 100.4–101.4 and 97.0–98.6 ppm are influenced by crystallinity, branching points, and preparation techniques.

The line intensity is often connected to conformational or crystalline defects, whereas the chemical shift of the latter line is linked more to the presence of $\alpha(1–6)$ linkages. In starch samples evaluated by solid-state C^{13} CPMAS/NMR the most frequently observed tendencies are changes in width peak with water sorption, which expresses a more homogeneous distribution of conformations and higher sensitivity to rehydration of more constrained conformations (freeze-dried samples).

5.3.3   Functional properties

Syneresis occurs during long-term storage, which is one of the disadvantages of unmodified starch as a gelling agent. The macromolecular reorganization has major effects on texture and nutritional value of foods containing starch. Starch gels are elastic although the yield values are low so that the gels are easily broken down. Amylopectin gels behave as Hookean solids at strains <0.1, and amylose gels at strains <0.2. Amylose gels exhibit a rapid rise in storage modulus whereas amylopectin gelation is a much slower process.

In addition to the amylose-to-amylopectin ratio, starch concentration in gels is the main parameter responsible for their functional properties.

There is a linear relationship between modulus and concentration at 10–25% w/w starch gel concentration after storage for 6 weeks at 1°C. At 10–20% w/w concentrations, storage modulus increases rapidly during the first 100 min and then the rate becomes slower; the final storage modulus value is only obtained 21 days later. With amylose–amylopectin mixtures, the higher the starch concentration is, the smaller the amount of amylose needed to obtain G' >> G" and higher values of moduli (Ortega-Ojedo, Larsson, and Eliasson, 2004).

Models used for the prediction of starch retrogradation kinetics are grounded in the Avrami equation, which was originally derived for polymer crystallization in melt and was observed to have a good fitting of the experimental values (Farhat and Blanshard, 2001). The equation has two parameters, the rate constant and the Avrami exponent, which depend on the type, nucleation, and the dimensions where growth takes place. Data treatments have led to predictions of the starch crystallization, taking into account both the presence of residual starch crystallites and the mechanism of crystal growth in freshly cooked durum wheat bread (Del Nobile et al., 2003).

Technological control of texture is possible by the dispersion level of starch granules, the amylose and water content, and the storage conditions. Whatever the macromolecular composition, gelation is faster at lower temperature. Rigidity development is enhanced significantly at lower temperature, mainly for amylopectin gelation. Crystallization presents a maximum in the range of starch concentration 50–60% w/w. Above 80% starch, crystallization is blocked by T_g. After pasting or drum drying of maize starch, amylose crystallizes independently from amylopectin by complex formation and retrogradation. This phenomenon is explained by the leaching of amylose during thermal treatment, subsequent phase separation, and then the crystallization within the amylose volume fraction. These crystallites, composed of linear chains with degree of polymerization (DP) range between 35–40, are acid-resistant and exhibit a high fusion temperature (above 100°C). In contrast, after extrusion cooking, amylose and amylopectin co-crystallize in the same manner as pure amylopectin.

Mungo starch vermicelli and rice flour noodles are based upon these starch networks (Mestres, Colonna, and Buleon, 1988). In both types of foods, an amylopectin-based structure is reinforced by amylose-based structures present mainly in the complexed (V-type diffraction pattern) and retrograded (B-type diffraction pattern) forms for rice flour noodles and mungo starch vermicelli, respectively. These crystallites are more resistant toward acid hydrolysis and their melting temperatures are between 82 and 119°C. Consequently, the cooking behavior of the glutenless noodles can be explained by these amylose networks.

Gelation is also involved in the formation of resistant starch (RS) fractions incompletely degraded in the upper gut (Colonna, Buléon, and Leloup,

1992) and fermented into the colon (Faisant et al., 1995). Details of RS types are discussed in Chapter 8. Englyst, Kingman, and Cummings (1992) have defined a type of RS, the RS3, consisting of retrograded starches formed during cooling and storage of gelatinized starches. For example, there are RS in bread and baked potato (Eerlingen and Delcour, 1995) and in gel gelation where, after cooling, amylose reassociates in networks that are strongly resistant to α-amylases (Rolland-Sabaté et al., 2004). However, the gelation process to obtain RS has an important role in the formulation of functional/nutraceutical food products used in low glycemic ingestion diets.

5.4   Reorganization of starch gel structure

5.4.1   Glass transition and plasticization

The increasing interest of food scientists in the concept of glass transition arises from an attempt to describe the complex solid systems arising from thermodynamic equilibrium. Relying on polymer science concepts transferred to hydrated biomaterials, the applicability of the kinetic description of amorphous glassy/rubbery systems has yielded limited dynamical views about the so-called "mobilities."

In amorphous and semi-crystalline polymers, at sufficiently low temperature, the mobility of polymer chains is strongly restricted, limited to very slow and local molecular motion. By applying thermal energy, molecular motion is initiated at a given temperature called the glass transition temperature. Above T_g, the materials become thermoplastic, rubbery, and flexible. At the glass transition, there is a discontinuous change in heat capacity and in the thermal expansion coefficient. T_g also decreases with increasing water content, which acts as a plasticizer of starchy products and increases with increasing crystallinity. This approach was applied to examine food systems, and especially the mechanical and storage properties of starch low-moisture foods, when water was shown to be an ubiquitous plasticizer of biopolymers (Franks, 1991). The concept was then amplified by numerous authors including Slade and Levine (1991a,b, 1993a,b, 1995), Blanshard and Lillford (1993), and Roos (1995).

Dynamical mechanical thermal analysis (DMTA) and differential scanning calorimetry (DSC) are the most common techniques to approach glass transition and plasticization. The advantage of DSC is the use of a confined sample whose composition and moisture content may be kept constant during temperature changes. The heat capacity change detected by this technique mainly probes the mobilization of translational and rotational degrees of freedom, allowing chain flow in the case of polymers, although specific motion in the structure cannot be precisely identified.

By comparing glass transition temperatures determined by DSC, Bizot et al. (1997) illustrated a relationship between structure and properties resulting from influence of molecular weight, degree of branching, and (1–4) versus (1–6) glycosidic linkage ratio upon the depression of T_g with water content. These results extended those of previous studies done by Kalichevsky et al. (1992) and Noel and Ring (1992).

The effect of water plasticization on amylose, amylopectin, phytoglycogen, pullulan, and different low molecular weight products issued from hydrolysis was described by Couchman and Karasz's correlation (1987). Linear chains appear to favor chain–chain interactions and induce partial crystallinity; branched molecules display lower T_g due to chain end effects as well as flexibility of branching points (Levine and Slade, 1988). The three dihedrals present in α (1–6) linkages seem to depress T_g in a similar fashion to internal plasticization. The example of linear α (1–4) amylose chains bearing (1–6) grafted fructose was also examined as a first step toward tailored structures designed to optimize mechanical properties and internal plasticization and inhibit recrystallization (Richards and Vandenburg, 1995).

5.4.2 Physical aging

Being far from thermodynamic equilibrium, glassy materials are subject to so-called "structural relaxation," a molecular rearrangement leading to lower states of energy. These alterations may be of significance for food products because they directly affect enthalpy, volume, mechanical, and diffusion properties.

The term "physical aging" was introduced by Struik (1978) to distinguish these effects from other aging processes such as chemical reactions, degradations, or changes in crystallinity. Physical aging has been widely studied by dielectric spectroscopy, being considered as related to the evolution of physical properties such as volume, enthalpy, relaxation times, and creep compliance (Simon et al., 1997).

Enthalpy relaxation is now clearly recognized as a significant phenomenon occurring during storage of food materials and a signature of physical aging. Evaluating DSC endotherms, enthalpy relaxation is attributed to corresponding endothermic peaks observed betweem 50°C and 60°C on fresh samples, which disappeared on second scans after heating samples. Its importance depends on ΔT (difference between T_g and T_{aging}) and on the time of aging (Shogren, 1992). This behavior is observed in a wide variety of products, including wheat gluten (Hoseney, Zeleznak, and Lai, 1986), breakfast cereals (Sauvageot and Blond, 1991), amylopectin (Kalichevsky et al., 1992), hydrated starch lintners (Le Bail, Bizot, and Buléon, 1993), and native rice starch (Seow and Teo, 1993).

Enthalpy relaxation of several hydrated polysaccharides after aging at different times and temperatures (15, 45, 65, and 100°C below T_g) was compared with synthetic polymers, such as polymethylmethacrylate (PMMA) and polyvinylpyrrolidone (PVP) by Borde et al. (2002). Structural relaxation during storage yielded either sub-T_g enthalpy recovery peaks or overshoots in the glass transition region, showing the width of the relaxation time spectrum. With an increase of the aging time, both amplitude and position on the temperature scale of the enthalpy recovery peak increase. The enthalpy recovery peak develops less markedly for the lower aging temperatures, but the shift on the temperature scale is larger. Extruded starch equilibrated apparently within the experimental time; however, amylopectin and phytoglycogen aged slightly faster than PMMA and hydrated PVP without completely reaching their equilibrium. In these samples, aging effects detected after short aging at very low temperatures (T_g 100°C) revealed the width of the relaxation time spectrum.

5.5   Interactions with other molecules during thermal transitions

Components of foods affect kinetic and characteristic temperatures of phase transitions, as well as the inherent pasting and gelling of starch. Consequently their contributions to the complete characteristics of complex foods may be significant. In the next sections the main interactions of some food components with starch and their influence on thermal transitions and rheological properties are discussed briefly.

5.5.1   Hydrocolloids and proteins

Starch pastes often present thermodynamic incompatibility with other biopolymers (Tolstoguzov, 2003). After their preparation, blends tend to separate into areas with a high concentration of polymer A and other areas with high concentrations of polymer B, known as phase separation. The phase boundary betweeen stable and unstable "polymer A–polymer B" is a key result, and its identification requires long observation. Easily detected with potato starch and gelatin (Abeysekera and Robards, 1995), the functional key is the phase separation threshold, that is, the minimal bulk concentration of biopolymers at which phase separation occurs. This threshold is rather high for mixtures of starch and proteins (Tolstoguzov, 2003).

Hydrocolloids modify the rheological properties of starch pastes and gels (Christianson et al., 1981; Doublier, 1987; Funami et al., 2005). Addition of hydrocolloids enhances or modifies the gelatinization and retrogradation behavior of starch and improves the water-holding capacity and the freeze–thaw stability of starch-based preparations. Incompatibility between leached amylose and hydrocolloid is responsible for a higher concentration of the hydrocolloid macromolecules in the suspension. In starch gels, whereas galactomannans increase only the rate of gelation without modifying the final rigidity, xanthan presence increases storage modulus values toward a solid-like behavior. These synergistic behaviors remain stable at room temperature. The starch–hydrocolloid system has to be considered as a suspension of starch ghosts dispersed in a solution of hydrocolloids (Closs et al., 1999).

Gluten is an important group of proteins that interacts with starch during cooking. Gluten in bread undergoes a transformation resulting in the release of water, which becomes absorbed by retrograding starch (Breaden and Wikkhoft, 1971). By evaluating a combination of DSC, x-ray, and NMR results, it was concluded that the presence of gluten has no effect on the kinetics, extent, or polymorphism of amylopectin retrograda- tion (Ottenhof and Farhat, 2004). However, on an industrial scale, gluten retarded water loss in the starch granule remnants (Wang, Choi, and Kerr, 2004). This could explain the decrease of retrogradation enthalpy as starch when increasing the level of gluten (Eliasson, 1983). Sevenou (2002), using the FTIR technique, demonstrated that starch gelatinization did not lead to gluten dehydration during the heating of the dough. The question of water transfers during baking has to be considered at different scales in order to separate rheological behavior from short distance interactions.

Interactions with starch and milk and its derivatives (mainly milk powder) have also been investigated. The presence of minor amounts of starch in milk increased the viscosity. Casein micelles strongly increased the modulus of the starch suspension certainly due to the inaccessibil- ity of swollen starch granules to the casein (Master and Steeneken, 1997). However, the effect of sodium caseinate on storage moduli of starch gels changes according to the starch botanic origin (Bertolini et al., 2005). In addition to origin, the effect of sodium caseinate addition on thermal and rheological behavior of starch gels is also affected by starch concentra- tion. In sodium caseinate mixtures with wheat, rice, corn, and cassava starches, sodium caseinate promoted an increase in the storage modulus and in the viscosity of the composite gel when compared with starch gels, but in potato starch the effect of sodium caseinate resulted in decrease of the storage modulus and in the viscosity. It was also reported that addi- tion of sodium caseinate resulted in an increase in the homogeneity in the matrices of cereal starch gels and promoted an increase in the onset temperature, gelatinization temperature, and end temperature, and a sig- nificant interaction between starch and sodium caseinate for the onset temperature, the peak temperature, and the end temperature (Bertolini et al., 2005).

The addition of sugars increases the apparent viscosity of starch– milk gels, with a higher influence of fructose rather than sucrose and glucose (Abu-Jdayil, Mohamed, and Eassa, 2004). When considering the effect of addition of starch on milk behavior (de Bont, van Kempen, and Vrecker, 2002), phase separation (binodal) is observed as with guar and locust gums. However, it occurs at a relatively high concentration

(>1.5%) of added amylopectin, with the formation of protein aggegates. The experimental results from confocal scanning light microscopy and image analysis show a remarkable agreement with Vrij's depletion theory (Vrij, 1976) for polymer and colloid mixtures. Amylopectin gives different microstructures in contrast to commonly studied gums. Using the experimental procedures of Sperry (1984) and the theoretical approach developed by Lekkerkerker et al. (1992), this difference is ascribed to a low value (<0.5) of the polymer–colloid size ratio ($\xi = \sigma_\pi/\sigma_\chi$) for milk protein–amylopectin systems related to the attraction range of polymers (diameter σ_π) and colloids (diameter σ_χ).

Systems of starch with proteins at low moisture content, such as in extrusion, also lead to phase separation, which can be easily detected by confocal light microscopy. At low zein levels (5–15% w/w db), starch gives a continuous phase with homogeneous distribution of zein particles (5–50 mm). Blends with higher zein content present a continuous zein phase with delimited amorphous starch domains (Chanvrier et al., 2005). An interesting feature is the behavioral difference between storage proteins and starch in the glassy state at similar T_g-ambient temperatures. Glassy starch with 12% moisture is ductile whereas storage proteins present an elastic limit with a fragile rupture observed for zein (Chanvrier et al., 2005) and gluten (Attenburrow et al., 1992). The fragility of the blends can be attributed to the phase separation between starch and zein, with a threshold close to 20–30% zein content. Above 50% zein behavior predominates thanks to the continuity of the protein. These results were obtained by Bertolini et al. (2005), where concentration of gels changed the polymer interactions and their effect on rheological behavior.

5.5.2 Sugars

Oligosaccharides are known to modify starch behavior during gelatinization by shifting gelatinization toward higher temperatures, increasing or decreasing the enthalpy change, decreasing the viscosity, and reducing swelling and solubility (Evans and Haiman, 1982; Eliasson, 1992; Ahmad and Williams, 1999; Crochet et al., 2005).

The general trend is that sugars delay gelatinization to higher temperatures (>90°C) and increase gelatinization enthalpy in the following order: water < ribose < fructose < glucose < maltose < sucrose. The swelling factors in the presence of sugar are higher compared to water alone for sugar concentration below 25% and reduced at sugar concentration >25% (Ahmad and Williams, 1999). Starch systems containing sugars present pronounced shear-thinning behavior. Mixing glucose and fructose gives results near those gained with equivalent amounts of one single sugar, although minor differences could suggest nonadditive effects (Sopade, Halley, and Junning, 2004).

The mechanism interpretation is based upon coupling DSC and x-ray diffractometry at large and small angles. The shifting of the endotherm upward reflects the shifting of the mobility threshold of amorphous growth ring regions of the starch granules (Perry and Donald, 2002). At this threshold, the mobility and the swelling of the amorphous growth regions reach the maximum level, beyond which native granular structure is irreversibly altered. This effect of sugars can be explained by the concept of antiplasticization, demonstrated with several plasticizers (Slade and Levine, 1991b; Gaudin et al., 1999, 2000). Sugar, in the presence of starch and water, behaves as a plasticizing cosolvent with water, and the combination of sugar–water leads to a plasticizing complex of higher molecular weight (and thereby lower free volume) than water alone. The plasticization of the amorphous regions, which immediatly precedes the gelatinization, is less pronounced than with water alone. Consequently the gelatinization temperature shifts to higher values to produce thermally induced mobility. Another interesting observation is that when a DSC scan is performed at conditions giving rise to a double endotherm for starch gelatinization, the double endotherm is shifted into one single endotherm. This observation is due to nonuniform moisture distribution among starch granules during gelatinization. The antiplasticization behavior of sugars is also present for amorphous starchy materials (Gaudin et al., 1999, 2000), where strain hardening is observed at low plasticizer content.

For gelatinized starches, and granular remains (ghosts), maltose and maltotriose induce a reduction of recrystallization (Smits et al., 2003). Threitol and xylitol give the most important reduction of recrystallization of potato starch.

5.5.3 Amylose complexation with small molecules

Amylose shows a unique capacity to form complexes with a variety of molecules. The best known interaction is with iodine, based upon the location of four dominant polyiodide chains I_9^{3-}, I_{11}^{3-}, I_{13}^{3-}, and I_{15}^{3-}. An interesting result is the absence of I_2 in the polyiodide chain (Yu, Houtman, and Atalla, 1996).

Monoacyl lipids and emulsifiers, as well as smaller ligands such as alcohols and flavor compounds, are able to induce the formation of left-handed amylose single helices, also known as V amylose. V amylose is a generic term for crystalline amyloses obtained as single helices co-crystallized with compounds such as iodine, dimethylsulfoxide, alcohols, or fatty acids. Although such compounds are required for formation of the V-type structure, they are not systematically included in the amylose helix. Resulting helical conformation and crystalline packing depend on the nature of the ligands and conditions for the formation of the complex. The specific interactions of amylose with lipids and aromas have a strong impact on food quality. Such complexation could have also important potential in the field of healthy foods when applied to protection and vectoring of vitamins and micronutrients.

5.5.3.1   Lipids

Cereal starches are characterized by the presence of monoacyl lipids, free fatty acids (FFAs), and lysophospholipids (LPLs) in amounts positively correlated to amylose content (Morrison et al., 1993). Wheat, barley, rye, and other triticale starches contain LPL almost exclusively, whereas other cereals contain mainly FFA together with a minority of lysophospholipids. Lysophosphatidylcholine is the major lipid found for both wheat and maize, with palmitic and linolenic acids. For barley starches, the fatty acid composition of the lipids becomes progressively more unsaturated as lipid content increases, but this pattern is less consistent in starches from maize. In starch from wheat and barley harvested at various stages of grain development, both amylose and LPL contents increase with maturity. The emerging picture is that of a large A-type starch granule displaying a gradient (from the hilum to the periphery) of increasing amylose and LPL.

Most waxy starches, as in starches from roots and tubers, contain negligible lipids. In the family of maize starches, by comparing different mutants, lipid content appears most directly correlated with long chain linear α–1,4–glucan (i.e., the backbone of amylose) revealed by enzymatic debranching.

Numerous studies have observed correlations between these monoacyl lipids and the functional properties of barley (Morrison, Milligan, and Azudin, 1984), oat (Wang and White, 1994), and wheat (Tester and Morrison, 1990) starches. Monoacyl lipids will induce the formation of amylose–lipid complexes during gelatinization. They will restrict swelling, dispersion of the starch granules, and solubilization of amylose, thus generating opaque pastes with reduced viscosity and increased pasting temperatures.

In food systems, polar lipids affect the viscous and gelling behavior of starch gels (Morrison, 1995). The technological importance of such interactions is clearly demonstrated when comparing cereal starches (excluding waxy genotypes) and lipid-free starches such as tuber starches. Differences in functional properties have been extensively studied and were most often analyzed on quantitative bases. However, ultrastructural aspects have not been reported due to issues in analyzing such complex molecules.

The intensity and nature of phase transitions (annealing, melting, polymorphic transitions, recrystallization, etc.) induced by hydrothermal treatments in crystalline structures are related to temperature and water content (Buleon et al., 1998). Despite its small concentration, the lipid phase

present mainly in cereal starches has a large influence on starch properties, particularly in complexing amylose. Many authors showed that complex formation occurs during heat/moisture treatments, especially during
gelatinization of starches naturally containing lipids (Kugimiya, Donovan,
and Wong, 1980; Kugimiya and Donovan, 1981), or when lipids are added
to defatted starches (Biliaderis et al., 1986; Biliaderis, Page, and Maurice,
1986), or pure amylose free of natural lipids (Biliaderis et al., 1985). Both
naturally occurring and heat-formed complexes present specific properties such as a decrease in amylose solubility or an increase in gelatinization temperatures (Eliasson et al., 1981; Morrison et al., 1993a). Polar lipids,
for example, fatty acids and their monoglyceride esters, are used for their
technological interest concerning the reduction of stickiness, the improvement of freeze–thaw stability (Mercier et al., 1980), and the retardation of
retrogradation. The foremost example is probably the use of fatty acids
and monoglycerides as anti-staling agents in bread and French cookies:
incorporation of such additives in the dough induces a slower crystallization (retrogradation) of the amylopectin fraction and therefore retards the
staling of bread (Krog, 1971; Champenois et. al, 1995).

Interaction of amylose with fatty acids and monoglycerides yields
more or less crystalline structures with the V_h crystalline type also
obtained with linear alcohols (Sarko and Zugenmaier, 1980; Rappenecker
and Zugenmaier, 1981; Whittam et al., 1989). In such a structure, the chain
conformation consists of six left-handed residues per helix turn with
0.792–0.805 nm (i.e., a rise per monomer between 0.132 and 0.136 nm).
In the case of amylose–lipid complexes, it is assumed that the aliphatic
part of the lipid is included inside the amylose helix, and the polar group
lies outside, as it is too large to be included (Godet et al., 1993; Morrison,
Law, and Snape, 1993). In the V_h-type, the most common form obtained
by complexation of amylose with lipids, the single helices are packed in
an orthorhombic unit cell (a = 1.37 nm, b = 2.37 nm, c = 0.805 nm) with
the space group P212121 and 16 water molecules within the unit cell.
Amylose–lipid complexes can be crystalline or amorphous, depending
on the temperature at which they are constituted (Biliaderis, 1992; Godet,
Bizot, and Buleon, 1995). The two forms cannot be found by DSC because
they yield very similar melting/decomplexing enthalpy and temperature.
Such behavior is apparently due to a major contribution of the intramolecular energy involved in the formation of a single helix relative to the total
energy of complex formation (packing energy); that is, interhelical bonding seems to contribute very little. The latter form consists of individual
complexed single helices, which are not involved in a crystalline packing.
The formation of V_h crystalline structures (Le Bail et al., 1999) can be also
observed by synchrotron x-ray diffraction in native maize starch heated at
intermediate and high moisture content (between 19 and 80% w/w). The
crystallization of amylose–lipid complexes can be shown upon heating

after gelatinization at 90 to 115°C in excess water; the second endotherm is attributed to lipid-containing starches and it can be assigned to the melting of crystalline amylose–lipid complexes formed upon heating. For intermediate water contents, mixed A + V_h (or B + V_h for high amylose starch) diffraction diagrams are observed and two mechanisms can be involved in amylose complexing: the first relating to crystallization of the amylose and lipid released during starch gelatinization, and the second to crystalline packing of separate complexed amylose chains (amorphous complexes) present in native cereal starches.

5.5.3.2 Alcohols, aromas, and flavors

Besides monoacyl lipids, amylose forms crystalline complexes with a variety of small ligands, resulting in V-amylose. Apart from very specific complexes obtained with iodine, the best documented complexes are obtained with alcohols. Linear alcohols yield the well-known V_h structure already mentioned for amylose–lipid complexes. Complexes induced by *n*-butanol yield a unit cell larger than that of the V_h type (Helbert and Chanzy, 1994). Similar observations were made on V-isopropanol amylose (Buléon et al., 1990) and, to a lesser extent, on V-glycerol complexes (Hullemann, Helbert, and Chanzy, 1996). For these structures, it is possible to transform crystals into the V_h type by solvent exchange. It is therefore hypothesized that amylose helices are also made of 6 D-glycosyl units per turn. However, the helix cavity may be too small to host and conceivably can be linked between helices. Amylose with a larger helical diameter is formed in presence of α-naphthol. In this case, the amylose helix consists of 8 D-glycosyl units per turn and the bulky molecules are included in the helical cavity as well as between helices (Yamashita and Monobe, 1971; Winter et al., 1998).

Knowledge of interactions between aroma and matrix is essential to optimize formulation in food processing. Complexing of amylose by aromas is frequently induced by food processing and is thought to be one of the possible mechanisms for aroma retention in food systems (Rutschmann and Solms, 1990; Nuessli et al., 1997). It could be approached through the determination by x-ray scattering of the crystalline type of complex, which influences the nature of inclusion (inter- and intra-helical), and through thermostability using DSC. In several cases, resulting complexes exhibit a different structure from that of the classical V_h amylose. Using both electron and x-ray diffraction techniques on lamellar crystals of pure amylose complexed with aromas such as fenchone, menthone, or geraniol, Nuessli et al. (2003) showed that a sevenfold helix corresponds to the V-isopropanol type, and a similar but larger unit to a sixfold V_h-type helix. Therefore, depending on the size of the ligand, it could be included within and/or between helices.

Complementary to x-ray diffraction, solid-state ^{13}C CPMAS/NMR is also a very efficient noninvasive technique to assess helical conformation or the crystalline structure of complexed starch. Gidley and Bociek (1988) demonstrated that C1 and C4 glycosidic sites were more sensitive to conformational changes than the C2, C3, or C5 carbons because C1 and C4 showed higher chemical shift dispersions under various conformations of the glycosidic linkage in α–(1–4) glucans (Marchessault et al., 1985; Gidley and Bociek, 1988). A similar approach was also applied to the different crystalline types of amylose–alcohol complexes (Kawada and Marchessault, 2004) and to linalol and menthone aromas (Rondeau-Mouro, Le Bail, and Buleon, 2004). Sequential washing of the powdered complexes with ethanol before and after desorption permitted probing intra- and inter-helical inclusions. High resolution magic angle spinning (HRMAS) recordings used to compare the chemical shifts of free and bound aroma revealed some hydrogen bonding involved in the amylose complexing. Moreover it shows that free aroma is completely removed by ethanol washings. Using cross polarization magic angle spinning (CPMAS) and x-ray scattering experiments, it was observed that the V-isopropanol type was retained for linalol, whatever the treatment used. Conversely, it shifted toward V_h type for menthone after ethanol washing before the desorption step, reflecting the disappearance of inter-helix associations between menthone and amylose. The stability of the complex prepared with linalol shows that this ligand is more strongly linked to amylose helices. The discrepancies observed in the chemical shifts attributed to carbons C1 and C4 in CPMAS spectra of V-isopropanol and V_h forms could be attributed either to a deformation of the single helix (with possible inclusion of the ligand inside) or to the presence of the ligand between helices (only water molecules are present in the V_h form).

5.6   Final remarks

Most starch transformations can now be described by physical and chemical factors based on a fundamental understanding. The final state of a food is a result of processing and storage paths, along with kinetic and thermodynamic factors.

With a wide range of available parameters, it is now possible for each technology to develop a state diagram representing a plot of composition and temperature, showing the boundaries of solid and liquid phases (states) in thermodynamic equilibrium.

This phase diagram concept has recently been extended (MacKenzie, 1977) to encompass thermodynamic states that are away from equilibrium, kinetically controlled metastable states. These supplemented phase diagrams (or dynamic phase diagrams) allow a qualitative analysis of food systems when mapping composition, time, and temperature.

In the frame of a general review on starch, all key factors necessary to build a dynamic phase diagram have been presented and should be adapted to each case. Data mining on applied publications should allow building complex dynamic phase diagrams. Glass transition has been considered in the last ten years as the central factor. A sucrose–water state diagram has revealed the relative locations of glass, solidus, liquid, and vapor curves, useful for explaining the cookie and cracker technologies (Slade and Levine, 1995). Time is a critical parameter when modeling kinetically controlled phenomena in the rubbery state. With starch, the question is very complex due to the large number of interactions in complex systems and the potential modifications of component thermal properties that might result from phase separation (Katelunc, 2003).

Assessing the structural level responsible for a property (texture, shelf life) will allow the use of state diagrams as a rational basis for the selection of the best technological paths (heat and mass transfers) or formulation (structure–function) for the final quality. A state diagram can also to be an important tool to elucidate some questions about starch structure, which are reflected in its thermorheological behavior.

References

Abeysekera, R.M. and Robards, A.W. 1995. Microscopy as an analytical tool in the study of phase separation of starch–gelatin binary mixtures. In: *Biopolymer Mixtures*, S.E. Harding, S.E. Hill, and J.R. Mitchell (Eds.). Lochborough, UK: Nottingham University Press, pp. 143–160.

Abu-Jdayil, B., Mohamed, H.A., and Eassa, A. 2004. Rheology of wheat starch-milk-sugar systems: Effect of starch concentration, sugar type and concentration and milk fat content. *J. Food Eng.* 64:207–212.

Ahmad, F.B. and Williams, P.A. 1999. Effect of sugars on the thermal and rheological properties of sago starch. *Biopolymers*. 50:401–412.

Attenburrow, G.E., Davies A.P., Goodband, R.M., and Ingman, S.J. 1992. The fracture behavior of starch and gluten in the glassy state. *J. Cereal Sci.* 16:1–12.

Bertolini, A.C., Eppink, M., Creamer, L., and Bolland, M. 2005. Some rheological properties of sodium caseinate–starch gels. *J. Agric. Food Chem.* 53:2248–2254.

Biliaderis, C.G. 1992. Structures and phase transitions of starch in food systems. *Food Technol.* 6:98–109.

Biliaderis, C.G., Page, C.M., and Maurice, T.J. 1986. On the multiple melting transitions of starch/monoglyceride systems. *Food Chem.* 22:279–295.

Biliaderis, C.G., Page, C.M., Maurice, T.J., and Juliano, B.O. 1986. Thermal characterization of rice starches: A polymeric approach to phase transitions of granular starch. *J. Agric. Food Chem.* 34:6–14.

Biliaderis, C.G., Page C.M., Slade, L., and Sirett, R.R. 1985. Thermal behavior of amylose-lipid complexes. *Carbohydrate Polym.* 5:367–389.

Bizot, H., Le Bail, P., Leroux, B., Davy, J., Roger, P., and Buléon, A. 1997. Calorimetric evaluation of the glass transition in hydrated, linear and branched polyanhydroglucose compounds. *Carbohydrate Polym.* 32:33–50.

Blanshard, J.M.V. 1987. Starch granule structure and function: A physicochemical approach. In: *Starch Properties and Potential, Critical Reports on Applied Chemistry*, 13, T. Gaillard (Ed.). New York: John Wiley & Sons, pp. 16–54.

Blanshard, J.M.V. and Lillford, P. 1993. *The Glassy State in Foods*, Loughborough, UK: Nottingham University Press.

Borde, B., Bizot, H., Vigier, B., and Buléon, A. 2002. Calorimetric analysis of the structural relxation in partially hydrated amorphous polysaccharides. II Phenomenological study of physical ageing. *Carbohydrate Polym.* 48:111–123.

Breaden, P.W. and Wikkhoft, E.M.A. 1971. Bread staling. III. Measurement of the redistribution of moisture in bread by gravimetry. *J. Sci. Food Agric.* 22:647–649.

Buléon, A., Delage, M.M., Brisson, J., and Chanzy, H. 1990. Single crystals of V amylose complexed with isopropanol and acetone. *Int. J. Biol. Macromol.* 12:25–33.

Buléon, A., Le Bail, P., Colonna, P., and Bizot, H. 1998. In: *The Properties of Water in Foods - ISOPOW* 6, D.S. Reid (Ed.). London: Blackie, pp. 160–178.

Champenois, Y., Colonna, P., Buléon, A., Dellavalle, G., and Renault, A. 1995. Gelatinisation et rétrogradation de l'amidon dans le pain de mie. *Sciences des Aliments* 15:593–614.

Chanvrier, H., Colonna, P., Della Valle, G., and Lourdin, D. 2005. Structure and mechanical behaviour of corn flour and starch–zein based materials in the glassy state. *Carbohydrate Polym.* 59:109–119.

Christianson, D.D., Hodge, J.E., Osborne, D., and Detroy, R.W. 1981. Gelatinization of wheat starch as modified by xanthan gum, guar gum and cellulose gum. *Cereal Chem.* 58:513–517.

Clark, A.H., Gidley, M.J., Richardson, P.K., and Ross Murphy, S.B. 1989. Rheological studies of aqueous amylose gels: The effects of chain length and concentration on gel modulus. *Macromolecules.* 22:346–351.

Closs, C.B., Conde-Petit, B., Roberts, I.D., Tolstoguzov, V.B., and Excher, F. 1999. Phase separation and rheology of aqueous starch-galactomannan systems. *Carbohydrate Polym.* 39:67–77.

Colonna, P., Buléon, A., and Leloup, V. 1992. Limiting factors of starch hydrolysis. *Eur. J. Clin. Nutr.* 46:S17–S32

Colonna, P., Buléon A., Mercier, C., and Lemaguer, M. 1982. *Pisum sativum* and *Vicia faba* carbohydrates. IV. Granular structure of wrinkled pea starch. *Carbohydrate Polym.* 2:43–51.

Couchman, P. and Karasz, F.E. 1987. A classical thermodynamic discussion of the effect of composition on glass transition temperatures. *Macromolecules.* 11:117–119.

Crochet, P., Beauxis-Lagrave, T., Nole, T.R., Parcker, R., and Ring, S.O. 2005. Starch crystal solubility and starch granule gelatinization. *Carbohydrate Res.* 340:107–113.

de Bont, P.W., van Kempen, G.M.P., and Vrecker, R. 2002. Phase separation in milk protein and amylopectin mixtures. *Food Hydrocolloids.* 16:127–138.

Del Nobile, M.A., Martoriella, T., Mocci, G., and La Notte, E. 2003. Modeling the starch retrogradation kinetic of durum wheat bread. *J. Food Eng.* 59:123–128.

Della Valle, G., Vergnes, B., Colonna, P., and Patria, A. 1997. Relations between rheological properties of molten starches and their expansion by extrusion. *J. Food Eng.* 31:277–296.

Donovan, J.W. 1979. Phase transitions of starch–water system. *Biopolymers*, 18:263–275.

Doublier, J.L. 1987. A rheological comparison of wheat, maize, fava bean, and smooth pea starches. *J. Cereal Sci.* 5:247–262.

Doublier, J.L. and Choplin, L. 1989. A rheological description of amylose gelation. *Carbohydrate Res.* 193:215–226.

Eerlingen, R.C. and Delcour, J.A. 1995. Formation, analysis, structure and properties of type III enzyme resistant starch. *J. Cereal Sci.* 22:129–138.

Eliasson, A.C. 1983. Differential scanning calorimetry studies on wheat starch-gluten mixtures. II. Effect of gluten and sodium stearoyl lactylate on starch crystallization during aging of wheat starch gels. *J. Cereal Sci.* 1:207–213.

Eliasson, A.C. 1992. A calorimetric investigation of the influence of sucrose on the gelatinisation of starch. *Carbohydrate Polym.* 18:131–138.

Eliasson, A.C, Carlson, T.L., Larsson, K., and Miezis, Y. 1981. Some effects of starch lipids on the thermal and rheological properties of wheat starch. *Stärke*. 33:130.

Englyst, H.N., Kingman, S.M., and Cummings, J.H. 1992. Classification and measurement of nutritional starch fractions. *Euro. J. Clin. Nutr.* 46:33–50.

Evans, I.D. and Haiman, D.R. 1982. The effect of solutes on the gelatization of potato starch. *Starch*. 34:224–231.

Faisant, N., Buléon, A., Colonna, P., Molis, C., Lartigue, S., Galmiche, J.P., and Champ, M. 1995. Resistant starch: A modified method adapted to high RS products. *Brit. J. Nutr.* 73:111–123.

Farhat, I.A. and Blanshard, J.M. 2001. Modeling the kinetics of starch retrogradation. In: *Bread Staling*, P. Chinachoti and J. Vodovotz (Eds.). Boca Raton, FL: CRC Press, pp. 163–172.

Flory, P.J. 1953. *Principles of Polymer Chemistry*. Ithaca: Cornell University Press.

Franks, F. 1991. Water sorption and glass transition behaviors of freeze-dried sucrose–dextran mixtures. In: *Water Relationships in Foods*, H. Levine and L. Slade (Eds.). New York: Plenum Press, pp. 1–19.

Funami, T., Kataoka, Y., Omoto, T., Goto, Y., Asai, I., and Nishimari, K. 2005. Food hydrolloids control the gelatinization and retrogradation behavior of starch. II. Functions of guar gums with different molecular weights on the gelatinization behavior of corn starch. *Food Hydrocolloids*. 19:15–24.

Garcia, V., Colonna, P., Lourdin, D., Buleon, A., Bizot, H., and Ollivon, M. 1996. Thermal transitions of cassava starch at intermediate moisture contents. *J. Thermal Anal.*. 47:1213–1228.

Gaudin, S., Lourdin, D., Forsell, P., and Colonna, P. 2000. Antiplasticisation and oxygen permeability of starch–sorbitol films. *Carbohydrate Polym.* 43:33–37.

Gaudin, S., Lourdin, D., Le Botlan, D., Ilari, J.L., and Colonna, P. 1999. Plasticization and mobility in starch–sorbitol films. *J. Cereal Sci.* 29:273–284.

Gidley, M. and Bulpin, P. 1989. Aggregation of amylose in aqueous systems: The effect of chain length on phase behaviour and aggregation kinetics. *Macromolecules*. 22:341–346.

Gidley, M.J. and Bociek, S.M. 1988. Carbon-13 CP/MAS NMR studies of amylose inclusion complexes, cyclodextrins, and the amorphous phase of starch granules: Relationships between glycosidic linkage conformation and solid-state carbon-13 chemical shifts. *J. Am. Chem. Soc.* 110:3820–3829.

Godet, M.C., Bizot, H., and Buleon, A. 1995. Crystallization of amylose-fatty acid complexes prepared with different amylose chain lengths. *Carbohydrate Polym.* 27:47.

Godet, M.C., Buléon, A., Tran, V., and Colonna, P. 1993. Structural features of fatty acids–amylose complexes. *Carbohydrate Polym.* 21:91–95.

Hanashiro, J., Abe, H., and Hizukuri, S. 1996. A periodic distribution of the chain length of amylopectin as revealed by high-performance anion-exchange chromatography. *Carbohydrate Res.* 283:151.

Helbert, W. and Chanzy, H. 1994. Single crystals of V amylose complexed with n-butanol or n-pentanol: Structural features and properties. *Int. J. Biol. Macromol.* 16:207–213.

Hizukuri, S. 1985. Relationship between the distribution of the chain length of amylopectin and the crystallite structure of starch granules. *Carbohydrate Res.* 141:295–305.

Hizukuri, S. 1986. Polymodal distribution of the chain length of amylopectin and its significance. *Carbohydrate Res.* 147:342–347.

Hoseney, R.C., Zeleznak, K., and Lai, C.S. 1986. Wheat gluten: A glassy polymer. *Cereal Chem.* 63:285–286.

Hullemann, S.H.D., Helbert, W., and Chanzy, H. 1996. Single crystals of V amylose complexed with glycerol. *Int. J. Biol. Macromolecules* 18:115–122.

Imberty, A., Chanzy, H., Perez, S., Buleon, A., and Tran, V.J. 1988. The double-helical nature of the crystalline part of A-starch. *Molec. Biol.* 201:365–378.

Inouchi, N., Glover, D.V., Sugimoto, Y., and Fuwa, H. 1991. DSC characteristics of gelatinization of starches of single, double, and triple mutants and their normal counterpart in the inbred Oh43 maize (*Zea mays* L.) background. *Starch/ Staerke.* 43:468–472.

Jenkins, P.J. and Donald, A.M. 1998. Gelatinisation of starch. A combined SAXS/ WAXS/DSC and SANS study. *Carbohydrate Res.* 308:133–147.

Kalichevsky, M. and Ring, S. 1987. Incompatibility of amylose and amylopectin in aqueous solution. *Carbohydrate Res.* 162:323–328.

Kalichevsky, M.T., Jaroszkiewicz, E.M., Ablett, S., Blanshard, J.M.V., and Lillford, P.J. 1992. The glass transition of amylopectin measured by DSC, DMTA, and NMR. *Carbohydrate Polym.* 18:77–88.

Katelunc, G. 2003. Construction of state diagrams. In: *Characterization of Cereals and Flours*, G. Kaletunc and K.J. Breslauer (Eds.). New York: Marcel Dekker, pp. 151–171.

Katelunc, G. and Breslauer, K.J. 1993. Glass transition of extrudates: Relationship with processing induced fragmetation and end-product attributes. *Cereal Chem.* 70:548–552.

Kawada, J. and Marchessault, R. 2004. Solid state NMR and X-ray studies on amylose complexes with small organic molecules. *Starch/Staerke.* 56:13–19.

Krog, N. 1971. Amylose complexing effect of food grade. *Starch/Staerke.* 22:206–210.

Kugimiya, M. and Donovan, J.W. 1981. Calorimetric determination of the amylose content of starches based on formation and melting of the amylose–lysolecithin complex. *J. Food Sci.* 46:765–777.

Kugimiya, M., Donovan, J.W., and Wong, R.Y. 1980. Phase transitions of amylose–lipid complexes in starches: A calorimetric study, *Starch/Stärke.* 32:265–270.

Le Bail, P., Bizot, H., and Buléon, A. 1993. 'B' to 'A' type phase transition in short amylose chains. *Carbohydrate Polym.* 21:99–104.

Le Bail, P., Bizot, H., Ollivon, M., Keller, G., Bourgaux, C., and Buleon, A. 1999. Monitoring the crystallization of amylose-lipid complexes during maize starch melting by synchrotron x-ray diffraction. *Biopolymers.* 5:99–110.

Lekkerkerker, H.N.W., Poon, W.C.K., Pusey, P., Stroobants, A., and Waren, P.B. 1992. Phase behaviour of colloid + polymer mixtures. *Europhys. Lett.* 20:559–564.

Leloup, V.M., Colonna, P., and Buleon, A. 1991. Influence of amylose–amylopectin ratio on gels properties. *J. Cereal Sci.* 13:1.

Leloup, V.M., Colonna, P., and Ring, S.G. 1990. Studies on probe diffusion and accessibility in amylose gels. *Macromolecules.* 23:862–866.

Leloup, V.M., Colonna, P., and Ring, S.G. 1992. Physicochemical aspects of resistant starch. *J. Cereal Sci.* 16:253–266

Levine, H. and Slade, L. 1988. Thermomechanical properties of small-carbohydrate glasses and "rubbers." Kinetically metastable systems at sub-zero temperatures. *J. Chem. Soc. Faraday Trans.* 84: 2619–2633.

Lourdin, D., Coignard, L., Bizot, H., and Colonna, P. 1997. Influence of equilibrium relative humidity and plasticizer concentration on the water content and glass transition temperature. *Polymer.* 38: 5401–5406.

Lourdin, D., Della Valle, G., and Colonna, P. 1995. Influence of amylose content on starch films and foams. *Carbohydrate Polym.* 27:261–270.

Marchant, J.L. and Blanshard, J.M.V. 1978. Studies of the dynamics of the gelatinization of starch granules employing a small angle light. *Starch/Staerke.* 30:257–264.

Marchessault, R.H., Taylor, M.G., Fyfe, C.A., and Veregin, R.P. 1985. Solid-state ^{13}C-CPMAS-NMR of starches. *Carbohydrate Res.* 144:C1–C5.

Master, A.M. and Steeneken, P.M. 1997. Rheological properties of highly cross-linked waxy maize starch in aqueous suspensions of skim milk components. *Carbohydrate Polym.* 32:297–305.

MacKenzie, A.P. 1977. Non-equilibrium freezing behaviour of aqueous systems. *Phil. Trans. R. Soc.* 278:167–189.

Mercier, C., Charbonniere, N., Grebaut, J., and de la Gueriviere, J.F. 1980. Formation of amylose-lipid complexes by twin-screw extrusion cooking of manioc starch. *Cereal Chem.* 57:4–9.

Mestres, C., Colonna, P., and Buleon, A. 1988. Gelation and crystallization of maize starch after pasting, drum-drying or extrusion cooking. *J. Cereal Sci.* 7:123–134.

Miles, M., Morris, V.J., and Ring, S.G. 1985. Gelation of amylose. *Carbohydrate Res.,* 135:257–269.

Miles, M.J., Morris, V.J., Orford, P.D., and Ring, S.G. 1985. The roles of amylose and amylopectin in the gelation and retrogradation of starch. *Carbohydrate Res.* 135:271–281.

Morrison, W., Law, R., and Snape, C. 1993. Evidence for inclusion complexes of lipids with V-amylose in maize, rice and oat starches. *J. Cereal Sci.* 18:107–109.

Morrison, W.R. 1995. Starch lipids and how they relate to starch granule structure and functionality. *Cereal Foods World* 40:437–446.

Morrison, W.R., Milligan, T.P., and Azudin, M.N. 1984. A relationship between the amylose and lipid contents of starches from diploid cereals. *J. Cereal Sci.* 2:257.

Morrison, W.R., Tester, R.F., Snape, C.E., Law, R., and Gidley, M.J. 1993. Swelling and gelatinisation of cereal starches. IV. Some effects of lipid-complexed amylose and free amylose in waxy and normal barley starches. *Cereal Chem.* 70:385–391.

Noel, T.R. and Ring, S.G. 1992. A study of the heat capacity of starch/water mixtures. *Carbohydrate Res.* 227:203–213.

Nuessli, J., Putaux, J.L., Le Bail, P., and Buleon, A. 2003. Crystal structure of amylose complexes with small ligands. *Int. J. Biol. Macromol.* 33:227–234.

Nuessli, J., Sig, B., Conde-Petit, B., and Escher, F. 1997. Characterization of amylose-flavor complexes by DSC and X-ray diffraction. *Food Hydrocolloids.* 11:27–34.

Ortega-Ojedo, F.E., Larsson, H., and Eliasson, A.C. 2004. Gel formation in mixtures of amylose and high amylopectin potato starch. *Carbohydrate Polym.* 57:55–66.

Ottenhof, M.A. and Farhat, I.A. 2004. The effect of gluten on the retrogradation of wheat starch. *J. Cereal Sci.* 40:269–274.

Paris, M., Bizot, H., Emery, J., Buzare, J.Y., and Buleon, A. 2001. NMR local range investigations in amorphous starchy substrates. 1- Structural heterogeneity probed by ^{13}C CP-Mas NMR. *Int. J. Biol. Macromol.* 29:127–136.

Perry, P.A. and Donald, A.M. 2002. The effect of sugars on the gelatinisation of starch. *Carbohydrate Polym.* 49:155–165.

Prasad, A. and Mandelkern, L.1989. Equilibrium dissolution temperature of low molecular weight polyethylene fractions in dilute solution. *Macromolecules.* 22:914–920.

Rappenecker, G. and Zugenmayer, P. 1981. Detailed refinement of the crystal structure of Vh-amylose. *Carbohydrate Res.* 89:11–19.

Richards, G.N. and Vandenburg, J.Y. 1995. Fructose-grafted amylose and amylopectin. *Carbohydrate Res.* 268:201–207.

Ring, S.G., Colonna, P., I'Anson, K.J., Miles, M.J., Morris V.J., and Orford, P. 1987. Gelation and crystallisation of amylopectin. *Carbohydrate Res.* 162:277–293.

Ring, S.G., I'Anson, K.J., and Morris, V.J. 1985. Static and dynamic light scattering studies of amylose solutions. *Macromolecules.* 18:182–188.

Rolland-Sabaté, A., Colonna, P., Potocki-Véronèse, G., Monsan, P., and Planchot, V. 2004. Elongation and insolubilization of α-glucans by action of *Neisseria polysaccharea* amylosucrase. *J. Cereal Sci.* 40:17–30.

Rondeau-Mouro, C., Le Bail, P.L., and Buleon, A. 2004. Structural investigation of amylose complexes with small ligands: Inter- or intra-helical associations? *Int. J. Biol. Macromol.*, 345:251–257.

Roos, Y.H. 1995. Food components and polymers. In: *Phase Transitions in Foods*, S.L. Taylor (Ed.). Academic Press, San Diego, pp.133–135.

Russel, P.L. 1987. Gelatinization of starches of different amylose/amylopectin content. A study by differential scanning calorimetry. *J. Cereal Sci.* 6:133–145.

Rutschmann, M. A. and Solms, J. 1990. Formation of inclusion complexes of starch with different organic compounds. *IV Lebensm Wiss u Technol*, 23:84–87.

Sarko, A. and Zugenmaier, P. 1980. In: *Fiber Diffraction Methods*, A.D. French and K.C. Gardner (Eds.). Washington: American Chemical Society, pp. 459–482.

Sauvageot, F. and Blond, G. 1991. Effect of water activity on crispness of breakfast cereals. *J. Texture Stud.* 22:423–442.

Seow, C.C. and Teo, C.H. 1993. Annealing of granular rice starches. Interpretation of the effect of phase transitions associated with gelatinization, *Starch/Stärke.* 45:345–351.

Sevenou, O. 2002. Starch: Its relevance to dough expansion during baking. Doctoral dissertation, University of Nottingham.

Shogren, R.L. 1992. Effect of moisture content on the melting and subsequent physical ageing of corn starch. *Carbohydrate Polymers.* 19:83–90.

Silverio, J.H., Fredriksson, R., Andersson, R., Eliasson, A-C., and Aman, P. 2000. The effect of temperature cycling on the amylopectin retrogradation of starches with different amylopectin unit-chain length distribution. *Carbohydrate Polym.* 42:175–184.

Simon, S.L., Plazek, D.J., Sobieski, J.W., and McGregor, E.T. 1997. Physical aging of a polyetherimide: Volume recovery and its comparison to creep and enthalpy measurements. *J. Polym. Sci. Part B Polym. Phys.* 356:929–936.

Slade, L. and Levine, H. 1987. Structural stability of of intermediate moisture foods: A new understanding. In: *Food Structure: Its Creation and Evaluation*, J.M.V. Blanshard and J.R. Mitchell (Eds.). London: Butterworths, pp. 115–147.

Slade, L. and Levine, H. 1988. Nonequilibrium of native granular starch. Part I: Temperature location of the glass transition temperature associated with gelatinization of a-type cereal starches. *Carbohdrate Polym.* 8:183–191.

Slade, L. and Levine, H. 1991a. In: *Water Relationships in Foods*, H. Levine and L. Slade (Eds.). New York, Plenum Press, pp. 29–101.

Slade, L. and Levine, H. 1991b. Beyond water activity: Recent advances based on an alternative approach to the assessment of food quality and safety. *Crit. Rev. Food Sci. Nutr.* 30:115–360.

Slade, L. and Levine, H 1993a. In: *The Glassy State in Foods*, J.M.V. Blanshard and P.J. Lillford (Eds.). Lochborough, UK: University of Nottingham Press, pp.35–101.

Slade, L. and Levine, H. 1993b. Water relationships in starch transitions. *Carbohydrate Polym.* 21:105–113.

Slade, L. and Levine, H. 1995. Water and the glass transition: Dependence of the glass transition on composition and chemical structure: Special implications for flour functionality in cookie baking. *J. Food Eng.* 24:431–509.

Smits, A.L.M., Kruiskamp, P.H., van Soest, J.J.G., and Vliegenthart, J.F.G. 2003. The influence of various small plasticizers and malto-oligosaccharides on the retrogradation of partly gelatinized starch. *Carbohydrate Polym.* 51:417–424.

Sopade, P.A., Halley, P.J., and Junning, L.L. 2004. Gelatinisation of starch in mixtures of sugars: Application of differential scanning calorimetry. *Carbohydrate Polym.* 58:311–321.

Sperry, P.R. 1984. Morphology and mechanism in latex flocculated by volume restriction. *J. Colloid Interface Sci.* 99:97–108.

Struik, L.C.E. 1978. *Physical Aging in Amorphous Polymers and Other Materials.* Amsterdam: Elsevier.

Tan, I., Wee, C.C., Sopade, P.A., and Halley, P.J. 2004. Investigation of the starch gelatinisation phenomena in water-glycerol systems: Application of modulated temperature differential scanning calorimetry. *Carbohydrate Polym.* 58:191–204.

Tester, F. and Morrison, W.R. 1990. Swelling and gelatinisation of cereal starches. I. Effects of amylopectin, amylose and lipids. *Cereal Chem.* 67:551–557.

Tolstoguzov, V. 2003. Thermodynamic considerations of starch functionality in foods. *Carbohydrate Polym.* 51:99–111.

Vergnes, B., Della Valle, G., and Colonna, P. 2003. Rheological properties of biopolymers and applications to cereal processing. In: *Characterization of Cereals and Flours*, G.L. Kaletunc and K.J. Breslauer (Eds.). New York: Marcel Dekker, pp. 209–265.

Vergnes, B., Villemaire, J-P., Colonna, P., and Tayeb, J. 1987. Interrelationships between thermomechanical treatment and macromolecular degradation of maize starch in a novel rheometer with preshearing. *J. Cereal Sci.* 5:189–202.

Vrij, A. 1976. Polymers at interfaces and the interactions in colloidal dispersions. *Pure Appl. Chem.* 44:471–483.

Waigh, T., Gidley, M., Komanshek, B., and Donald, A. 2000b. The phase transformations in starch during gelatinisation: A liquid crystalline approach. *Carbohydrate Res.* 328:165–176.

Waigh, T., Kato, L., Donald, A., Gidley, M., Clarke, C., and Riekel, C. 2000a. Side-chain liquid crystalline model for starch. *Starch/Staerke.* 52:450–460.

Waigh, T., Perry, P., Riekel, C., Gidley, M., and Donald, A. 1998. Chiral side-chain liquid-crystalline polymeric properties of starch. *Macromolecules.* 31:7980–7984.

Wang, L.Z. and White, P.J. 1994. Structure and properties of amylose, amylopectin and intermediate material of oat starches. *Cereal Chem.* 71:263–268.

Wang, X., Choi, S-G., and Kerr, W. 2004. Effect of gluten content on recrystallisation kinetics and water mobility in wheat starch gels. *J. Sci. Food Agric.* 84:371–379.

Whittam, M.A., Noel, T.R., and Ring, S.G. 1990. Melting behaviour of A- and B-type starches. *Int. J. Biol. Macromol.* 12:359–362.

Whittam, M.A., Noel, T.R., and Ring, S. 1991. Melting and glass/rubber transition of starch polysaccharides. *Food Polymers, Gels and Colloids,* Royal Society of Chemistry.

Whittam, M.A., Orford, P.D., Ring, S.G., Clark, S.A., Parker, M.L., Cairns, P., and Miles, M.J. 1989. Aqueous dissolution of crystalline and amorphous amylose-alcohol complexes. *Int. J. Biol. Macromol.* 11:339–344.

Winter, W.T., Chanzy, H., Putaux, J.L., and Helbert, W. 1998. Inclusion compounds of amylose. *Polym. Prep.* 39:703.

Yamashita, Y. and Monobe, K. 1971. Single crystals of amylose V complexes. III. Crystal with 81 helical. *J. Polym. Sci.,* Part A-2, 9:1471–1481.

Yu, X., Houtman, C., and Atalla, R.H. 1996. The complex of amylose and iodine. *Carbohydrate Res.* 292:129–141.

Zeleznak, K.J. and Hoseney, R.C. 1987. The glass transition of starch. *Cereal Chem.* 64:121–124.

chapter six

Starch-based plastics

Rossana Mara da Silva Moreira Thiré
PEMM/COPPE, Universidade Federal do Rio de Janeiro

Contents

6.1 Introduction

In recent years, there has been increasing concern over the harmful effects of synthetic plastic materials in the environment. A strong increase in the research and development of new biodegradable materials, especially for single-use plastic items, was prompted by growing ecological awareness among the general public, politics, industry, and consumers' changing political conditions. The biodegradable plastics, whose components are derived entirely or partially from renewable raw materials, are designated as bioplastics (Stevens, 2001). In general, a bioplastic contains one or more biopolymers (polysaccharides, protein, peptides, etc.) as essential components, plasticizers, and other additives.

New bioplastics are still more expensive than their petroleum-based relatives, but this gap is decreasing. Rising oil prices and ecologically motivated political support have been leading to price advantages for bio-plastics, especially with regard to raw materials and disposal. Thus, as the production of some bioplastics has reached an industrial scale, the remaining economic disadvantages due to limited production capac-ity can be compensated and bioplastics are becoming more competitive compared to conventional plastics, especially in the packaging industry. Meanwhile, efforts are being made to retain the conventional processing methods used for petrochemical polymers, applying them to natural raw materials (Endres, Siebert, and Kaneva, 2007).

Among the bioplastics available in the global market, the most impor-tant are those composed of starch–starch-blends, polylactide (PLA) and polyester–polyester-blends (Endres, Siebert, and Kaneva, 2007). According to Degli Innocentini and Bastioli (2002), 75–80% of the global market for biopolymers corresponds to starch–starch blends. It is expected that the worldwide manufacturing capacity of starch-based polymers will increase from 100,000 ton/year to approximately 350,000 ton/year until 2010 (Endres, Siebert, and Kaneva, 2007).

The commercial starch-based polymers used for bioplastic production correspond to thermoplastic materials in the form of pellets obtained by extrusion, by sequential steps of extrusion and blending, or by a combined extrusion/blending (reactive blending) step. The pellets can be converted into a finished product on slightly modified standard thermoplastic resin machinery. Conversion technologies in use include film blowing, extru-sion, thermoforming, injection molding, and foaming (Crank et al., 2005). In addition, granular starch can also be compounded into plastics in the form of biodegradable filler (Griffin, 1977; Yavuz and Babaç, 2003).

This chapter outlines some properties of starch-containing plastic films and sheets and the way they are affected by starch composition, processing parameters, and storage conditions. It deals with the structure and properties of starch-based plastic films and sheets, including crys-tallinity, morphology, mechanical properties, and aging. The influence of starch composition, processing parameters, and storage conditions on thermoplastic starch (TPS) properties is discussed. Finally, some strate-gies usually adopted to overcome the water sensitivity and postprocess-ing aging of TPS are reviewed.

6.2   Thermoplastic starch (TPS)

6.2.1   Definitions

Except when starch is used as filler to produce reinforced plastics, its appli-cation as a bioplastic requires the transformation of the semicrystalline starch granules into a homogeneous, essentially amorphous matrix, in order to enhance the processability as compared to granular starch. The disruption of the molecular order within the granules can be accomplished by thermal or mechanical energy input associated with the addition of plasticizers such as water. "Destructurized starch" (DS) and "thermoplastic starch" are terms commonly employed to designate starch in this form.

Early patents (Stepto, Tomka, and Beat, 1988; Stepto and Beat, 1989) described destructurized starch as a material obtained by thermoplastic melt formation, by heating starch and water, with moisture content in the range of 5 to 40% (w/w), in a closed volume. The heating proceeds until all endothermic transitions have occurred, including the final narrow one, just prior to the endothermic change characteristic of oxidative and thermal degradation of starch. The starch is heated above the melting and glass transition temperatures of its components so that they undergo an endothermic transition. As a consequence, a melting and disordering of the molecular structure of the starch granule takes place, so that a destructurized starch is obtained. The applicability of DS is limited because of the degradation of starch due to water loss at elevated temperatures. Hence, the materials can only be processed by the addition of water, other plasticizers, or melt flow accelerators after the granular disruption step.

In order to overcome these problems, some patents (Tomka, 1994; Jurgen, Winfred, and Harald, 2001) described a starch-based composition, made by mixing starch with an appropriated additive or plasticizing agent, such as glycerol and sorbitol, under conditions that yield a thermoplastic starch melt after thermomechanical transformations. The additive or plasticizer is a substance that, when mixed with starch, reduces its melting point, bringing it below the decomposition temperature. This homogeneous amorphous material was claimed as a thermoplastically processable starch or thermoplastic starch in the patents.

As intermediate forms of starch are difficult to distinguish as DS or TPS from the descriptions above, van Soest (1996) suggested an empirical distinction of these starch materials. Accordingly, DS is essentially an amorphous starch made from granular starch of which the native order and crystallinity were completely lost by thermomechanical processing irrespective of the type of additives or storage conditions encountered after processing. When DS is obtained by any conventional thermoplastic process, such as extrusion and injection molding into shaped forms without the addition of extra plasticizers or melt flow accelerators, it can also be designated TPS.

Recently, Avérous (2004) introduced other terms to designate starch products as a function of water content and destructuring level of granules. Most starch applications require water dispersion and partial or complete gelatinization. Details of starch thermal transitions are discussed in Chapter 5. According to this classification, when material is obtained with the combination of water (high content) and heat, it is called gelatinized starch. Conversely, thermoplastic starches (or plasticized starches) are described as starch materials that were obtained with both low water content (less than 20 wt.%) and a high level of destructuration, corroborating the concept suggested by van Soest (1996).

6.2.2 Production

Disruption or melting of granular starch can be accomplished by a process predominantly regulated by thermal energy input, such as film casting (essentially heat application with low mechanical input), or regulated by the combination of thermal and mechanical energy input, such as conventional plastic processing techniques (e.g., extrusion, kneading, and injection molding). Extrusion is the most common industrial process used. Because melting temperature (220–240°C) is higher than the degradation temperature range of starch (200–220°C, approximately), some plasticizer is necessary in order to process granular starch (Shogren, 1993). Plasticizers are low molecular weight substances used to enhance flexibility and processability of a polymeric compound by decreasing the hydrogen bonding of the polymeric chains, which leads to an increasing free volume or molecular mobility of polymers. Hence, the properties of TPS mainly rely on the hydrogen bond-forming abilities between plasticizers and starch molecules. The amount and type of plasticizer strongly influence the physical properties of the processed starch by controlling its destructuration and depolymerization and affecting the final properties of the material, such as its glass transition temperature (T_g) and elastic modulus (Avérous, 2004).

Water is a primary plasticizer of starch. In addition, the most commonly used plasticizers are polyols; mono-, di-, or oligosaccharides; fatty acids; lipids; and derivatives. Many studies have demonstrated the plasticization effect of water on starches and also the various techniques for analyzing T_g have been compared (Zeleznak and Hoseney, 1987; Kalichevsky et al., 1992; Shogren, 1993; Mathew and Dufresne, 2002). On a molecular level, moisture-induced plasticization of a polymer leads to increased intermolecular distances (free volume), decreased local viscosity, and increased backbone chain segmental mobility (Slade and Levine, 1991). The addition of low molecular weight plasticizers to an amorphous matrix has basically the same effect as increased temperature on molecular mobility. Due to the hydrophilic nature of starch, the water content of starch-based materials is dependent on air relative humidity (RH) during processing and storage, which, in turn, directly affect their physical properties. The water content of starch increases from 7% to 24% (w/w), when the relative humidity increases from 20% to 90% (w/w), whereas a decrease in material T_g from 140°C to 18°C was observed (Shogren, 1993).

Similar behavior was observed in relation to T_m; its value dropped from 126°C to 72°C, as the water content of corn starch increased from 10% to 60% (Souza and Andrade, 2002).

Glycerol has also been extensively employed as a plasticizer for obtaining TPS materials. The presence of small amounts of glycerol can promote a classic antiplasticizing effect on starch films (Forssell et al., 1997; Avérous et al., 2000; Altskär et al., 2008). Due to strong interaction between glycerol and starch, a hydrogen-bonding network is formed and a reinforced material is obtained. As the plasticizer contents increase, interactions between plasticizer and starch become stronger, resulting in swelling and a plasticization effect (Lourdin, Bizot, and Colonna, 1997; Chang, Karima, and Seow, 2006). It was also observed that 20% (w/w) glycerol seems to be the maximum that can act as plasticizer. Above this percentage, phase separation occurs and, because glycerol is hygroscopic, the amount of adsorbed water increases as it binds to starch film as well as to "free" glycerol (Godbillot et al., 2006). At very high plasticizer content (above 50% w/w), starch materials become soft and behave more like a gel or paste.

Several other compounds have also been shown to be useful as plasticizers. In general, monohydroxyl alcohols and high molecular weight glycols failed to plasticize starch, whereas shorter glycols and sorbitol were effective (Gaudin et al., 1999; Mathew and Dufresne, 2002; Da Róz et al., 2006). The antiplasticizing effect is observed in compress molding of corn starch sheets plasticized with 1,4-butanediol, diethyleneoxide glycol, and sorbitol (Da Róz et al., 2006). Studies demonstrated that starch films obtained with plasticizers containing amide groups are less sensitive to aging effects than those plasticized with glycols. However, the former films present little internal flexibility, which negatively affects their mechanical properties (Ma and Yu, 2004).

The main difference between the conversion of starch granules into TPS by casting or by other conventional thermomechanical processes is the amount of water or plasticizer utilized during the gelatinization or melting of the granular starches.

A casting process is often employed to produce edible films for packaging and coating of food products (Liu, 2005). When a casting technique is used, starch is processed by heating under moderate shear in the presence of excess water or other plasticizer, which causes irreversible swelling of the granules. According to the classical gelatinization model of starch, swelling is accompanied by disruption of the native crystalline structure and solubilization of amylose. The structural changes that take place during gelatinization include simultaneous crystallite melting and double-helix unwinding, absorption of water in the amorphous growth ring, changes in shape and size of granules, dispersion of blocklet-like structures, and leaching of amylose from the granule. Details of this subject can be found in Chapters 3 and 5. In the last decade, due to the development of experimental techniques, several researchers have reported many studies on the gelatinization of starch and have put forward the classical starch gelatinization model (Jenkins and Donald, 1998; Vermeylen et al., 2006; Peng, Zhongdong, and Kennedy, 2007).

The mechanism of starch film formation depends, among several factors, on the solid concentration and amylose content. Generally, aggregation and packing of swollen granules dominate film formation of starch dispersions with a relatively high solid concentration. In relation to dilute starch solutions, the film formation follows the order of helical formation, aggregation or reorganization of aggregates; the former is primarily driven by cooling and the latter by dehydration (Liu and Han, 2005). Preparation conditions, such as temperature, relative humidity, heating period, shear rate, drying procedure, and composition, starch source, and plasticizers are important to the film structure.

In the processing of a TPS by extrusion, a common plastic processing technique, the temperature and shear stress are higher compared to those employed in a casting technique. In most cases, water content is lower than 20% w/w and this kind of processing leads to differences in starch melting due to water content, morphology of the materials, and differences in aggregation and crystallization behavior of the amylose and amylopectin compared to cast films.

During thermomechanical processing by extrusion, the first step is the compaction and heating of the starch granules. Then, granules are partially transformed at the same time by heat (internal granular disorganization) and by mechanical processing (granular fragmentation). Fragmentation itself induces only partial loss of crystallinity. Small fragments finally are melted due to a local temperature increase, contributing to interparticle friction. Susceptibility to fragmentation is related to the mechanical properties of starch granules associated with their architecture and botanical origin. This difference could be related to differences in granular organization, that is, the presence of brittle areas or defects (Barron et al., 2001). Under temperature and shearing, starch is destructured, plasticized, and melted, but also partially depolymerized (Avérous, 2004). The transformations of starch that occur during extrusion are influenced by extruder geometry as well as by processing variables such as extrusion temperature, screw speed, feed rate, and moisture content (Lai and Kokini, 1991; Souza and Andrade, 2002).

In general, the processing of starch into TPS is accompanied by a decrease in molecular weight, although the depolymerization is significantly less pronounced for the cast films (Altskär et al., 2008). It is accepted that, in the heating–shearing process, starch degradation increases as water content decreases and temperature or rotation speed increases (Chinnaswamy and Hanna, 1990; Silva et al., 2004). Nevertheless, some studies indicated that an optimum in breakdown occurs at a certain moisture content, leading to an inversion of the expected behavior (Govindasamy, Campanella, and Oates, 1996; van den Einde et al., 2004a). Although only the mechanical contribution associated with specific mechanical energy has been accounted for in the evaluation of starch molecular breakdown during thermomechanical processing (Della Valle et al., 1995; Martin, Averous, and Della Valle, 2003; van den Einde et al. 2004b), it was suggested that thermal breakdown also plays an important role in this process (van den Einde et al. 2004b).

6.2.3 Microstructure

6.2.3.1 Morphology of TPS films

The morphology of amylose, amylopectin, and starch films has been evaluated by several microscopy techniques, such as light microscopy, scanning electron microscopy (SEM), transmission electron microscopy (TEM), and atomic force microscopy (AFM). Chapter 2 discusses AFM on starches in detail.

In general, the surface of starch films has a network structure, which has been suggested to be built up by crystalline or co-crystallized amylose and amylopectin strands. The formation of such a network requires high moisture content or a slow cooling rate of melt (as in the case of compression-molded films) during processing in order to provide sufficient molecular mobility (Stading, Rindlav-Westling, and Gatenholm, 2001; Rindlav-Westling, Stading, and Gatenholm, 2002; Thiré, Simão, and Andrade, 2003a; Altskär et al., 2008).

Similar to crystallinity, morphology of starch films is also influenced by film-forming conditions. Under certain conditions, the starch films could form amylose-rich and amylopectin-rich phases, which results in a phase-separated structure (Rindlav-Westling, Stading, and Gatenholm, 2002; Rindlav-Westling and Gatenholm, 2003). It was suggested that amylose–amylopectin phase separation is related to amylose ability in forming a continuous network faster than phases could separate. It was observed that amylose inherent in starch has a lower ability to form a continuous network than amylose-free starch. Hence, phase separation in native starch films was observed, even though the blended films of amylose–amylopectin in the same proportion as native starch did not show this characteristic (Rindlav-Westling, Stading, and Gatenholm 2002). The phase separation induces the heterogeneity of structure and affects the properties of starch films. Because there is less contact between amylose and amylopectin in the heterogeneous regions of films, there are few possibilities for co-crystallization to take place, leading to a lower crystallinity as compared to non-phase-separated starch films. In addition to this phenomenon, starch films containing glycerol can also exhibit a separation into glycerol-rich and glycerol-poor regions (Forssell et al., 1997; Souza and Andrade, 2002; Thiré, Simão, and Andrade, 2003a).

Some specific features were also visualized on the surface of certain films. Cast native potato starch films showed rounded protrusions of 15–35 nm in diameter and 1–4 nm high on the outermost surface. These protrusions were attributed to protein aggregates, which were released along the starch polymers during granule disruption (Rindlav-Westling and Gatenholm, 2003). In cast cassava starch films, some granule remnants (ghosts) were identified (Yakimets et al., 2007) and maintenance of the network is attributed to the presence of these granules.

It is worth noting that ghosts are defined as the remnants of the envelope after structural collapse during gelatinization, where the majority of the starch polymers have been released (Atkin, Abeysekera, and Robards, 1998). They are primarily composed of amylopectin and exhibit elastic/plastic properties that could influence the properties of the films. These structures could be identified at a certain point of granule gelatinization despite the starch source. Figure 6.1 shows a polygonal-shaped ghost on the surface of corn starch film. This film was obtained by casting, using water and glycerol as plasticizers.

The heating time and addition of glycerol as plasticizer affect the granular disruption level and consequently, the morphology of cast corn

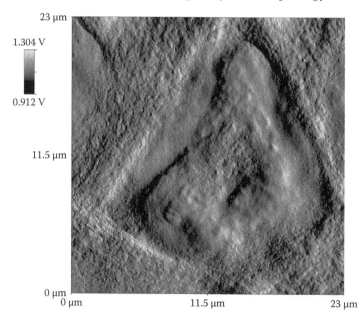

Figure 6.1 AFM topographic image of a ghost on the surface of corn starch film.

starch films (Thiré et al., 2003a). Glycerol retards the gelatinization of starch granules (van Soest et al., 1996a; Tan et al., 2004) and regardless of the preparation conditions, crystalline domains resulting from recrystallization of amylose and glycerol-rich and glycerol-poor regions can be identified in a continuous matrix of glycerol-plasticized films by AFM phase contrast technique (Thiré, Simão, and Andrade, 2003a).

The surface morphology of compression-molding hydroxypropylated and oxidized potato starch, plasticized with water and glycerol, seems to be very similar to corresponding cast films. Fewer ghosts were observed on the surface of film-blown material as compared to compression-molded films, probably due to higher shear rates in the extruder and more extensive mixing. Considering the mechanical properties, the existence of residual granules most likely has a negative effect on the film-blowing properties, as they are considered weak spots in the material. These weak spots could cause the failure of the biaxially stretched bubble and therefore restrict the extent of stretching allowed (Altskär et al., 2008).

6.2.3.2   Crystallinity

The final degree of crystallinity in any polymer sample depends on the ability of the chains to form crystals as well as the mobility of the chains during the crystallization process. In the case of starch films and sheets, the development of crystallinity is affected by the composition content (e.g., starch source and type and amount of plasticizers), applied processing conditions (e.g., temperature, relative humidity, and screw speed), and also storage conditions. Different types of crystallinity can be observed in thermoplastic starches, which can be divided into three groups (van Soest et al., 1996b; Avérous 2004):

- Residual crystallinity: Related to native crystallinity of starch granules that were incompletely melted during processing (A, B, and C types)
- Crystallinity induced by processing: Related to fast recrystallization of amylose molecules into single-helical structure (V_H, V_A, and E_H types)
- Crystallinity induced by post processing aging: Related to the B-type crystallinity developed after processing when TPS are stored at temperatures above T_g

The V- and E-type structures are related to the complex formed between amylose and lipids from starch granules or between amylose and plasticizers during processing (Yang, Yu, and Ma, 2006; van Soest and Vliegenthart, 1997). The single helical structure E_H is more stable at low moisture content and is transformed into the V_H form by increasing water content of the film. Differences between the two morphologies are attributed to different interaxial distances between helices. Hence, the change from E_H to V_H is suggested to be a polymorphous change rather than an increase in crystallinity (Shogren, Fanta, and Doane, 1993; van Soest et al., 1996a).

Whenever TPS materials are stored above T_g, retrogradation or recrystallization of starch molecules into a B-type double helix structure is observed, which reflects recrystallization of amylose and of amylopectin (van Soest and Knooren, 1997; Hulleman et al., 1999; Rindlav-Westling, Stading, and Gatenholm, 2002). It has been suggested that co-crystallization between amylose and amylopectin occurs (Gudmundsson and Eliasson, 1990; Hulleman et al., 1999; Rindlav-Westling, Stading, and Gatenholm, 2002). Crystallization of amylose is fast and takes place within hours, whereas the crystallization rate of amylopectin is slow and takes a few weeks. Then, during storage of rubbery starch materials, a slower development of the crystallinity occurs. This process has been essentially attributed to amylopectin (Hulleman et al., 1999). Retrogradation of amylose and amylopectin occurs according to a different kinetic process. Nevertheless, some stages of retrogradation, such as chain interactions, double-helix formation, crystallization, and formation of interconnected networks, are similar with both molecules (Putaux Buléon, and Chanzy, 2000).

On compression molding of extruded potato starch, an increase in screw speed during extrusion led to an increase in the formation of single-helical amylose type crystallinity (van Soest et al., 1996a). A variation in screw speed probably affects the amount of applied shear stress on the material. The processing temperature also influences the formation of the type of crystallinity; below 180°C, the V_H structure is formed, whereas above this temperature, E_H structure formation is favored. Temperature, relative humidity, and time of the drying procedure affect the crystallinity of cast starch films. B-type crystallinity develops upon film formation, varying from almost none to 23% (Rindlav-Westling, Hulleman, and Gatenholm, 1997). A slow drying procedure, in which cast films are dried in high relative humidity or at low temperatures, promotes crystallization of the starch polymers (Rindlav-Westling et al., 1998) and affects phase separation (Rindlav-Westling, Stading, and Gatenholm, 2002). A lower drying rate gives the chains a longer time to arrange into more favorable conformations (i.e., in the form of crystals), and a higher crystallinity is thereby achieved for films.

The effect of plasticizer on crystallinity is complex and mainly involves two opposite factors. When the plasticizing effect increases, the mobility of amorphous chains increases and the T_g value decreases. In this case, the mobility of polymer chains favors the formation of crystalline domains. Considering this phenomenon, the crystallization of starch should decrease with increasing the molecular weight of the plasticizer. At the same time, decreasing the molecular weight of the plasticizer leads to an increase of the amount of amylopectin–plasticizer interactions and therefore to a decrease of the possible interactions among the amylopectin chains. These restricted interactions hinder the formation of crystalline domains from amorphous amylopectin chains. Considering the latter phenomenon, the crystallization of starch should decrease with decreasing the molecular weight of the plasticizer. It seems that the latter phenomenon is predominant in a dry atmosphere (Mathew and Dufresne, 2002).

Potato amylose films with or without the addition of glycerol as a plasticizer exhibit a relatively high degree of B-type crystallinity, despite the rate of the drying procedure. Nonplasticized potato amylopectin films are amorphous even when dried slowly in 90% relative humidity, whereas glycerol-plasticized amylopectin results in B-type crystallinity, depending on the relative humidity conditions during film formation. It seems that amylopectin is more sensitive than amylose to plasticization caused by glycerol (Rindlav-Westling et al., 1998).

An increasing crystallinity with increasing amylose concentration is observed in potato starch materials obtained by casting (Rindlav-Westling, Stading, and Gatenholm, 2002) and by extrusion (van Soest and Essers, 1997). No crystallinity is observed in casting cassava film plasticized with water and dried at 20°C, even at high relative humidity, probably due to the low content of amylose in cassava starch (Yakimets et al., 2007). The amount of V_H-type crystallinity has been shown to be linearly dependent on the amylose content of TPS (Forssell et al., 1999).

The amount and type of plasticizer did not appreciably affect the crystallinity of compress molding corn starch sheets plasticized with several polyols. However, propylene glycol- and ethylene glycol-plasticized sheets showed an increase in B-type and V_H-type crystallinity as a function of plasticizer content; this increase was associated with the progressively higher chain mobility promoted by higher plasticizer concentration (Da Róz et al., 2006). Nevertheless, the increasing amount of glycerol or water led to a decrease in E_H-type crystallinity in extruded potato starch sheets (van Soest et al., 1996a). This effect was attributed to change in melt viscosity, which caused a change of shear stress forces and hence changed the orientation of amylose molecules during processing.

However, the presence of ghost granules may influence starch retrogradation, depending on the nature of the plasticizer used. Ghosts can act as nuclei for crystallization when no plasticizer or small plasticizers up to four hydroxyl groups are added. Otherwise, when plasticizers containing more hydroxyl groups are used, they may interact with the ghost structures and obstruct interaction of the polysaccharide chains, thus preventing recrystallization (Smits et al., 2003).

By changing the processing parameters, such as temperature and water content during molding, it is possible to vary the amount of B-type crystallinity and thus the properties of compression-molded, glycerol-plasticized potato starches. Increasing the water content during molding at 160°C led to an increase in B-type crystallinity for both regular potato starch and amylopectin obtained from potato starch, indicating a major contribution of amylopectin to the overall crystallinity. At high water content during molding, mainly amylopectin recrystallization takes place (Hulleman et al., 1999). Conversely, molding at low water content led to residual B-type crystallinity originated from amylopectin, which is caused by incomplete melting of potato starch granules.

In addition to B-type structure formation, changes from E_H to V_H can occur during storage (Rindlav-Westling, Hulleman, and Gatenholm, 1997; van Soest and Knooren, 1997; Forssell et al. 1999). However, changes in crystallinity types are suggested to be polymorphous rather than an increase in crystallinity (Shogren, Fanta, and Doane, 1993; van Soest et al., 1996a).

6.2.4   Mechanical properties

The mechanical properties of starch films depend, among other things, on the ratio of amylose to amylopectin, plasticizer, water content, and storage conditions. Moreover, these properties are also affected by corresponding changes in glass transition temperature related to the mobility of macro-molecular chains in the amorphous phase and the degree of crystallinity of films. Crystals behave as fillers and physical cross-linkers, which tend to strengthen and stiffen the films.

For example, extruded potato starch sheets containing less than 9% (w/w) water are glassy (T_g above 40°C) with an elastic modulus between 500–1000 MPa. For rubbery materials (with 13–15% water), elastic modulus drops to a value in the range of 0–100 MPa. Above a water content of 15% w/w, the sheets are weak and soft, with low elongation and low energy to breakpoint. At higher water content (30% w/w approximately), the interactions of starch molecules, especially hydrogen bondings, are weakened to such an extent that the material loses its strength (van Soest, Benes, and de Wit, 1996).

Higher strength, stiffness, and toughness are observed in ther-moplastic starch materials with high amylose content (Lourdin, Della Valle, and Colonna, 1995; van Soest and Essers, 1997; Thunwall, Boldizar, and Rigdahl, 2006). Unplasticized cast starch films show a continuous increase in tensile strength (40–70 MPa) and elongation (4–6%) when amylose increases from 0 to 100%. For plasticized films, influence of amy-lose is masked by the presence of glycerol. For 40–100% amylose, no plas-ticization of the material is apparent and the maximum elongation and tensile strength are 5% and 25 MPa, respectively (Lourdin, Della Valle, and Colonna, 1995). Glycerol-plasticized, compression-molded sheets of high-amylose potato starch and of normal potato starch present a tensile modulus of 160 MPa and 120 MPa, respectively, after conditioning at 53%

relative humidity and 23°C. The stress at breakpoint seems to be substantially higher for high-amylose potato starch as compared to normal starch, 9.8 and 4.7 MPa, respectively (Thunwall, Boldizar, and Rigdahl, 2006).

Several factors may contribute to the differences in mechanical performance due to amylose content, such as (i) higher T_g of materials containing high levels of amylose, (ii) a more entangled network attributed to a greater amount of amylose, and (iii) residual crystalline structures or recrystallized structures associated with amylose, to a lesser extent.

Addition of plasticizers tends to decrease elastic modulus and tensile strength and increase elongation, resulting in starch films that are softer and more flexible, but weaker. This behavior may be explained by the T_g decrease of the films (Lourdin, Della Valle, and Colonna, 1997; Mathew and Dufresne, 2002). However, starch films plasticized by low amounts of polyols, such as glycerol and sorbitol, can show decreased elongation with increased plasticizer content due to the antiplasticization effect (Lourdin, Della Valle, and Colonna, 1997; Gaudin et al., 1999). The molecular weight of the plasticizer also affects mechanical properties of TPS films. Use of polyols with high molecular weight as plasticizers produces TPS starch films and sheets with relatively high modulus and elongation at break (Mathew and Dufresne, 2002). This occurs because the degree of crystallinity and water uptake of moist samples tends to decrease with the molecular weight of the plasticizer. Therefore, the cohesion between amorphous domains and crystalline zones remains sufficient to ensure good mechanical properties.

6.2.5 Aging

The term "physical aging" was introduced by Struik (1978) to distinguish these effects from other aging processes, such as chemical reactions, degradations, and changes in crystallinity. These alterations directly affect enthalpy, volume, and mechanical and diffusion properties. Hence, physical aging may be quantitatively measured as the changes in specific volume or relaxation enthalpy.

The mechanical behavior of low-moisture glassy starch shows a progressive embrittlement upon storage. The mechanical relaxation time increases and the material becomes stiffer. These progressive changes indicate that starch chains become less mobile and the matrix is densified during aging (Lourdin et al., 2002). Data from differential scanning calorimetry and dynamic mechanical analysis suggested that the embrittlement can occur as a result of free volume relaxation during sub-T_g aging as well as a result of plasticizer loss by evaporation (Shogren, 1992). It also seems that amylose retards the relaxation of amorphous regions toward equilibrium state (Chung and Lim, 2004).

After processing, TPS compounds are not usually at equilibrium and exhibit time-dependent changes in structure and macroscopic properties that are associated with aging (Hulleman et al., 1999; Rindlav-Westling, Stading, and Gatenholm, 2002; Smits et al., 2003; Avérous, 2004; Thiré, Andrade, and Simão, 2005). Two different aging behaviors can be detected, depending on water content and glass temperature of films:

- Materials with intermediate to high moisture content (room temperature above T_g) incur plasticizer–starch phase separation and crystallization of starch chains (retrogradation) (van Soest and Knooren, 1997).
- Materials with low moisture content, that is, with water content below 30% of the total wet weight (sub-T_g domain) show physical aging or structural relaxation (Lourdin et al., 2002; Shogren, 1992).

Glassy materials display mechanical and physical properties similar to those of crystalline solids, while keeping a molecular arrangement more characteristic of a liquid. Being far from the thermodynamic equilibrium, glassy materials stored at a temperature below their glass transition temperature are subject to molecular rearrangements, leading to lower states of energy. Thus, physical aging can be considered a molecular rearrangement (structural relaxation) toward an equilibrium as a function of storage time and temperature (Borde et al., 2002; Chung and Lim, 2003).

For rubbery starch materials, at room temperature above T_g, aging occurs mainly due to the formation of an intermolecular double helix and B-type crystallinity structure: the retrogradation process. As amylopectin retrogradation is slow, its crystallization has an important role in TPS aging (van Soest et al., 1996c; Forssell et al., 1999; Hulleman et al., 1999). However, the crystallization of amylose also contributes to time-dependent changes in properties, especially when high content of plasticizers is used (Shi et al., 2007). The V-type crystals, which are formed at the initial storage period, also contribute to changing the mechanical properties during aging and act as physical cross-linking points in the material.

Mechanical properties of rubbery starch materials are strongly influenced by retrogradation. Similar to synthetic polymers, an increase in the amount of B-type crystallinity results in an increase of the elastic modulus and hardness of material (van Soest et al., 1996c; van Soest and Knooren, 1997). Intermolecular crystallinity leads to a reinforcement of the network by the formation of physical cross-links. Intramolecular crystallization of the amylopectin decreases the mobility of the amylopectin and increases the stress in the material at highly crystalline junction zones. At the same time, the elongation between the points is restricted, which results in the decrease of the elongation at breakpoint. At crystalline regions, TPS materials may spontaneously break as a result of internal stress and cracks generated by crystals (van Soest, de Wit, and Vliegenthart, 1996).

Aging effects depend, among other issues, on the botanic origin of the starch. Variations in surface roughness on extruded oat and barley starch films aged 5 weeks, with phase separation in the surface, can be observed by atomic force microscopy on both films (Kuutti et al., 1998), whereas no changes are observed in general morphology and roughness on the surface of cast corn starch films, even after 270 days of storage (Thiré, Andrade, and Simão, 2005). However, an increasing number of ordered domains at the surface of aged films was attributed to amylose recrystallization.

In films from corn starch plasticized with glycerol and stored for 90 days, the water content (3.5 to 13.9%) and the amount of B-type crystallinity (20 to 39%) increased, as the surrounding relative humidity increased from 30 to 90% (Thiré et al., 2003b). It seems that the increase in the surrounding relative humidity led the network of amylose in the matrix and the amylopectin molecules of the granular region to take up water and swell. Hence the plasticization of the starch by increased water content led to higher mobility of the starch molecules and subsequent higher crystallinity of the material.

6.3   TPS drawbacks and strategies

In general, starch films have high tensile strength, but are brittle and exhibit almost no elongation at break (Briassoulis, 2004). For instance, at 15% water content, extruded starch has poor mechanical properties and is inadequate for film applications (initial tensile strength 20–30 MPa and elongation at break 10–15%) (Imam et al., 1995). Moreover, these properties are affected by relative humidity and, thus, by the corresponding changes in T_g. At low relative humidity, there are problems with brittleness, whereas at high relative humidity, with softness. In addition to moisture sensitivity, TPS materials change their time-dependent mechanical properties (also called "postprocessing aging"). In order to overcome these limitations, different strategies are often adopted, such as: (i) chemically modified starch in formulations; (ii) suitable plasticizers; (iii) association of biodegradable, renewable, or synthetic polymers to thermoplastic starch; and (iv) surface modifications of starch plastics.

6.3.1   TPS from modified starches

Starch-based materials have been developed through chemical modification by well-known organic reactions, to obtain material with improved water sensitivity and mechanical performance (Fringant, Desbrières, and Rinaudo, 1996; Demirgöz et al., 2000; Wilpiszewska and Spychaj, 2007). Modified starches are discussed further in Chapter 8.

Starch esterification by acetylation has been shown to be an efficient chemical modification for obtaining TPS-based materials with reinforced hydrophobicity and higher thermal stability (Fringant, Desbrières, and Rinaudo, 1996); better properties are achieved with degrees of substitution (DS) higher than 1.7. The properties of starches modified by chemical derivatization are also some options for TPS materials; they depend greatly on DS and the length of the introduced alkyl chain; the higher the DS and the longer the aliphatic chain the better are the hydrophobic properties (Wilpiszewska and Spychaj, 2007) and blends with cross-linking modified starch by hydroxyl groups and tri-sodium tri-meta phosphate (Demirgöz et al., 2000).

However, according to Avérous (2004), the strategy of chemical modification is strongly limited by: (i) the toxicity and the diversity of byproducts obtained during the chemical reactions, (ii) the high cost of modification and product purification stages, and (iii) the unfavorable alterations in TPS mechanical properties caused by chemical reactions.

6.3.2   *Plasticizers and biodegradable polymers in TPS*

Both aging through crystallization and water sensitivity are affected by the choice of plasticizer. Thus, it is very important for the application and development of starch materials to choose a plasticizer that is able to impart flexibility to TPS materials and to suppress retrogradation during the aging time. In order to achieve this objective, several plasticizers have been incorporated with pure thermoplastic starch, such as water, glycol, sorbitol, urea, amide, sugars, and quaternary amine. Recently, it was suggested that the use of high glycerol content or of plasticizers with other polar groups that have stronger interaction with the starch molecular structure than glycerol should be an effective approach to restrain aging (Shi et al., 2007). The addition of glyceryl monostearate (GMS) as surfactant to TPS films decreased their retrogradation tendencies (Mondragón, Arroyo, and Romero-García, 2008) and this effect is attributed to the formation of amylose–GMS complexes which is favored over the amylose retrogradation.

Starch has been also widely blended with biodegradable polymers, including polycaprolactone (Avérous et al., 2000), polylactic acid (Zhang and Sun, 2004), polyhydroxybutyrate (Thiré, Ribeiro, and Andrade, 2006), and poly(hydroxyester ether) (Willet and Doane, 2002). Thermoplastic starch blended with thermoplastic polyesters, such as polycaprolactone (PCL), is the most popular base film (Bastioli, 2000). Commercially, TPS is used alone mainly in soluble compostable foams, such as loose fillers, and other expanded items as a replacement for polystyrene. These films are biodegradable, compostable, and fulfill the requirements of different application fields, such as packaging and agriculture. Thus, the association of TPS with other biodegradable compounds is a promising strategy to obtain compostable multiphase materials with improved properties. The relationships among structure, process, and properties of biodegradable multiphase materials (i.e., blends, composites, and multilayers) based on TPS have been reviewed recently by Avérous (2004). Details about biodegradability of starch-based materials are discussed in Chapter 10.

Biodegradable blends of TPS and waterborne polyurethane (PU) obtained from castor oil presented improved mechanical properties, surface and bulk hydrophobicity, and water resistance, as compared to pure TPS films (Lu et al., 2005). Glycerol-plasticized blended films of wheat TPS and polycaprolactone obtained by extrusion and injection molding (Avérous, et al. 2000) show phase separation with significant improvement of stability and hydrophobicity due to the presence of polycaprolactone.

6.3.3 Effects of multilayer structures on TPS constitution

In addition to blends, the association of starch and other polymers can be obtained by multilayer structures. These structures are composed of an internal layer of TPS and two external layers of hydrophobic polymer. Compared to blends, in general, multilayer structures present higher water resistance, because moisture sensitivity is not fully addressed in a blend as a result of starch phase distribution close to the surface (Avérous, 2004).

Starch-based materials coated with water-resistant biodegradable polyesters are viable for a broader range of uses, including those typical of food preparation, packaging, and consumption (Shogren and Lawton, 1998). This method involves: (i) the application of a natural resin layer, such as shellac or pine rosin, to the surface of a starch-based article and (ii) subsequent application of a continuous layer of a water-resistant polyester, such as poly(beta-hydroxybutyrate-co-valerate) (PHBV) and poly(lactic acid) (PLA).

For plasticized starch–polyethylene multilayer systems, the compatibility of starch and polyethylene is achieved through maleic anhydride functionalized polyethylene (PEg) (Dole et al., 2005). The use of the starch–multilayer structure (PE/PEg/starch/PEg/PE) improves gas barrier properties at high relative humidity as compared to existing commercial multilayer systems (PE/PEg/EVOH/PEg/PE). The higher quantity of water sorbed by TPS and the specific water sorption isotherm of starch significantly extended the water equilibration time in the hydrophilic inner layer (Dole et al., 2005).

Manufacturing starch composites with reinforced natural fibers is another way to achieve TPS-based multilayer materials with improved properties (Dufresne and Vignon, 1998; Curvelo, de Carvalho, and Agnelli, 2001; Avérous and Boquillon, 2004). Improvements of TPS-reinforced film performance can be attributed to usual matrix reinforcement and to the interrelations of the fiber matrix (Avérous, 2004). According to Dufresne and Vignon (1998), the water resistance of starch-based plastics can be improved by adding a small amount of commercial cellulose fibers up to 15% w/w. Extruded and injection-molded TPS sheets reinforced by cellulose and lignocellulose fillers exhibit improved thermal resistance compared to nonreinforced sheets (Avérous and Boquillon, 2004). Mechanical behavior of these reinforced biocomposites changes according to the filler content, fiber nature, and fiber length. Composites prepared with regular corn starch plasticized with glycerol and reinforced with short cellulosic fibers (16% w/w) from bleached pulp (Curvelo, de Carvalho, and Agnelli, 2001), increased 100% in tensile strength and more than 50% in modulus in relation to nonreinforced films.

Recently, nanoscale fillers have been incorporated into starch blends to enhance thermal, mechanical, and processing properties of the products. Two kinds of nanoparticles have been tested in TPS composites: whiskers obtained from cellulose (Samir, Alloin, and Dufresne, 2005; Angellier et al., 2006) and organoclays (McGlashan and Halley, 2003; Huang, Yu, and Ma, 2004). Details about the use of cellulose in starch-based nanocomposities are discussed in Chapter 9.

6.3.4   Surface modification

Surface modification is a suitable approach to alter superficial physical and chemical properties of starch films, such as hydrophilicity and water uptake, without affecting their bulk properties. It can be accomplished, for instance, by chemical reactions between an external reagent and superficial hydroxylic groups of starch or by depositing a protective layer over the starch film or sheet surface (coating process). The surface of compress-molded corn starch films is modified by reaction with coupling agents (isocyanate, phenol, blocked isocyanate, epoxy and acid chloride moieties, i.e., small molecules and polymeric ones) (Carvalho, Curvelo, and Gandini, 2005) and the polymeric reagents coupled simultaneously with OH groups, immobilizing the latter species. Dipping followed by heating proved to be an adequate process for surface modification, which led to a decrease of the hydrophilic character of TPS films. Decreasing water uptake and improving mechanical properties of glycerol–cassava starch films are achieved by coating a chitosan solution on starch film (Bangyekan, Aht-Ong, and Srikulkit, 2006). An increase in chitosan coating concentration results in a significant increase in tensile stress at maximum load and tensile modulus and a decrease in elongation at break. The combination of (i) hydrogen bonding, (ii) opposite charge attraction between chitosan cations and negatively charged starch film surface, (iii) hydrophilicity, and (iv) compatible water activities provides a good adherence between the two layers.

Another approach to overcome starch limitation is the protection of starch materials from humidity changes with a thin polymeric layer, deposited by plasma polymerization (or glow discharge polymerization). Plasma technology has become an important industrial process, used to alter chemical and physical properties of polymeric surfaces without affecting their bulk properties. Glow discharge polymerization is a specific type of plasma chemistry, which consists of complex reactions between charged and neutral plasma species, between plasma and surface species, and between surface species (Chan, Ko, and Hiraoka, 1996; Yasuda, 1981). Unlike conventional polymers, plasma polymers do not consist of chains with a regular repeating unit, but tend to form cross-linked networks. Deposition of thin films by plasma polymerization offers several advantages over conventional coating techniques: (i) the process may occur in a single reaction step, (ii) a nanometer-thick coating is obtained, (iii) good adhesion between film and substrate is generally achieved, (iv) the deposition of the film is fairly uniform over the entire surface, and (v) problems with residual solvents are avoided (Chan, Ko, and Hiraoka, 1996). The process involved in macromolecular surface modification in cold plasma environments was recently reviewed by Denes and Manolache (2004).

Ultra-thin, glass-like SiOx coating (20–50 nm thick coating) plasma-deposited on different starch-based films is proposed as a technical solution to improve their barrier properties against water vapor and gases (Johansson, 2000). Starch foils coated in a radiofrequency glow discharge with two different monomers, hexamethyldisilazane (HMDSN) and hexamethyldisiloxane (HMDSO). Argon feeds at radiofrequency discharge powers of 5 and 2 W improved their water-repellent properties without affecting the biodegradability of foils (Behnisch et al., 1998).

Similar results are obtained with cast corn starch films coated by low-pressure plasma polymerization in a 1-butene atmosphere (Andrade, Simão, and Thiré, 2005). For partially gelatinized coated films, an average reduction of 52% in water absorption is observed, independent of coating thickness and for completely gelatinized coated films, a maximum reduction of 90% in water absorption is observed for coatings 80 nm in thickness. When 1,3-butadiene is used as monomer to plasma polymerization, the water absorption of partially gelatinized coated films is reduced up to 80% (Simão et al., 2006a). Unmodified and 1-butene plasma-modified corn starch films can be seen in Figure 6.2. The unmodified film folded due to water absorption, whereas the plasma-modified film remained visually intact, illustrating the protection against water imparted by plasma coating.

The process based on plasma polymerization in a methane atmosphere followed by sulfur hexafluoride (SF$_6$) plasma treatment can also improve water repellence of corn starch films (Simão et al., 2006b). The

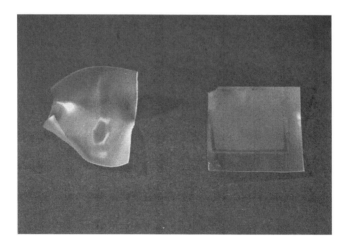

Figure 6.2 Photograph of unmodified (left) and 1-butene plasma-modified (right) corn starch films after contact with a drop of water.

increase of surface hydrophobicity may be related to the resulting topography originated from SF_6 etching of the homogeneous coating obtained by previous methane plasma treatment.

6.4   Final remarks

The twenty-first century will be highlighted by progress toward the intelligent use of natural resources and climate protection. The road toward sustainable development will be built on innovation. Starch-based materials present a great potential to play a major role toward achieving these goals. Due to its low cost, worldwide availability from a large number of crops, and high biodegradability, interest is growing in the use of starch as a raw material for bioplastics production.

Packaging is the dominant application area for starch-based plastics, in special starch blends. They are used for packaging film, shopping bags, strings, straws, tapes, loose-fill foams, laminated paper, fast food and fresh produce containers, and so on. Applications in agriculture include mulch film, planters, and planting pots. Further novel applications include materials for encapsulation and slow release of agrochemicals (Crank et al., 2005). Recently, nanodroplets of a complexed starch have been used as biopolymeric filler in tires, partially replacing traditional fillers (carbon black, diatomite, and silica).

Although there are several starch-based products already available on the global market, starch-based plastics have some drawbacks that limit their application, including limited long-term stability caused by water absorption, postprocessing aging, and poor mechanical properties. Over recent decades, different approaches have been used to overcome these limitations and increase the number of sectors in which starch-based plastics can be used. Hence, innumerable studies have been carried out to develop a completely biodegradable starch-based plastic, with mechanical properties similar to conventional polymers, that are not affected by moisture and postprocessing aging.

References

Altskär, A., Andersson, R., Boldizar, A., Koch, K., Stading, M., Rigdahl, M., and Thunwall, M. 2008. Some effects of processing on the molecular structure and morphology of thermoplastic starch. *Carbohydrate Polym.* 71:591–597.

Andrade, C.T., Simão, R.A., and Thiré, R.M.S.M. 2005. Surface modification of maize starch films by low-pressure glow 1-butene plasma. *Carbohydrate Polym.* 61:407–413.

Angellier, H., Molina-Boisseau, S., Dole, P., and Dufresne, A. 2006. Thermoplastic starch-waxy maize starch nanocrystal nanocomposites. *Biomacromolecules.* 7:531–539.

Atkin, N.J., Abeysekera, R.M., and Robards, A.W. 1998. The events leading to the formation of ghost remnants from the starch granule surface and the contribution of the granule surface to the gelatinization endotherm. *Carbohydrate Polym.* 36:193–204.

Avérous, L. 2004. Biodegradable multiphase systems based on plasticized starch: A review. *J. Macromol. Sci. Part C Polym. Rev.* C 44:231–274.

Avérous, L. and Boquillon, N. 2004. Biocomposites based on plasticized starch: Thermal and mechanical behaviours. *Carbohydrate Polym.* 56:111–122.

Avérous, L., Moro, L., Dole, P., and Fringant, C. 2000. Properties of thermoplastic blends: Starch–polycaprolactone. *Polymer.* 41:4157–4167.

Bangyekan, C., Aht-Ong, D., and Srikulkit, K. 2006. Preparation and properties evaluation of chitosan-coated cassava starch films. *Carbohydrate Polym.* 63:61–71.

Barron, C., Bouchet, B., Della Valle, G., Gallant, D.J., and Planchot, V. 2001. Microscopical study of the destructuring of waxy maize and smooth pea starches by shear and heat at low hydration. *J. Cereal Sci.* 33:289–300.

Bastioli, C. 2000. Global status of the production of biobased packaging materials. In: *Conference Proceedings of the Food Biopack Conference*, J.W. Claus, Ed., Copenhagen, pp. 2–7.

Behnisch, J., Tyczkowski, J., Gazicki, M., Pela, I., Holländer, A., and Ledzion, R. 1998. Formation of hydrophobic layers on biologically degradable polymeric foils by plasma polymerization. *Surface Coatings Technol.* 98:872–874.

Borde, B., Bizot, H., Vigier, G., and Buléon, A. 2002. Calorimetric analysis of the structural relaxation in partially hydrated amorphous polysaccharides. II. Phenomenological study of physical ageing. *Polymer Rev.* 48:111–123.

Briassoulis, D. 2004. An overview on the mechanical behaviour of biodegradable agricultural films. *J. Polym. Environ.* 12:65–81.

Carvalho, A.J.F., Curvelo, A.A.S., and Gandini, A. 2005. Surface chemical modification of thermoplastic starch: Reaction with isocyanates, epoxy functions and stearoyl chloride. *Indust. Crops Products.* 21:331–336.

Chan, C.M., Ko, T.M., and Hiraoka, H. 1996. Polymer surface modification by plasmas and photons. *Surface Sci. Rep.* 24:1–54.

Chang, Y.P., Karima, A.A., and Seow, C.C. 2006. Interactive plasticizing–antiplasticizing effects of water and glycerol on the tensile properties of tapioca starch films. *Food Hydrocolloids* 20:1–8.

Chinnaswamy, R. and Hanna, M.A. 1990. Macromolecular and functional properties of native and extrusion cooked starch. *Cereal Chem.* 67:490–499.

Chung, H.J. and Lim, S.T. 2003. Physical aging of glassy normal and waxy rice starches: Effect of aging time on glass transition and enthalpy relaxation. *Food Hydrocolloids* 17:855–861.

Chung, H.J. and Lim, S.T. 2004. Physical aging of glassy normal and waxy rice starches: Thermal and mechanical characterization. *Carbohydrate Polym.* 57:15–21.

Crank, M., Patel M., Marscheider-Weidermann, F., Schleich, J., et al. 2005. Techno-economic feasibility of large-scale production of bio-based polymers in Europe. EUR 22103 EN. http://www.biomatnet.org/publications/1944rep. pdf (accessed January, 2008).

Curvelo, A.A.S., de Carvalho, A.J.F., and Agnelli, J.A.M. 2001. Thermoplastic starch–cellulosic fibers composites: Preliminary results. *Carbohydrate Polym.* 45:183–188.

Da Róz, A.L., Carvalho, A.J.F., Gandini, A., and Curvelo, A.A.S. 2006. The effect of plasticizers on thermoplastic starch compositions obtained by melt processing. *Carbohydrate Polym.* 63:417–424.

Degli Innocenti, F. and Bastioli, B. 2002. Starch-based biodegradable polymeric materials and plastics: History of a decade of activity. United Nations Industrial Development Organization, Trieste. http://www.ics.trieste.it/chemistry/plastics/ egm-edp2002.htm (accessed January, 2008).

Della Valle, G., Boché, Y., Colonna, P., and Vergnes, B. 1995. The extrusion behaviour of potato starch. *Carbohydrate Polym.* 28:255–264.

Demirgöz, D., Elvira, C., Mano, J.F., Cunha, A.M., Piskin, E., and Reis, R.L. 2000. Chemical modification of starch based biodegradable polymeric blends: Effects on water uptake, degradation behaviour and mechanical properties. *Polymer Degradation Stability* 70:161–170.

Denes, F.S. and Manolache, S. 2004. Macromolecular plasma chemistry: An emerging field of polymer science. *Progress Polym. Sci.* 29:815–885.

Dole, P., Avérous, L., Joly, C., Della Valle, G., and Bliard, C. 2005. Evaluation of starch-PE multilayers: Processing and properties. *Polym. Eng. Sci.* 45:217–224.

Dufresne, A. and Vignon, M.R. 1998. Improvement of starch film performances using cellulose microfibrils. *Macromolecules.* 31:2693–2696.

Endres, H.J., Siebert, A., and Kaneva, Y. 2007. Overview of the current biopolymers market situation. *Bioplastics Mag.* 2:31–33.

Forssell, P.M., Hulleman S.H.D., Myllärinen, P.J., and Moates, G.K. 1999. Ageing of rubbery thermoplastic barley and oat starches. *Carbohydrate Polym.* 39:43–51.

Forssell, P.M., Mikkilä, J.M., Moates, G.K., and Parker, R. 1997. Phase and glass transition behaviour of concentrated barley starch-glycerol-water mixtures, a model for thermoplastic starch. *Carbohydrate Polym.* 34:275–282.

Fringant, C., Desbrières, J., and Rinaudo, M., 1996. Physical properties of acetylated starch-based materials: Relation with their molecular characteristics, *Polymer.* 37:2663–2673.

Gaudin, S., Lourdin, D., Le Botlan, D., Ilari, J.L., and Colonna, P. 1999. Plasticisation and mobility in starch-sorbitol films. *J. Cereal Sci.* 29:273–284.

Godbillot, L., Dole, P., Joly, C., Rogé, B., and Mathlouthi, M. 2006. Analysis of water binding in starch plasticized films. *Food Chem.* 96:380–386.

Govindasamy, S., Campanella, O.H., and Oates, C.G. 1996. High moisture twin-screw extrusion of sago starch: 1. Influence on granule morphology and structure. *Carbohydrate Polym.* 30:275–286.

Griffin, G.J.L. 1977. Biodegradable synthetic resin sheet material containing starch and a fatty material. Patent No. US 4,016,117.

Gudmundsson, M. and Eliasson, A.-C. 1990. Retrogradation of amylopectin and the effects of amylose and added surfactants/emulsifiers. *Carbohydrate Polym.* 13:295–315.

Huang, M.F., Yu, J.G., and Ma, X.F. 2004. Studies on properties of montmorillonite-reinforced thermoplastic starch. *Polymer.* 45:7017–7023.

Hulleman, S.H.D., Kalisvaart, M.G., Janssen, F.H.P., Feil, H., and Vliegenthart, J.F.G. 1999. Origins of B-type crystallinity in glycerol-plasticised, compression-moulded potato starches. *Carbohydrate Polym.* 39:351–360.

Imam, S.H., Gordon, S.H., Shogren, R.L., and Greene, R.V. 1995. Biodegradation of starch-poly(β-hydroxybutyrate-co-valerate) composites in municipal activated sludge. *J. Environ. Polym. Degrad.* 3:205–213.

Jenkins, P.J. and A. M. Donald. 1998. Gelatinisation of starch: A combined SAXS/WAXS/DSC and SANS study. *Carbohydrate Polym.* 308:133–147.

Johansson, K.S. 2000. Improved barrier properties of renewable and biodegradable polymers by means of plasma deposition of glass-like SiO_x coatings. In: *Proceedings of the Food Biopack Conference*, J.W. Claus (Ed.), 110. Copenhagen.

Jurgen, L., Winfred, P., and Harald, S. 2001. Compositions and methods for manufacturing thermoplastic starch blends. US Patent 6,235,816.

Kalichevsky, M.T., Jaroszkiewicz, E.M., Ablett, S., Blanshard, J.M.V., and Lillford, P.J. 1992. The glass transition of amylopectin measured by DSC, DMTA and NMR. *Carbohydrate Polym.* 18:77–88.

Kuutti, L., Peltonen, J., Myllärinen, P., Teleman, O., and Forssell, P. 1998. AFM in studies of thermoplastic starches during ageing. *Carbohydrate Polym.* 37:7–12.

Lai, L.S. and Kokini, J.L. 1991. Physicochemical changes and rheological properties of starch during extrusion (a review). *Biotechnol. Prog.* 7:251–266.

Liu, Z. 2005. Edible films and coatings from starches. In: *Innovations in Food Packaging*, J.H. Han (Ed.). Amsterdam: Elsevier, pp. 318–337.

Liu, Z. and Han, J.H. 2005. Film-forming characteristics of starches. *J. Food Sci.* 70:E32–36.

Lourdin, D., Bizot, H., and Colonna, P. 1997. Antiplasticization in starch–glycerol films. *J. Appl. Polym. Sci.* 63:1047–1053.

Lourdin, D., Colonna, P., Brownsey, G.J., Noel, T.R., and Ring, S.G. 2002. Structural relaxation and physical ageing of starchy materials. *Carbohydrate Polym.* 337:827–833.

Lourdin, D., Della Valle, G., and Colonna, P. 1995. Influence of amylose content on starch films and foams. *Carbohydrate Res.* 27:261–270.

Lu, Y., Tighzert, L., Dole, P., and Erre, D. 2005. Preparation and properties of starch thermoplastics modified with waterborne polyurethane from renewable resources. *Polymer.* 46:9863–9870.

Ma, X. and Yu, J. 2004. The plasticizers containing amide groups for thermoplastic starch. *Carbohydrate Polym.* 57:197–203.

Martin, O., Averous, L., and Della Valle, G. 2003. In-line determination of plasticized wheat starch viscoelastic behavior: Impact of processing. *Carbohydrate Polym.* 53:169–182.

Mathew, A.P. and Dufresne, A. 2002. Plasticized waxy maize starch: Effect of polyols and relative humidity on material properties. *Biomacromolecules.* 3:1101–1108.

McGlashan, S.A. and Halley, P. J. 2003. Preparation and characterisation of biodegradable starch-based nanocomposite materials. *Polymer Int.* 52:1767–1773.

Mondragón, M., Arroyo, K., and Romero-García, J. 2008. Biocomposites of thermoplastic starch with surfactant. *Carbohydrate Polym.* 74:201–208.

Peng, L., Zhongdong, L., and Kennedy, J.F. 2007. The study of starch nano-unit chains in the gelatinization process. *Carbohydrate Polym.* 68:360–366.

Putaux, J.L., Buléon, A., and Chanzy, H. 2000. Network formation in dilute amylose and amylopectin studied by TEM. *Macromolecules.* 33:6416–6422.

Rindlav-Westling, Å. and Gatenholm, P. 2003. Surface composition and morphology of starch, amylose and amylopectin films. *Biomacromolecules.* 4:166–172.

Rindlav-Westling, Å., Hulleman, S.H.D., and Gatenholm, P. 1997. Formation of starch films with varying crystallinity. *Carbohydrate Polym.* 34:25–30.

Rindlav-Westling, Å., Stading, M., and Gatenholm, P. 2002. Crystallinity and morphology in films of starch, amylose and amylopectin blends. *Biomacromolecules.* 3:84–91.

Rindlav-Westling, Å., Stading, M., Hermansson, A.M., and Gatenholm, P. 1998. Structure, mechanical and barrier properties of amylose and amylopectin films. *Carbohydrate Polym.* 36:217–224.

Samir, M.A.S.A., Alloin, F., and Dufresne, A. 2005. Review of recent research into cellulosic whiskers, their properties and their application in nanocomposite field. *Biomacromolecules.* 6:612–626.

Shi, R., Liu, Q., Ding, T., Han Y., Zhang, L., Chen, D., and Tian, W. 2007. Ageing of soft thermoplastic starch with high glycerol content. *J. Appl. Polym. Sci.* 103:574–586.

Shogren, R.L. 1992. Effect of moisture content on the melting and subsequent physical aging of cornstarch. *Carbohydrate Polym.* 19:83–90.

Shogren, R.L. 1993. Effects of moisture and various plasticizers on the mechanical properties of extruded starch. In: *Biodegradable Polymers and Packaging*, C. Ching, D.L. Kaplan, and E.L. Thomas (Eds.). Lancaster: Technomic, pp. 141–150

Shogren, R.L. and Lawton, J.W. 1998. Enhanced water resistance of starch-based materials. US Patent 5,756,194.

Shogren R.L., Fanta, G.F., and Doane, W.M. 1993. Development of starch based plastics: A reexamination of selected polymer systems in historical perspective. *Starch/Stärke.* 45:276–280.

Silva, M.C., Thiré, R.M.S.M., Pita, V.J.R.R., Carvalho, C.W.P., and Andrade, C.T. 2004. Processamento de amido de milho em câmara de mistura. *Ciência e Tecnologia de Alimentos* 24:303–310.

Simão, R.A., da Silva, M.L.V., Martins, M., Thiré, R.M.S.M., and Andrade, C.T. 2006b. Sulphur hexafluoride plasma treatment to enhance the hydrophobicity of CVD carbon coatings produced. *Macromol. Symp.* 245–246:519–524.

Simão, R.A., Thiré, R.M.S.M., Coutinho, P.R., de Araújo, P.J.G., Achete, C.A., and Andrade, C.T. 2006a. Application of glow discharge butadiene coatings on plasticized cornstarch substrates. *Thin Solid Films.* 515:1714–1720.

Slade, L. and H. Levine. 1991. Beyond water activity: Recent advances based on an alternative approach to the assessment of food quality and safety. *Crit. Rev. Food Sci. Nutr.* 30:115–360.

Smits, A.L.M., Kruiskamp, P.H., van Soest, J.J.G., and Vliegenthart, J.F.G. 2003. The influence of various small plasticisers and malto-oligosaccharides on the retrogradation of (partly) gelatinised starch. *Carbohydrate Polym.* 51:417–424.

Souza, R.C.R. and C.T. Andrade. 2002. Investigation of the gelatinization and extrusion processes of corn starch. *Adv. Polym. Technol.* 21:1–8.

Stading, M., Rindlav-Westling, Å., and Gatenholm, P. 2001. Humidity-induced structural transitions in amylose and amylopectin films. *Carbohydrate Polym.* 45:209–217.

Stepto, R.F.T. and Beat, D. 1989. Method of producing destructurized starch. European Patent EP 0,326,517.

Stepto, R.F.T., Tomka, I., and Beat, D. 1988. Destructurized starch and method for making same. European Patent EP 0,282,451.

Stevens, E.S. 2001. *Green Plastics: An Introduction to the New Science of Biodegradable Plastics.* Princeton, NJ: Princeton University Press.

Struik, L.C.E. 1978. *Physical Aging in Amorphous Polymers and Other Materials.* Amsterdam: Elsevier.

Tan, I., Wee, C.C., Sopade, P.A., and Halley, P.J. 2004. Investigation of the starch gelatinisation phenomena in water–glycerol systems: Application of modulated temperature differential scanning calorimetry. *Carbohydrate Polym.* 58:191–204.

Thiré, R.M.S.M., Andrade, C.T., and Simão, R.A. 2005. Effect of aging on the microstructure of plasticized cornstarch films. *Polímeros: Ciência e Tecnologia.* 15:130–133

Thiré, R.M.S.M., Pizzorno, B.S., Simão, R.A., and Andrade, C.T. 2003b. Investigation of humidity-induced microstructural changes in cornstarch films by atomic force microscopy. *Acta Microscópica.* 12:96–99.

Thiré, R.M.S.M., Ribeiro, T.A.A., and Andrade, C.T. 2006. Effect of starch addition on compression-molded poly(3-hydroxybutyrate)/starch blends. *J. Appl. Polym. Sci.* 100:4338–4347.

Thiré, R.M.S.M., Simão, R.A., and Andrade, C.T. 2003a. High resolution imaging of the microstructure of maize starch films. *Carbohydrate Polym.* 54:149–158.

Thunwall, M., Boldizar, A., and Rigdahl, M. 2006. Compression molding and tensile properties of thermoplastic potato starch materials. *Biomacromolecules.* 7:981–986.

Tomka, I. 1994. Thermoplastically processable starch and a method of making it. US Patent 5,362,777.

van den Einde, R.M., Akkermans, C., van der Goot, A.J., and Boom, R.M. 2004a. Molecular breakdown of corn starch by thermal and mechanical effects. *Carbohydrate Polym.* 56:415–422.

van den Einde, R.M., Bolsius, A., van Soest, J.J.G., Janssen, L.P.B.M., van der Goot, A.J., and Boom, R.M. 2004b. The effect of thermomechanical treatment on starch breakdown and the consequences for process design. *Carbohydrate Polym.* 55:57–63.

van Soest, J.J.G. 1996. Starch plastics: Structure-properties relationships. PhD Thesis, Utrecht University.

van Soest, J.J.G. and Essers, P. 1997. Influence of amylose-amylopectin ratio on properties of extruded starch plastic sheets. *J. Macromol. Sci.: Pure Appl. Chem.* A34:1665–1689.

van Soest, J.J.G. and Knooren, N. 1997. Influence of glycerol and water content on the structure and properties of extruded starch plastics sheets during aging. *J. Appl. Polym. Sci.* 64:1411–1422.

van Soest, J.J.G. and Vliegenthart, J.F.G. 1997. Crystallinity in starch plastics: Consequences for material properties. *Trends Biotechnol.*15:208–213.

van Soest, J.J.G., Benes, K., and de Wit, D. 1996. The influence of starch molecular mass on the properties of thermoplastic starch. *Polymer.* 37:3543–3552.

van Soest, J.J.G., Bezemer, R.C., de Wit, D., and Vliegenthart, J.F.G. 1996a. Influence of glycerol on the melting of potato starch. *Industr. Crops. Prod.* 5:1–9.

van Soest, J.J.G., de Wit, D., and Vliegenthart, J.F.G. 1996. Mechanical properties of thermoplastic waxy maize starch. *J. Appl. Polym. Sci.* 61:1927–1937.

van Soest, J.J.G., Hulleman, S.H.D., de Wit, D., and Vliegenthart, J.F.G. 1996c. Changes in the mechanical properties of thermoplastic potato starch in relation with changes in B-type crystallinity. *Carbohydrate Polym.* 29:225–232.

van Soest, J.J.G., Hulleman, S.H.D., de Wit, D., and Vliegenthart, J.F.G. 1996b. Crystallinity in starch bioplastics. *Industr. Crops Prod.* 5:11–22.

Vermeylen, R., Derycke, V., Delcour, J.A., Goderis, B., Reynaers, H., and Koch, M.H.J. 2006. Gelatinization of starch in excess water: Beyond the melting of lamellar crystallites. A combined wide- and small-angle x-ray scattering study. *Biomacromolecules.* 7:2624–2630.

Willet, J.L. and Doane, W.M. 2002. Effect of moisture content on tensile properties of starch/poly(hydroxyester ether) composite materials. *Polymer.* 43:4413–4420.

Wilpiszewska, K. and Spychaj T. 2007. Chemical modification of starch with hexamethylene diisocyanate derivatives. *Carbohydrate Polym.* 70:334–340.

Yakimets, I., Paes, S.S., Wellner, N., Smith, A.C., Wilson, R.H., and Mitchell, J.R. 2007. Effect of water content on the structural reorganization and elastic properties of biopolymer films: A comparative study. *Biomacromolecules.* 8:1710–1722.

Yang, J., Yu, J., and Ma, X. 2006. Study on the properties of ethylenebisformamide and sorbitol plasticized corn starch (ESPTPS). *Carbohydrate Polym.* 66:110–116.

Yasuda, H. 1981. Glow discharge polimerisation. *J. Polym. Sci.: Macromol. Rev.* 16:199–293.

Yavuz, H. and Babaç, C. 2003. Preparation and biodegradation of starch/polycaprolactone films. *J. Polym. Environ.* 11:107–113.

Zeleznak, K.J. and Hoseney, R.C. 1987. The glass transition in starch. *Cereal Chem.* 64:121–124.

Zhang, J.F. and Sun, X. 2004. Mechanical properties of poly(lactic acid)/starch composites compatibilized by maleic anhydride. *Biomacromolecules.* 5:1446–1451.

chapter seven

Trends in microbial amylases

Marcia Nitschke
Instituto de Química de São Carlos, Universidade de São Paulo

Contents

7.1   Introduction

The history of the industrial production of enzymes dates back to 1894, when Dr. J. Takamine began the production of digestive enzyme by wheat bran koji culture of *Aspergillus oryzae.* Enzymes have proven to be of great value over the last 15–20 years in the starch industry. In the 1950s, fungal amylase was used in the manufacture of specific types of syrup, that is, those containing a range of sugars, which could not be produced by conventional acid hydrolysis. The real turning point was reached early in the 1960s when the glucoamylase enzyme was first used to completely break down starch into glucose. Within a few years, almost all glucose production was reorganized and enzyme hydrolysis was used instead of acid hydrolysis due to benefits such as greater yield, higher degree of purity, and easier crystallization. Therefore, in recent decades, there has been a shift from acid hydrolysis of starch to the use of starch-converting enzymes in the production of maltodextrins, modified starches, glucose, and fructose syrups.

The starch-processing industry is the second largest segment of the industrial enzyme market, representing approximately 30% of world enzyme production. The major enzymes are amylases, glucoamylases, and glucose isomerases which are used for starch liquefaction, saccharification, and isomerization.

For many years, industry has been using starch hydrolysis to convert starch-bearing raw materials to starches, starch derivatives, and starch saccharification products, which have several applications in food processing. Initially, the hydrolysis of starch utilized inorganic acids, but in recent decades, the use of enzymes has successfully replaced the chemical process, offering a number of advantages associated with improved yields and favorable economics. Enzymatic hydrolysis allows greater control and high specificity of reaction, better stability of generated products, lower energy requirements, and the elimination of neutralization steps. Moreover, the milder enzymatic catalysis conditions reduce unwanted side reactions and the formation of off-flavor and off-color compounds.

Amylases for food use are traditionally obtained from microbial sources, mainly fungi and bacteria. An advantage of using microorganisms for the production of amylases is the economical bulk production capacity and the ease of manipulation to obtain enzymes of desired characteristics. The conditions prevailing in industrial applications of amylases are generally extreme, especially with respect to temperature and pH. For that reason, there is a continuing demand to improve stability of enzymes and thus meet the requirements set by specific applications. Exploring their natural diversity (finding amylases with special characteristics), gene cloning, and protein engineering are some attempts that have been made to develop novel starch-degrading enzymes that fit industrial needs.

This chapter discusses those enzymes acting on starch and focuses on recent developments involving enzymes obtained from extremophilic bacteria and new molecular and protein engineering approaches for improving enzyme properties to fit industrial needs. Some industrial applications of enzymatic modified starches are also presented.

7.2   *Microbial amylase producers*

Amylases are important industrial enzymes and have a wide range of applications from conversion of starch to sugar syrups to the production of cyclodextrins. Amylases are starch-degrading enzymes that catalyze the hydrolysis of glycosidic bonds of amylose and amylopectin moieties of the polysaccharide. Most amylases are metalloenzymes that require calcium ions for their activity, structural integrity, and stability. They belong to the family 13 of the glycoside hydrolase group of enzymes (Bordbar, Omidiyan, and Hosseinzadeh, 2005) comprising hydrolases, transferases, and isomerases with nearly 30 different specificities. The amylase family can be divided basically into five groups (van der Maarel et al., 2002):

1. Endoamylases: Cleavage of internal α 1,4 bonds resulting in α-anomeric products; the α-amylase (E.C. 3.2.1.1) is the major known endoamylase. Final products of α-amylase action are α- oligosaccharides of varied chain lengths and α-limited dextrins which constitute branched oligosaccharides.

2. Exoamylases: Cleavage of α 1,4 and α 1,6 bonds from external (nonreducing end) glucose residues of amylose and amylopectin resulting in α- or β-anomeric products. β-amylase (E.C. 3.2.1.2) removes one maltose unit at a time by exclusively hydrolyzing α 1,4 bonds whereas glucoamylase (E.C. 3.2.1.3) and α-glucosidase (E.C. 3.2.1.20) are able to hydrolyze both types of glycosidic bonds. The difference between glucoamylase and α-glycosidase is the preferential substrate: α-glycosidase acts preferentially on short malto-oligosaccharide chains, whereas glucoamylase acts on long-chain polysaccharides. The resulting products of exoamylase action on starch are glucose or maltose and β-limited dextrin molecules.

3. Debranching enzymes: Isoamylase (E.C. 3.2.1.68) and pullulanase I (E.C. 3.2.1.41) hydrolyze α 1,6 bonds exclusively, leaving long linear polysaccharides. Pullulanases are able to cleave pullulan and amylopectin, whereas isoamylase acts only on amylopectin molecules. Another group of pullulanases are also able to act on α 1,4 bonds and are called pullulanase II or amilopullulanases which generate maltose and maltotriose as final products.

4. Transferases: Cleavage of an α 1,4 glycosidic bond of the donor molecule and transfer of part of the donor to a glycosidic acceptor forming a new glycosidic bond. When cyclodextrin glycosyltransferase (CGTase, E.C. 2.4.1.19) acts on starch or a starch derivative, the product is a circular ring of glucose units linked by α 1,4 bonds. The enzyme catalyzes an intramolecular transglycosylation reaction, cleaves the bonds within a starch substrate to create a linear chain of glucose units, and rejoins the ends to form the ring structure. CGTases generate cyclic oligosaccharides containing six (α-cyclodextrin), seven (β-cyclodextrin), or eight (γ-cyclodextrin) glucose units and branching dextrins of high molecular weight.

5. Amylomaltases (E.C. 2.4.1.25): Similar to CGTases with respect to the enzymatic reaction, however, the final products are linear oligosaccharides. The branching enzymes (E.C. 2.4.1.18) are involved in glycogen synthesis, forming α 1,6 bonds in lateral chains of glycogen molecules.

The nomenclature of the enzymes used commercially for starch hydrolysis is somewhat confusing and the EC numbers sometimes group enzymes with subtly different activities. For example, α-amylases may be subclassified as liquefying or saccharifying amylases but even this classification is inadequate to encompass all the enzymes that are used in commercial starch hydrolysis. One reason for the confusion about this nomenclature is the use of the anomeric form of the released reducing group in the product rather than that of the bond being hydrolyzed; the products of bacterial and fungal α-amylases are in the α-configuration and the products of β-amylases are in the β-configuration, although all these enzymes cleave between α–1,4-linked glucose residues.

Industrial amylases exhibit a variety of different specificities that act on starch molecules, promoting the hydrolysis of α–1–4, α–1–6, glycosidic bonds in endo- or exo-acting mode. Endoamylases catalyze hydrolysis in a random manner in the interiors of the starch macromolecules producing linear and branched oligosaccharides of various chain lengths. Exoamylases act from the nonreducing end, successively resulting in short end products (Pandey et al., 2000; Gupta et al., 2003); the most important industrial compounds are α-amylases, β-amylases, glucoamylases, isoamylases, and pullulanases (Van der Maarel et al., 2002).

Several microorganisms have been suggested as efficient production sources of these enzymes, but only a few selected strains of fungi and bacteria meet the criteria for commercial production. An example of the wide diversity of microorganisms involved in amylase production is shown in Table 7.1. The industrial production of α-amylases commonly employs *Bacillus* species such as *B. amyloliquefaciens, B. licheniformis, B. stearothermophilus,* and *Aspergillus* species. β-amylases are generally produced by bacterial strains belonging to *Bacillus,* however, *Pseudomonas, Clostridium, Streptomyces,* and fungal strains from *Rhizopus* sp. are also involved. Glucoamylases are obtained from filamentous fungi belonging to *Aspergillus, Penicillium,* and *Rhizopus* genera. Isoamylases are produced from *Cytophaga* sp., *Escherichia coli, Saccharomyces cerevisiae,* and *Pseudomonas* strains. Pullulanase production involves essentially *Bacillus* species.

7.3   Industrial needs

The enzymatic conversion of starch to high fructose corn syrup is a well-established process in which the consecutive use of several enzymes is necessary. The first step in the process is the conversion of starch to oligomaltodextrins by the action of α-amylase. The concomitant injection of steam puts extreme demands on the thermostability of the enzyme. Although commercial thermostable α-amylase is available, in general the pH has to be adjusted to an undesirable high level and ions must be added to stabilize the enzyme (Crab and Mitchinson, 1997).

α-Amylases are Ca^{+2}-dependent enzymes; Ca^{+2} is typically needed for activity and stability. Ideally, the enzyme should be active and stable at a low pH (~4.5) and not demand calcium for stability. One of the concerns of the starch industry is the calcium requirement of the α-amylases and

Table 7.1 Diversity of Microbial Amylase Producers

Enzyme	Microorganisms involved	Enzyme	Microorganisms involved
α-amylases	*Aspergillus awamori*	Glucoamylases	*Aspergillus* sp.
	A. flavus		*A. awamori*
	A. fumigatus		*A. candidus*
	A. niger		*A. niger*
	A. oryzae		*A. oryzae*
	A. usanii		*A. phoenicus*
	Bacillus brevis		*A. terreus*
	B. licheniformis		*Bacillus firmus*
	B. stearothermophilus		*B. stearothermophilus*
	B. amyloliquefaciens		*Clostridium acetobutylicum*
	Clostridium acetobutylicum		*C. thermosaccharoliticum*
	C. thermosulfurogenes		*Flavobacterium* sp.
	C. thermohydrosulfuricum		*Mucor rouxianus*
	Lactobacillus brevis		*Penicillium italicum*
	Micrococcus luteus		*P. oxalicum*
	Mucor pusillus		*Rhizoctonia solanii*
	Nocardia asteroides		*Rhizopus* sp.
	Penicillium brunneum		*R. javanicus*
	Pseudomonas stutzeri		*R. oligospora*
	Rhizopus sp.		*R. oryzae*
	Thermococcus profundus		*Trichoderma reesei*
	Thermonospora viridis		*T. viride*
	Aspergillus sp.		
β-amylases	*Bacillus cereus*	Pullulanases	*Bacillus* sp.
	B. circulans		*B. acidopolluliticus*
	B. polymyxa		*B. cereus*
	B. megaterium		*B. circulans*
	Clostridium thermosulfurogenes		*B. macerans*
	C. thermocellum		*C. thermohydrosulfuricum*
	Pseudomonas sp.		*C. thermosaccharoliticum*
	Rhizopus japonicus		*Klebsiella aerogenes*
	Streptomyces sp.		*K. oxytoca*
			K. pneumoniae

(continued)

Table 7.1 Diversity of Microbial Amylase Producers (Continued)

Enzyme	Microorganisms involved	Enzyme	Microorganisms involved
Isoamylases	*Cytophaga* sp.		*Micrococcus* sp.
	Saccharomyces cerevisiae		*Pyrococcus woesei*
	Escherichia coli		*Thermatoga maritima*
	Pseudomonas amyloderamosa		*Thermococcus aggregans*
			Thermoanaerobacter ethanolicus
			Thermus aquaticus
			T. caldophilus

Source: Adapted from Pandey et al. (2000) and Doman-Pytka and Bardowski (2004).

the formation of calcium oxalate, a substance that may block process pipes and heat exchangers. In addition, such accumulations in certain products such as beer are not acceptable. Its precipitation can be reduced through a decrease in the calcium ion requirement of the enzymes and lowering the pH of the production process. The calcium requirement is not desirable in starch-based sweetener production; Ca^{+2} is known to inhibit glucose isomerase, the enzyme involved in isomerization of glucose to fructose (Smith, Rangarajan, and Hartley, 1991). Calcium-free amylases are of great importance for the detergent industry. The presence of chelating agents, usually found in these products, can inhibit the majority of available alkaline amylases often used as additives to detergents for washing machines and automatic dishwashers (Nielsen and Borchert, 2000). However, the activities of a number of starch-hydrolyzing enzymes are calcium-dependent (Haki and Rakshit, 2003).

Saccharification involves hydrolysis of remaining oligosaccharides (8–12 glucose units) into either maltose syrup by β-amylase or glucose syrups by glucoamylase (Crab and Mitchinson, 1997). The process is run at pH 4.2–4.5 and 60°C, at which temperature the currently used *Aspergillus niger* glucoamylase is stable. Still, the temperature has to be reduced after liquefaction and the pH has to be adjusted in order for the glucoamylase to act. More economically feasible would be use of an enzyme that was active at the same pH and temperature range as the liquefaction enzymes.

The water content in the starch slurry is generally high (35% w/w or higher) and reduction of the moisture content could be more economical. This has been shown to be possible when including a shearing treatment (Van der Veen et al., 2006) followed by an increase of isomaltose production. Increasing temperatures would also require enzymes with very high thermostability.

New amylases with optimized properties, such as enhanced thermal stability, acid tolerance, and the ability to function without the addition of calcium, are required and will offer obvious benefits to the industry.

7.4   Approaches to enzyme improvement

7.4.1   Isolation of new microbial strains

Natural microorganisms have over the years been a great source of enzyme diversity. An interesting approach is the screening for novel microorganism strains from extreme environments (Veille and Zeikus, 2001). As enzymatic liquefaction and saccharification of starch are performed at high temperatures (100–110°C), thermostable and thermoactive amylolytic enzymes from extremophiles are of great interest for the development of valuable products such as glucose, crystalline dextrose, dextrose syrup, maltose, and maltodextrins. Currently, thermostable amylases of *Bacillus stearothermophilus* or *Bacillus licheniformis* are used in starch processing industries (Crabb and Mitchinson, 1997). Thus, the highly thermostable and thermoactive amylolytic enzymes from extreme thermophiles are still being sought for the improvement of the starch hydrolysis process.

Many extreme and hyperthermophilic bacteria and archaea, as *B. stearothermophilus, Clostridium thermosulfurogenes, Clostridium thermocellum, Desulfurococcus mucosus, Fervidobacterium pennavorans, Pyrococcus furiosus, Pyrococcus woesei, Thermoanaerobacter* sp., *Thermococcus aggregans, Thermococcus celer, Thermococcus guaymasensis, Thermococcus profundus, Thermococcus hydrothermalis, Thermococcus litoralis, Thermatoga maritima,* and *Rhodothermus marinus* produce amylases and/or pullulanase utilizing starch and related carbohydrates (Bertoldo and Antranikian, 2002; Gomes, Gomes, and Steiner, 2003).

For the most part, thermophilic archaeal α-amylases do not differ from their mesophilic counterparts in molecular weight and amino acid composition. With the exception of *Pyrococcus furiosus* intracellular α-amylase, which is a homodimer with a subunit molecular weight of 66 kDa (Laderman et al., 1993), all thermophilic archaeal α-amylases are monomeric enzymes with a molecular weight between 42 and 68 kDa. When expressed in the mesophilic eubacteria *Escherichia coli* or *Bacillus subtilis*, the recombinant archaeal α-amylases retained all the biochemical properties of the native enzymes (Jorgensen, Vorgias, and Antranikian, 1997).

Although Ca^{+2} has often been reported as an activator of archaeal α-amylases, two α-amylases do not require Ca^{+2}: archaeal *P. furiosus* and eubacterial *Thermus* sp. extracellular α-amylases (Koch et al., 1990; Shaw, Bott, and Day, 1999). The effects of other metals on thermophilic archaeal α-amylase activities are metal-specific and depend on the enzyme source.

Three results have been observed: inhibition, activation of the enzymatic activity, or no significant effect (Leveque et al., 2000).

Most of the pullulanases produced by different microorganisms are not suitable for operation under conditions prevailing in the industry because of the low stability to pH and temperature (Haki and Rakshit, 2003), but many thermophilic and mesophilic bacteria are rich sources of these enzymes (Doman-Pytka and Bardowski, 2004). Thermophilic archaeal pullulanases were detected on *P. furiosus, P. woesei, T. aggregans, T. celer, T. guaymasensis, T. hydrothermalis, Thermococcus litoralis*, and numerous other thermococcal strains. The archaeal pullulanases are typically optimally active between 80 and 100°C and between pH 5.5 and 6.5 (Leveque et al., 2000).

Screening of mesophilic and thermophilic microorganisms showing amylase activity could facilitate the discovery of enzymes suitable for new industrial applications. Recent reports show a new isolated strain of *B. subtilis* JS-2004 producing high levels of thermostable α-amylase suitable for applications in starch processing (Asgher et al., 2007). A new strain of *Bacillus* sp. I-3 was able to produce a thermostable α-amylase that can be used for the direct hydrolysis of raw potato starch (Goyal, Gupta, and Soni, 2005), lowering the cost of the final product because the starch gelatinization step can be suppressed or reduced.

The starch industry largely uses hydrolysis by enzymatic processes to produce maltodextrins which serve as coating agents, viscosity providers, flavor carriers, moisture controllers, crystallization inhibitors, fat substitutes, and many other functions. The efficiency of each property depends strongly on the degree of polymerization (DP) of the individual maltosaccharide. There is growing interest in producing maltodextrins with a narrow DP distribution, appropriate for a specific application and more adequate for a possible subsequent purification. An example is the formulation of sports drinks, consumed during physical activities, that provide a burst of energy and water to compensate for fluid loss from perspiration. A balanced carbohydrate composition is essential to obtain such a result. An optimal drink would contain short linear oligosaccharides with DP of 3 to 6, because they are absorbed at the highest rate and keep the osmolality at moderate levels, thus preventing fluid loss and possible side effects such as diarrhea and cramps (Marchal, Beeftink, and Tramper, 1999). An atypical amylase produced by *B. subtilis* US116 is capable of generating a mixture of 30% maltohexaose (DP6) and 20% maltoheptaose (DP7) from starch (Messaoud et al., 2004).

Research on thermostable amylases has concentrated on the enzymes of thermophiles and extreme thermophiles. Facultative thermophiles are a largely unexplored microbial group capable of growing at mesophilic temperatures, but can grow well at higher temperatures. However, not much is known about the processes for enzyme production involving such organisms (Pandey et al., 2000; Haki and Rakshit, 2003). Relatively little information is available regarding the mechanism involved in thermal stability. Saboto et al. (1999) reported that thermophilic enzymes are more rigid proteins than their mesophilic counterparts, whereas Fitter et al. (2001) determined that a mechanism characterized by entropic stabilization could be responsible for the higher thermostability of thermophilic enzymes. Comparative analysis of the structures and physicochemical properties of mesophilic and thermophilic microbial amylases may provide information about the molecular basis of high-temperature tolerance ability of thermostable amylases.

Cultivation of thermophiles at high temperature is technically and economically interesting as it reduces the risk of contamination and also reduces viscosity, making mixing easier, and leads to a high degree of substrate solubility. However, compared to their mesophilic counterparts, the biomass achieved by these organisms is usually disappointingly low. The low cell yield is undesirable for both large- and small-scale production, which makes extensive studies of their enzymes very difficult (Turner, Mamo, and Karlsson, 2007). To increase the thermophile cell yield, several media compositions and culture optimizations have been proposed (Krahe, Antranikian, and Markel, 1996) and special equipment and specific processes have been developed to improve fermentation of thermophiles and hyperthermophiles (Schiraldi and De Rosa, 2002). However, due to factors such as the requirement of complex and expensive media, low solubility of gas at high temperature, low specific growth rates, and product inhibition, large-scale commercial cultivation of thermophiles for enzyme production remains an economic challenge. These examples demonstrate the importance of a continuous search for novel environmental microorganisms showing starch degrading activities.

7.4.2   Gene cloning and protein engineering

Reducing the production cost of thermophilic enzymes is fundamental for their breakthrough on a large scale. Recombinant technology has been proposed as an alternative to reduce production costs and increase the yield. Industrial production of recombinant enzymes is preceded by an extensive research and development phase that culminates in the construction of a successful production strain. This process typically involves the following development of the host (recipient) strain, construction of the expression vector, transformation of the host strain, identification of the best recombinant strain, additional improvements, and characterization of the production strain (Olempska-Beer et al., 2006).

A great deal of work has been done on the construction of recombinant microorganisms, mostly *Escherichia coli* and *Saccharomyces cerevisiae* to improve enzyme yields and properties, and to reduce the levels of secondary toxic metabolites (Pandey et al., 2000; Olempska-Beer et al., 2006). In addition to genetic stability, an indispensable characteristic of the mutant strains, there is great concern regarding the safety of recombinant microorganisms for food applications. Usually the recombinant genes are inserted on GRAS (generally recognized as safe) organisms traditionally used in food enzyme production such as *B. subtilis*, *E. coli* K12, *Aspergillus oryzae*, and *A. niger*. However, some fungal strains produce toxic secondary metabolites under certain conditions and genetic manipulation can reduce or eliminate their expression (Olempska-Beer et al., 2006).

Recombinant production strains can be further improved using classical mutagenesis. Expression vectors may integrate into the host genome at different loci and various copy numbers. Consequently, the transformation procedure yields a population of transformants, which produce different levels of the intended enzyme. These transformants are subsequently grown under different conditions and assessed for enzyme expression and other characteristics (Olempska-Beer et al., 2006). Once a satisfactory transformant is identified, it can be subjected to mutagenesis using either a chemical mutagen, UV, or ionizing radiation. Subsequently, the population of mutants is screened for enzyme yield to identify the best performer.

Gene cloning has been performed extensively for the molecular study of proteins, their hyperproduction, and protein engineering. Cloning of amylases is an important tool for studying their sequence, characteristics, production, and expression. A gene encoding a new α-amylase of *B. licheniformis* was cloned and expressed in *E. coli*; genomic DNA of *B. licheniformis* was digested with EcoRI and BamHI and linked to the pBR322 vector. The transformed *E. coli* carried the recombinant plasmid pIJ322 containing a 3.5 kb fragment of *B. licheniformis* DNA. The purified enzyme encoded by pIJ322 was capable of hydrolyzing pullulan and cyclodextrin as well as starch (Kim et al., 1992), indicating its simultaneous action as α-amylase and pullulanase, and offering the advantage of acting on liquefaction and saccharification steps during syrup production. This work constitutes a good example of enzyme cloning.

Differences in codon usage or improper folding of the proteins can result in reduced enzyme activity or low level of expression (Duffner et al., 2000). Moreover, many complex enzymes, such as hetero-oligomers or those requiring covalently bound co-factors, can be produced with difficulty in mesophilic hosts. Because of this, a search for genetic tools for the overexpression of such enzymes in thermophilic host systems has been pursued (Turner, Mamo, and Karlsson, 2007), but use of novel thermophilic expression systems is, however, still at the research level and more work remains before exploitation on a large or industrial scale can be considered.

Enzyme properties may also be adapted to specific use conditions by using modern genetic techniques. Site-specific mutagenesis can be used to introduce specific changes in the amino acid sequence of an enzyme. Site-specific mutagenesis is most effective when the three-dimensional structure of the enzyme is known and the relationships between structure and enzyme properties have been elucidated (Olempska-Beer et al., 2006). Hashida and Bisgaard-Frantzen (2000) have reported a high calcium-independent and acid-stable α-amylase produced via site-directed mutagenesis. Protoplast fusion and mutagenesis have been widely used in protein engineering to achieve strains with high enzyme productivity or specific characteristics (Pandey et al., 2000).

The process of molecular evolution is a powerful approach for improving enzyme properties and consists of several steps, often performed in an interrelated manner (Tobin, Gustafsson, and Huisman, 2000). In the first step, one or several parent genes are chosen and, if several genes are used, they are often derived from diversified sources to provide sequence diversity. These genes are subsequently mutagenized in a random manner using techniques such as an error-prone PCR mutagenesis, sequential random mutagenesis, or gene shuffling to create a large number of gene variants. A library of altered genes is then constructed in a suitable host microorganism. The clones are screened using high-throughput methods to identify those expressing improved enzymes. Genes encoding these enzymes are isolated, sequenced, and usually recycled through the process until an enzyme with the desired characteristics is identified (Bessler et al., 2003).

A thermostable α-amylase, active at low pH, that does not require added calcium when used for starch hydrolysis was developed using molecular evolution from three amylases discovered in nature. The microorganisms that produce these amylases belong to the order Thermococcales within Archaea. The hybrid α-amylase is produced from *P. fluorescens* biovar I and is intended for use in starch processing and fermentation of ethanol for alcoholic beverages (Landry et al., 2003).

Enzyme engineering is known to be a promising technique to find new and potentially interesting enzymes. It consists of integrating desired properties in the appropriate gene, which may include high thermostability, wide pH profile, calcium ion independence, raw starch degrading ability, activity at a high concentration of starch, protease resistance, insensibility to catabolic repression, and hyperproduction (Sivaramakrishnan et al., 2006). Engineering α-amylase for changed pH activity profiles also would add to the stability of the enzymes (Nielsen and Borchert, 2000) and engineering of commercially available enzymes is used with great success (Van Der Maarel et al., 2002). Introduction of three mutations— Asn172 → Arg, His156 → Tyr, and Ala181 → Thr—was reported to lead to a fivefold increase in the thermostability of α-amylase of *B. licheniformis* (Declerck et al., 2003). Thermostability at low pH was achieved by substitutions of Met15 → Thr and Asn188 → Ser (Shaw, Bott, and Day, 1999).

Stabilization of proteins has been done by insertion of prolines in loop regions (Matthews, Nicholson, and Bechtel, 1987). It has also been shown that incorporation of hydrophobic residues at the surface of *B. licheniformis* α-amylase resulted in increased resistance to high temperature (Machius et al., 2003).

A comparison of the active sites and the surroundings of different α-amylase activity at medium and high temperatures identified regions that could be important in the functioning of *B. licheniformis* α-amylase at medium temperatures. The regions 181–195, 141–149, and 456–463, the individual amino acids at positions 311, 346, and 385, and mutations therein or deletions contribute to improved pH stability from 8 to 10.5, improved Ca^{2+} stability at pH 8 to 10.5, or increased specific activity at 30 to 40°C (Borchert et al., 1999). Introduction of disulfide bonds in enzymes and changes in the oxidized amino acids prone to an unaffected amino acid by oxidative agents (Barnett et al., 1998) leads to improved stability of the enzyme.

Developments in bioinformatics and the availability of sequence data have greatly increased the possibility of isolating an efficient gene from nature. Protein engineering offers the possibility of structural, biochemical, and biophysical property changes, pointing to a valuable new tool for enzyme optimization. Recently, various attempts at understanding the important parameters in directed evolution have emerged and successful examples of combining rational engineering with directed evolution have been reported (Kirk, Borchert, and Fuglsang, 2002).

7.5   Applications of enzymatic modified starches

Recent advances in protein engineering and molecular biology related to various enzymes involved in the degradation and biosynthesis of starch are leading to a new way to create structured and modified starches and, consequently, to improve their range of commercial applications.

One of the most desired properties is freeze–thaw stability of gelatinized starch used in the food industry, which improves the quality of frozen and chilled products due to their low starch retrogradation. A thermostable 4-α-glucanotransferase (also called D-enzyme) originated from *Thermus scotoductus* is used to modify the physical and structural properties of starch, and is known to catalyze the disproportionation of added malto-oligosaccharides by intermolecular transglycosylation. It promotes the reduction of amylose chains and modification of amylopectin lateral chains, improving freeze–thaw stability of the enzyme-treated rice starch gel (Lee et al., 2006).

Amylomaltase from the hyperthermophilic bacterium *Thermus thermophilus* HB8 overproduced in *Escherichia coli* promoted hydrolyses in both short and long chains of potato starch, resulting in a thermoreversible gel at concentrations of 3% (w/v) or more, which suggested it would be a good substitute for gelatin in vegetarian food products (van der Maarel et al., 2005).

Esterification of starch with long-chain fatty acids, such as palmitic acid, yields thermoplastic starch now widely used in the plastics industry and in pharmaceutical and biomedical applications such as materials for bone fixation and carriers for controlled release of drugs and other bioactive agents. Unlike chemical esterification, enzymatic esterification is ecofriendly and avoids the use of toxic solvents. A lipase from *Candida rugosa* was proposed as a catalyst for starch esterification. The modification of the starch OH groups by esterification imparts thermoplasticity and water resistance to the starch ester over the unmodified starch (Rajan, Sudha, and Abraham, 2008).

A neopullulanase from *Bacillus stearothermophilus* promotes selective amylose hydrolysis, barely cutting amylopectin chains. Dextrin obtained from this hydrolyzed starch exhibits low retrogradation and lower viscosity than gelatinized starch. Thus this enzyme can be employed to obtain low-amylose or amylose-free starch (Kamasaka et al., 2002).

Designed biosynthetic starches obtained by genetic modification are new tools employed in starch-based products. Activity reduction of starch synthase isozymes produced by gene inhibition on potato transgenic plants resulted in an amylase-free starch with short-chain amylopectin, which has high freeze–thaw stability and almost no syneresis (Jobling et al., 2002). Even if construction of a tailor-made starch structure by enzymatic transgenic modifications is an exciting way to improve starch properties and increase the range of industrial applications, much research is still needed to obtain sufficient knowledge.

7.6 Final remarks

Enzymes found in nature have been widely used in the production of fermented starchy foods and hydrolyzed starch products used in the food, textile, paper, and pharmaceutical industries. Research to develop new amylases adapted to industrial needs is often focused on thermotolerant, pH-tolerant, and calcium-independent amylases, which can be obtained by new isolated microorganisms or by genetic and protein engineering techniques. As advances in molecular genetics allowed reshaping enzyme production, it became possible to clone genes encoding enzymes and express them in host microorganisms that are well adapted to large-scale industrial fermentation. Enzyme yield could be substantially increased by using efficient promoters and introducing multiple copies of the enzyme-encoding gene. It has also become possible to tailor enzyme properties to food processing conditions, such as temperature or pH, by modifying the amino acid sequence of the enzyme using either rational design or molecular evolution.

Amylases are important industrial enzymes and efforts to improve their performance should be reinforced by a combination of traditional microbiology and new molecular research strategies. Enzyme stability, high-scale production, and purification, as well as the safety of engineered proteins are some of the challenges that will be faced by the scientific and industrial communities in introducing new successful commercial amylases. Knowledge of amylolytic enzyme action and detailed studies about mechanisms involving modification of starch structure will allow us to obtain new starches with the desired and appropriate properties.

References

Asgher, M., Asad, M.J., Rahman, S.U., and Legge, R.L. 2007. A thermostable α-amylase from a moderately thermophilic *Bacillus subtilis* strain for starch processing. *J. Food Eng.* 79:950–955.

Barnett, C.C., Mitchinson, C., Power, S.D., and Roquat, C.A. 1998. Oxidatively stable α-amylases. US Patent 5,824,532.

Bertoldo, C. and Antranikian, G. 2002. Starch-hydrolyzing enzymes from thermophilic archaea and bacteria. *Curr. Opin. Chem. Biol.* 6:151–160.

Bessler, C., Schmitt, J., Maurer, K., and Schmid, R.D. 2003. Directed evolution of a bacterial α-amylase: Toward enhanced pH-performance and higher specific activity. *Protein Sci.* 12:2141–149.

Borchert, T.V., Andersen, C., Nissen, T., and Kjaerulff, S. 1999. α-Amylase mutants. International Patent WO 99/23211.

Bordbar, A.K., Omidiyan, K., and Hosseinzadeh, R. 2005. Study on interaction of α-amylase from Bacillus subtilis with cetyl trimethylammonium bromide. *Colloids Surf.* B 40: 67–71.

Crabb, W.D. and Mitchinson, C. 1997. Enzymes involved in the processing of starch to sugars. *Trends Biotechnol.* 15: 349–352.

Declerck, N., Machius, M., Joyet, P., Wiegand, G., Huber, R., and Gaillardin, C. 2003. Hyperthermostabilization of *Bacillus liqueniformis* α-amylase and modulation of its stability over 50°C temperature range. *Protein Eng.* 16:287–293.

Doman-Pytka, M. and Bardowski, J. 2004. Pullulan degrading enzymes of bacterial origin. *Crit. Rev. Microbiol.* 30:107–121.

Duffner, F., Bertoldo, C., Andersen, J.T., Wagner, K., and Antranikian, G. 2000. A new thermoactive pullulanase from *Desulfurococcus mucosus*: Cloning, sequencing, purification and characterization of the recombinant enzyme after expression in *Bacillus subtilis*. *J. Bacteriol.* 182:6331–6338.

Fitter, J., Herrmann, R., Hauss, T., Lechner, R.E., and Dencher, N.A. 2001. Dynamical properties of α-amylase in the folded and unfolded state: The rate of thermal equilibrium fluctuations for conformational entropy and protein stabilization. *Physica B.* 301:1–7.

Gomes, I., Gomes, J., and Steiner, W. 2003. Highly thermostable amylase and pullulanase of the extreme thermophilic eubacterium *Rhodothermus marinus*: Production and partial characterization. *Biores. Technol.* 90:207–214.

Goyal, N., Gupta, J.K., and Soni, S.K. 2005. A novel raw starch digesting thermo-stable α-amylase from *Bacillus* sp. I-3 and its use in the direct hydrolysis of raw potato starch. *Enz. Microb. Tech.* 37:723–734.

Gupta, R., Gigras, P., Mohapatra, H., Goswami, V.K., and Chauhan, B. 2003. Microbial α-amylases: A biotechnological perspective. *Process Biochem.* 38:1599–1616.

Haki, G.D. and Rakshit, S.K. 2003. Developments in industrially important ther-mostable enzymes: A review. *Bioresource Technol.* 89:17–34.

Hashida, M. and Bisgaard-Frantzen, H. 2000. Protein engineering of new indus-trial amylases. *Trends Glycosci. Glyc.* 12:389–401.

Jobling, S.A., Westcott, R.J., Tayal, A., Jeffcoat R., and Schwall, G.P. 2002. Production of a freeze-thaw-stable potato starch by antisense inhibition of three starch synthase genes. *Nature Biotechnol.* 20:295–299.

Jorgensen, S., Vorgias, C.E., and Antranikian, G. 1997. Cloning, sequencing, charac-terization, and expression of an extracellular α-amylase from the hyperther-mophilic archaeon *Pyrococcus furiosus* in *Escherichia coli* and *Bacillus subtilis. J. Biol. Chem.* 272:16335–16342.

Kamasaka, H., Sugimoto, K., Takata, H., Nishimura, T., and Kuriki, T. 2002. *Bacillus stearothermophilus* neopullulanase selective hydrolysis of amylose to maltose in the presence of amylopectin. *Appl. Environ. Microbiol.* 68:1658–1664.

Kim, I.C., Cha, J.H., Kim, J.R. , S.Y., Seo, B.C., Cheong, T.K., Lee, D.S., Choi, Y.D., and Park, K.H. 1992. Catalytic properties of the cloned amylase from *Bacillus licheniformis. J. Biol. Chem.* 267:22108–22114.

Kirk, O., Borchert, T.V., and Fuglsang, C.C. 2002. Industrial enzyme applications. *Curr. Opin. Biotechnol.* 13:345–351.

Koch, R., Zablowski, P., Spreinat, A., and Antranikian, G. 1990. Extremely ther-mostable amylolytic enzyme from the archaebacterium *Pyrococcus furiosus. FEMS Microbiol. Lett.* 71:21–26.

Krahe, M., Antranikian, G., and Markel, H. 1996. Fermentation of extremophilic microorganisms. *FEMS Microbiol. Rev.* 18:271–285.

Laderman, K., Davis, B., Krutzsch, H. Lewis, M.S., Griko, Y.V., Privalov, P.K., and Anfinsen, C.B. 1993. The purification and characterization of an extremely thermostable α-amylase from the hyperthermophilic archaebacterium *Pyrococcus furiosus. J. Biol. Chem.* 268:24394–24401.

Landry, T.D., Chew, L., Davis, J.W., Frawley, N., Foley, H.H., Stelman, S.J., Thomas, J., Wolt, J., and Hanselman, D.S. 2003. Safety evaluation of an α-amylase enzyme preparation derived from the archaeal order Thermococcales as expressed in *Pseudomonas fluorescens* biovar I. *Regul. Toxicol. Pharm.* 37:149–168.

Lee, K.Y., Kim, Y., Park, K.H., and Lee, H.G. 2006. Effects of α-glucanotransferase treatment on the thermo-reversibility and freeze-thaw stability of a rice starch gel. *Carbohydr. Polym.* 63:347–354.

Leveque, E., Janecekc, S., Hayea, B., and Belarbi, A. 2000. Thermophilic archaeal amylolytic enzymes. *Enz. Microb. Tech.* 26:3–14.

Machius, M., Declerck, N., Huber, R., and Wiegand, G. 2003. Kinetic stabilization of *Bacillus liqueniformis* α-amylase through introduction of hydrophobic resi-dues at the surface. *J. Biol. Chem.* 278:11546–11553.

Marchal, L.M., Beeftink, H.H., and Tramper, J. 1999. Towards a rational design of commercial maltodextrins. *Trends Food Sci. Tech.* 10:345–355.

Matthews, B.W., Nicholson, H., and Bechtel, W.J. 1987. Enhanced protein thermo-stability from site-directed mutations that decrease the entropy of unfolding. *Proc. Natl. Acad. Sci. USA.* 84:6663–6667.

Messaoud, E.B., Ben Ali, M., Elleuch, N., Fourati Masmoudi, N., and Bejar, S. 2004. Purification and properties of a maltoheptaose- and maltohexaose-forming amylase produced by *Bacillus subtilis* US116. *Enz. Microb. Tech.* 34:662–666.

Nielsen, J.E. and Borchert, T.V. 2000. Protein engineering of bacterial α-amylases. *Biochim. Biophys. Acta.* 1543:253–274.

Olempska-Beer, Z.S., Merker, R.I., Ditto, M.D., and DiNovi, M.J. 2006. Food-processing enzymes from recombinant microorganisms: A review. *Regul. Toxicol. Pharm.* 45:144–158.

Pandey, A., Nigam, N., Soccol, C.R., Soccol, V.T., Singh, D., and Mohan, R. 2000. Advances in microbial amylases. *Biotechnol. Appl. Biochem.* 31:135–152.

Rajan, A., Sudha, J.D., and Abraham, E. T. 2008. Enzymatic modification of cassava starch by fungal lipase. *Ind. Crop. Prod.* 27:50–59.

Saboto, D., Nucci, R., Rossi, M., Gryczynski, I., Gryczynski, Z., and Lakowicz, J. 1999. The α-glycosidase from the hyperthermophilic archaeon *Sulfolobus solfataricus*. Enzyme activity and conformational dynamics at temperatures above 100°C. *Biophys. Chem.* 81:23–31.

Schiraldi, C. and De Rosa, M. 2002. The production of biocatalysts and biomolecules from extremophiles. *Trends Biotechnol.* 20:515–521.

Shaw, A., Bott, R., and Day, A.G. 1999. Protein engineering of α-amylase for low pH performance. *Curr. Opin. Biotechnol.* 10:349–352.

Sivaramakrishnan, S., Gangadharan, D., Nampoothiri, K.M., Soccol, C.R., and Pandey, A. 2006. α-Amylases from microbial sources: An overview on recent developments. *Food Tech. Biotechnol.* 44:173–184.

Smith, A., Rangarajan, M., and Hartley, S. 1991. D-xylose (D-glucose) isomerase from *Arthrobacter* strain NRRL B 3728. *Biochem. J.* 277:255–261.

Tobin, M.B., Gustafsson, C., and Huisman, G.W. 2000. Evolution: The rational basis for irrational design. *Curr. Opin. Struc. Biol.* 10:421–427.

Turner, P., Mamo, G., and Karlsson, E. N. 2007. Potential and utilization of thermophiles and thermostable enzymes in biorefining. *Micro. Cell Fact.* 6:1–23.

Van der Maarel, M.J.E.C., Capron, I., Euverink, G.W., Bos, H.T., Kaper, T., Binnema, D.J., and Steeneken, P.A.M. 2005. A novel thermoreversible gelling product made by enzymatic modification of starch. *Starch.* 57:465–472.

Van der Maarel, M.J.E.C., Van der Veen, B., Uitdehaag, J.C.M, Leemhuis, H., and Dijkhuizen, L. 2002. Properties and applications of starch-converting enzymes of the α-amylase family. *J. Biotechnol.* 94:137–155.

Van der Veen, M.J.E.C., Veelaert, S., Van der Goot, A.J., and Boom, R.M. 2006. Starch hydrolysis under low water conditions: A conceptual process design. *J. Food Eng.* 75:178–186.

Veille, C. and Zeikus, G.J. 2001. Hyperthermophilic enzymes: Sources, use, and molecular mechanisms for thermostability. *Microbiol. Mol. Biol. Rev.* 65:1–43.

chapter eight

Modified starch
Chemistry and properties

Kerry C. Huber
School of Food Science, University of Idaho

James N. BeMiller
Whistler Center for Carbohydrate Research, Purdue University

Contents

8.1 Introduction

Granular starch is utilized in many food and industrial applications due to its universal abundance, relatively low cost (Whistler, 1984), and ability to impart a broad range of functional properties to food and non-food products (Jane, 1995; BeMiller, 2007). Current annual production (in millions of tons) for the primary starch sources follows the order: corn (46.1), cassava (9.1), wheat (5.15), and potato (2.45) (Röper and Elvers, 2008). However, most starches in their native form have limitations that make them less than ideal for the diversity of desired applications. For this reason, most granular starch utilized as a food or industrial ingredient is first modified (chemically and/or physically), while yet in the granular form, to alter and improve the physical properties of starch polymers in accordance with the intended end use. Starch, including native and particularly modified types, accounts for more than 85% of all hydrocolloids used in food systems worldwide (Wanous, 2004). Some of the major industrial applications for various modified food starches are summarized in Table 8.1. Focused and comprehensive discussions of traditional commercial starch modification practices and modified starch applications are provided by Rutenberg and Solarek (1984) and Wurzburg (1986a), and specific reagents and reagent levels allowed for production of modified starches for use in food are summarized by Xie, Liu, and Cui (2005). The present chapter provides a synopsis of the current level of understanding of starch reactivity (granular and molecular aspects), advances in starch modification processes, and an overview of the physical changes that result from starch modification, with emphasis on recent scientific literature.

Table 8.1 Some Major Applications of Modified Starches

Industry	Modified starch types
Adhesive	British gums, yellow dextrins
Food	Distarch phosphates, distarch adipates, hydroxypropyl starches, starch acetates, starch octenylsuccinates, monostarch phosphates, hydroxypropylated distarch phosphates, phosphorylated distarch phosphates, acetylated distarch phosphates, acetylated distarch adipates, thinned (via use of acid or alkaline hypochlorite solution) versions of the above, oxidized starches, pregelatinized versions of the above, white dextrins
Glass fiber	Cationic starches, benzylated starches
Papermaking	Hydroxyethyl starches, cationic starches, acid-thinned versions of the above, oxidatively thinned starches, graft copolymer products
Pharmaceutical	Pregelatinized starches, sodium starch glycolate (sodium carboxymethyl starch)
Textile	Hydroxyethyl starches, carboxymethyl starches, starch acetates, distarch phosphates, acid-modified (thin-boiling) starches and acid-modified versions of the above, oxidized starches

8.2   Reasons for modification of starch

Starches are modified chemically or physically or both to accentuate their positive characteristics, diminish their undesirable qualities, or add new attributes. Common limitations associated with native normal starches are excessive viscosity at low solids content (difficulty in handling, lack of body), high susceptibility to retrogradation (gel opacity, syneresis, and lack of freeze–thaw stability), and lack of process tolerance. By proper modification, changes can be made in one or more of the following attributes.

- Ability to act as an emulsifying agent
- Ability to act as an emulsion stabilizer
- Ability to encapsulate
- Charges on starch molecules (via adding positive or negative charges)
- Cold-water swellability
- Cooking characteristics (degree of breakdown, degree of setback—retrogradation, energy required to cook, gelatinization/pasting temperature, hot-paste viscosity)
- Digestibility
- Film formation
- Flowability
- Interactions with other substances
- Paste and gel characteristics (adhesiveness, clarity, freeze–thaw stability, gel strength, rate and extent of syneresis, retrogradation stability, sheen, viscoelasticity, viscosity)
- Process tolerance (pH tolerance, shear tolerance, temperature tolerance)
- Solubility in hot and room-temperature water
- Stability in high-salt environments
- Water resistance of films (water resistance or water-holding capacity)

8.3   Starch modification

8.3.1   Chemically modified starches

Chemical modification, which is the most common industrial means of enhancing starch properties, entails reaction or treatment of starch with chemical reagents to introduce new chemical substituent groups, effect molecular scission, or promote oxidation or molecular rearrangements (Wurzburg, 1986b). Although chemical modification of granular starch is practiced industrially, a complete understanding of starch granule reactivity has not yet been realized, due in part to the lack of understanding of the relationship between starch granular and molecular structural regimes and their relationships to modification patterns. Starch modification patterns can be chemically tracked and understood at the universal (overall extent of substitution or modification), granular (pattern of modification within the granule), and molecular (pattern of reaction on individual starch molecules) levels. Each of these levels provides useful fundamental insight and understanding into starch modification practices.

At the universal level, the extent of modification for reactions that introduce new chemical substituent groups is most commonly quantified by the degree of substitution (DS). The DS, which has a maximum value of three (based on the average potential number of hydroxyl groups available for reaction on each glucosyl unit), represents the average number of modified hydroxyl groups per glucosyl unit. For modifications in which it is possible for the introduced chemical derivative group to undergo further reaction with the reagent (e.g., introduced hydroxyalkyl groups), the extent of the reaction is defined in relation to molar substitution (MS, average number of moles of reactant per glucosyl unit). Methods for determination of starch DS and MS vary according to the type of modifying agent.

Starch granular reaction patterns (distribution of reagent within granules) have been directly visualized with varying degrees of certainty using transmission electron microscopy (TEM), scanning electron microscropy (SEM, via x-ray microanalysis or backscattered electron imaging), light microscopy, and confocal laser scanning microscopy (CLSM) methods (Whistler and Spencer, 1960; Berghofer and Klaushofer, 1977; Vihervaara et al., 1990; Hamunen, 1995; Whistler et al., 1998; Huber and BeMiller, 2001; Gray and BeMiller, 2004). Details of microscopic techniques used in the study of starch structure are discussed in Chapter 2. Granular reaction sites of modified starches were effectively visualized by Huber and BeMiller (2001) using SEM. This technique involved conversion of anionic starch derivatives to heavy metal salts and visualization of granular reaction sites using SEM compositional backscattered electron imaging. This general approach was simplified by Gray and BeMiller (2004), who modified the method to effect visualization of granular reaction patterns via optical sectioning using reflectance confocal laser scanning microscopy (R-CLSM). The limitation of the latter method is that it will only detect sites in which several anionic substituent groups are located in close proximity to one another; it is not capable of detecting single reaction sites.

Determination of reaction patterns on starch molecules is traditionally accomplished indirectly by systematic enzyme- and/or acid-catalyzed hydrolysis of starch derivatives and characterization of their respective hydrolytic products (Hood and Mercier, 1978; Turneport, Salomonsson, and Theander, 1990; Steeneken and Woortman, 1994; Zhu and Bertoft, 1997; Kavitha and BeMiller, 1998; van der Burgt et al., 1998; Zhu et al., 1998; Manelius et al., 2000; Manelius, Nurmi, and Bertoft, 2000; Shi and

BeMiller, 2000; Richardson et al., 2003). Such methods are based upon the rationale that introduced chemical substituent groups interfere with the normal pattern of enzyme-catalyzed hydrolysis, and utilize deviations from normal hydrolysis patterns (based on the unmodified starch) to pinpoint molecular reaction sites on modified starch chains. More recently, Higley (2005) developed a model reaction system utilizing a fluorescent probe that directly facilitates elucidation of reaction patterns on intact AM and AP starch polymers and AP branch chains. For this method, starches derivatized with a fluorescent reagent may be analyzed as either intact polymers or debranched chains via size exclusion chromatography aided by fluorescence detection. This latter technique has the advantage of directly elucidating molecular reaction patterns on intact starch polymers and AP branch chains without destructive hydrolysis of the molecular chains themselves.

8.3.1.1   Factors influencing chemical modification
A primary underlying principle defining starch granule reactivity is that some degree of granule swelling is generally necessary for efficient derivatization to occur (Hauber, BeMiller, and Fannon, 1992; Kweon, Bhirud, and Sosulski, 1996). Thus, starch attributes or derivatization conditions that facilitate or promote the swelling of starch granules (without incurring pasting) and maximize the accessibility of starch molecules within granules to reagent generally enhance overall reactivity (Gray and BeMiller, 2005). These factors can be classified as either intrinsic, that is, inherent to the starch material itself, or extrinsic, related to the conditions imposed by the reaction system, as previously set forth by Han et al. (2006a) (Table 8.2).

Table 8.2 Some Factors Affecting Granular Starch Reactions

General characteristic	*Specific factors*
Intrinsic	
Granule composition	AM:AP ratio, AM and AP chemical fine structure, presence of nonstarch constituents
Granule structure	Granule architecture, nature of granule surface, presence of pores and channels, granule size
Extrinsic	
Reagent parameters	Reagent type (nature of reactive group, molecular size and physical properties) and concentration
Reaction medium conditions	pH, type and concentration of swelling-inhibiting salt, solvent system comprising reaction medium, reaction temperature, reaction time

8.3.1.1.1 Intrinsic factors The granule architecture exhibits a marked effect on both starch molecular and granular patterns of reaction. One of the more compelling arguments in support of this notion is based on the differential molecular reaction patterns observed for granular and nongranular starches subjected to identical derivatization conditions (Steeneken and Smith, 1991; Jane et al., 1992; Steeneken and Woortman, 1994; Wilke and Mischnik, 1997; Richardson et al., 2003). For nongranular reactions, derivatization patterns on starch macromolecules are largely homogeneous or random in nature. In contrast, substituent groups on starch polymers modified in granular form are unevenly distributed among starch macromolecules and over their length; both observations have led to the hypothesis that molecular reaction sites within granular starch occur in clusters or blocks (Hood and Mercier, 1978; Steeneken and Smith, 1991; Steeneken and Woortman, 1994; Kavitha and BeMiller, 1998; Shi and BeMiller, 2000; Chen, Schols, and Voragen, 2004a).

For granular reactions, this uneven or heterogeneous molecular derivatization pattern is derived from the reality that reaction is preferentially and almost exclusively confined to granule amorphous regions (Hood and Mercier, 1978; Steeneken and Smith, 1991; Manelius et al., 2000a) and to the outer lamellae of crystalline regions (Chen, Schols, and Voragen, 2004). In this scenario, the amorphous regions of the granule, which generally contain AM molecules, branching regions of AP, and some nonreducing ends of AP branch chains are the favored loci for reaction (Hood and Mercier, 1978; Steeneken and Woortman, 1994; Kavitha and BeMiller, 1998). For most granular starch derivatives, AM is on average more heavily derivatized than AP (Steeneken and Woortman, 1994; Kavitha and BeMiller, 1998; Shi and BeMiller, 2000; Chen, Schols, and Voragen, 2004; Kaur, Singh, and Singh, 2004), although the reverse has also been reported (Salomonsson, Fransson, and Theander, 1991). Molecular reaction patterns of wheat, potato, and cassava starches (derivatized with a fluorescent probe) directly revealed the relative reaction densities for intact starch macromolecules (AM, AP) and AP branch chains (Figure 8.1) (Higley, 2005). Medium-length chains of AP (B1 and A chains) were least reacted, most likely due to their extensive involvement in the granule crystalline structure. Others have observed more than 40% of AP branch chains of DP ≈ 15 to be completely unreacted for the same reason cited above (Hood and Mercier, 1978; Kavitha and BeMiller, 1998). Reaction densities of long (B2, B3, and possibly some C chains) and very short AP branch chains (possibly too short for participation in the crystalline structure) were more than threefold higher than those of medium branch chains, and were equivalent to or greater than those of AM (Higley, 2005).

Although starches of various botanical sources share some fundamental similarities, it is essential to recognize that they differ in the specific details of their compositions and fine structures (AM:AP ratios, molecular

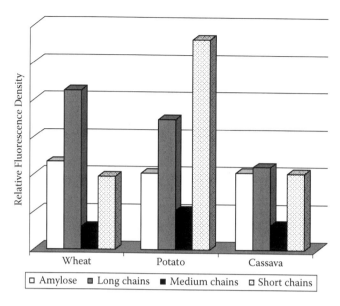

Figure 8.1 Relative reaction densities of AM and AP branch chains of wheat, potato, and cassava starches subjected to derivatization with a fluorescent reagent. AM reaction densities are relatively consistent for the three starch botanical sources studied, whereas AP long and medium branch chains are the most and least densely reacted, respectively, for all three evaluated modified starches (adapted from Higley, 2005).

structures, nonstarch constituents, etc.) and that these differences affect granular structure and, subsequently, reactivity. Heterogeneity among the various botanical sources and genotypes with respect to starch granule composition results in differential starch granule reactivities and reaction patterns (Mussulman and Wagoner, 1968; Singh, Ali, and Divakar, 1993; Azemi and Wootton, 1995; Bhattacharya, Singhal, and Kulkarni, 1995; Huber and BeMiller, 2001; Bertolini et al., 2003; Gray and BeMiller, 2005). In short, it is virtually certain that the granule structure affects both starch granular (crystalline vs. amorphous regions) and molecular reaction patterns, and that variation in structure, and thus reactivity, occurs not only among differing botanical sources of starch, but also within a given population of granules of a single source (Allen, Hood, and Parthasarathy, 1976; Ji et al., 2004).

Granule microstructure has also been shown to affect starch granule reactivity. Pores at the surfaces of corn, sorghum, millet, wheat, rye, and barley starch granules (Hall and Sayre, 1970; Fannon, Hauber, and BeMiller, 1992; Fannon, Schull, and BeMiller, 1993; Li et al., 2004; Waduge et al., 2006; Kim and Huber, 2008) were demonstrated to be openings to channels leading to the granule interior (Fannon, Schull, and BeMiller, 1993; Huber and

BeMiller, 1997; Kim and Huber, 2008). Their characterization and influence on chemical modifications have been reviewed elsewhere (Fannon et al., 2003, 2004; Han et al., 2006a). For granules possessing channels, it was observed that the diffusion of dye (simulation of reagent flow) or reagent solution into the granule matrix occurred most readily through channels (laterally) and interior cavities (from the granule interior cavity outward into the matrix) (Huber and BeMiller, 2000, 2001). More recently, channels of corn and wheat starch granules are reported to be lined with protein (Han et al., 2005), which could further affect the reactivity or permeability of starch granules. In the absence of channels, reagent may enter the granule matrix via external granule surfaces (BeMiller, 1997), although, in the case of corn and sorghum starch granules, the rate of reagent entry by this mode appeared to be somewhat impeded by an outer shell, probably composed of tightly packed molecules (Huber and BeMiller, 2001). Consequently, for granules possessing pores and channels, these features facilitate reagent delivery into the granule matrix and significantly affect granular reaction patterns (Huber and BeMiller, 2001; Gray and BeMiller, 2004).

Starch derivatization is also reportedly influenced by starch granule size (Boruch, 1985; Bertolini et al., 2003; Stapley and BeMiller, 2003; Chen, Schols, and Voragen, 2004; Singh and Kaur, 2004; Van Hung and Morita, 2005a,b), although this effect has not been consistently observed in all cases (van der Burgt et al., 2000; Bertolini et al., 2003; Stapley and BeMiller, 2003). Small granules contribute a greater collective surface area for reaction on an equal starch weight basis when compared with large ones; however, the extent of starch derivatization does not always coincide with specific surface area measures. This observation is likely explained by the fact that differences in granule size are almost always confounded by other accompanying compositional or structural differences that themselves may also have an impact on granule reactivity. Thus, the effect of granule size is not always clear-cut, and can be reagent-specific (Bertolini et al., 2003).

Nonstarch constituents present in granules can also affect starch granule reactions. For example, waxy wheat starch was observed to exhibit a higher apparent reactivity relative to normal wheat starch in reactions with phosphoryl chloride (Bertolini et al., 2003). This observation is likely explained by the lack of a significant AM–lipid complex and a higher swelling power for waxy relative to normal wheat starch (Van Hung, Maeda, and Morita, 2007). The specific impact of AM–lipid complexes on granule reactivity has not been fully investigated, although Gunaratne and Corke (2007a) reported that these complexes are at least partially disrupted by alkaline conditions typical of starch reactions. In any event, the presence and stability of AM–lipid complexes have the potential to influence molecular mobility and reactivity of the granule amorphous regions during reaction. Additionally, proteins are also anticipated substrates in starch reactions, due to their broad distribution at granule surfaces, inside channels (i.e., channel-associated), and within the granule matrix (i.e., granule-bound), and based on the fact that they contain nucleophilic groups capable of reacting with starch derivatizing reagents. The likelihood for protein to be located in accessible or amorphous regions of the granule further increases its potential for reaction. The potential accessibility and reactivity of the various starch granule proteins to a fluorescent probe have been demonstrated (Han and Hamaker, 2002; Han et al., 2005), although protein has been rarely, if at all, investigated as a factor in starch reactions involving industrial reagents.

8.3.1.1.2 Extrinsic factors Of the various extrinsic factors, Gray and BeMiller (2005) identified pH as the most significant factor influencing starch reaction efficiency in the reaction with propylene oxide. For the majority of chemical modification schemes in which starch functions as a nucleophile, at least some degree of starch hydroxyl group ionization (formation of the alkoxide anion) is essential to "activate" starch for reaction. With a pK_a ~12.6, starch hydroxyl groups would be expected to become increasingly ionized with an escalating pH. In general, an increased pH results in higher reaction efficiency (Wu and Seib, 1990; Lim and Seib, 1993; van Warners, Stamhuis, and Beenakers, 1994; Gray and BeMiller, 2005; Han and BeMiller, 2006), which is attributed to increased granule swelling (due to electrostatic repulsion between ionized starch hydroxyl groups) and/or increased starch alkoxide ion levels (BeMiller and Pratt, 1981; Gray and BeMiller, 2005). However, the effect of pH can vary with reagent type. An exception to the aforementioned trend may occur when a highly anionic reagent (e.g., sodium tripolyphosphate) becomes excessively ionized under alkaline conditions, inducing electrostatic repulsion between starch and reagent molecules to reduce the extent of reaction (Lim and Seib, 1993). In such cases, moderation of the reaction pH results in improved reaction efficiency.

The nature of the reaction medium greatly influences starch derivatization. Swelling–inhibiting salts (e.g., sodium sulfate, sodium chloride), which are frequently incorporated into the reaction medium to control granule swelling and prevent starch pasting, influence both the extent of derivatization (Shi and BeMiller, 2000; Villwock and BeMiller, 2005) and granular patterns of reaction (Gray and BeMiller, 2005). Approximately 1.8 times as much reagent was needed to achieve the same MS level on corn starch when a highly effective granule swelling inhibitor (potassium citrate) was used in the reaction medium in place of the usual sodium sulfate (Shi and BeMiller, 2000). Both the type and concentration of swelling-inhibiting salt are important considerations, as reaction medium conditions that promote a greater degree of granule swelling generally increase reagent access to the granule and the overall extent of derivatization (Shi and BeMiller, 2000; Gray and BeMiller, 2005; Villwock and

BeMiller, 2005). Kweon, Bhirud, and Sosulski (1996), using an aqueous ethanol reaction system, observed that the extent of reaction was enhanced as the proportion of water in the reaction medium increased. In other cases, an increased ionic strength has been observed to enhance derivatization efficiency for reactions involving anionic reagents or reagent intermediates (Wu and Seib, 1990), possibly by shielding like charges between reagents and starch alkoxide anions (Woo and Seib, 1997).

For starch reacted as a granular slurry, the temperature of the reaction medium can be increased to enhance reactivity (Wu and Seib, 1990; Lammers, Stamhuis, and Beenackers, 1993; Woo and Seib, 1997), although temperature increases are generally limited to the starch subgelatinization range (typically $\leq 50^{\circ}C$) to prevent pasting of granules. Possible explanations for reactivity enhancement with increasing temperature include: (1) a decrease in the pK_a of starch hydroxyl groups corresponding to an increased starch alkoxide ion concentration (Lammers, Stamhuis, and Beenackers, 1993), (2) a reduction of the Donnan potential to grant catalyzing hydroxide anions greater access to starch granules (Oosten, 1982), or (3) increased granule swelling due to electrostatic repulsion among neighboring starch alkoxide anions (Donovan, 1979; Gray and BeMiller, 2005). Nevertheless, reactivity enhancement with increasing temperature has its limits. Wu and Seib (1990) reported that temperature was not a factor in cross-linking reactions conducted at pH values above 10.75, and Gray and BeMiller (2005) observed no significant temperature effect over the range of $44-54^{\circ}C$ for substitution reactions. It would seem that once a specific degree of granule swelling or alkoxide level is achieved (via elevated pH or temperature), additional temperature increases offer no further benefit. Aside from starch–slurry reactions noted above, high temperature ($120-170^{\circ}C$) is also used to drive derivatization of starch impregnated with inorganic (e.g., orthophosphate) salts using a semi-dry reaction process (Kerr and Cleveland, 1962). The heating of starch at reduced levels of moisture prevents gelatinization of starch granules.

Reagent type affects starch granular reaction patterns. Fast-acting reagents (e.g., phosphoryl chloride) have been shown to be more likely to react at granule surfaces (first point of granular contact) to produce heterogeneous reaction patterns (Gluck-Hirsch and Kokini, 1997; Huber and BeMiller, 2001; Gray and BeMiller, 2004). In contrast, slow-reacting reagents generally produce more homogeneous granular reaction patterns, as the reagent has a longer timeframe to diffuse into and disperse throughout the granule matrix prior to reacting (Huber and BeMiller, 2001; Gray and BeMiller, 2004). Furthermore, the order of reagent addition during preparation of dual-modified starches affects overall derivatization levels and reaction patterns (Yeh and Yeh, 1993; Wang and Wang, 2000), implying that various reagents impart differential effects on granule reactivity. Initial cross-linking reduces the consequent extent of substitution, whereas initial substitution actually increases the extent of subsequent cross-linking (Yeh and Yeh, 1993). This finding is congruent with the general belief that the reagent reacts with the less-ordered (more amorphous) areas of granules first, causing these regions to swell. The swelling not only allows the amorphous regions to react more, but may also disrupt some of the crystalline regions, increasing the starch available for reaction (Gray and BeMiller, 2005). However, this general belief was not corroborated in starch reactions with propylene oxide (Han and BeMiller, 2005). It could be that this phenomenon is reagent-specific, as it is known to be possible to substitute starch granules (especially with anionic reagents) to such high levels that ambient temperature pasting of granules occurs as a function of extensive derivatization.

Although the appropriate length of reaction varies according to the reagent, most granular slurry reactions are thought to experience an initial stage of relatively high (reagent-specific) rate of reaction, in which the most accessible granular sites (amorphous regions) are reacted by reagent, followed by a decline or leveling off in reaction rate as the principal reaction sites become derivatized or reagent concentration declines (Han and BeMiller, 2005). The length of reaction can also potentially alter granule composition. For starch substitution reactions with propylene oxide, an increasing degree of leaching of starch material from granules was observed as the length of reaction increased (0–30 h). Molar substitution values for leached starch increased over the course of the reaction, and consistently exceeded those of the corresponding granular starch for all evaluated time intervals within the reaction period.

8.3.1.2   Types of chemical modification

8.3.1.2.1   Conversion   In starch science, the term "conversion" refers to depolymerization. Products that are partially depolymerized are called "converted starches." There are three basic processes for making converted starches: using acid, using an oxidant in an alkaline system, and application of heat (Section 8.3.1.2.2).

To obtain the products variously known as "acid-modified," "acid-converted," or "thin-boiling" starches, a mineral acid is added to a stirred slurry of starch granules (36–40% solids) held at a temperature below the gelatinization temperature of the starch (usually 40–60°C). When the desired degree of limited hydrolysis or conversion is reached, the suspension is neutralized and the granules are recovered by centrifugation or filtration, washed, and dried. Normal acid-catalyzed hydrolysis of the glycosidic linkages occurs, but the hydrolysis is not random with respect to either hydrolysis of linkages or hydrolysis of polymer molecules. Experimental evidence indicates that (1–6) linkages are hydrolyzed preferentially to (1–4) linkages, probably because the (1–6) linkages are found in the amorphous regions containing the AP branch points. However, in solution, the preference for hydrolysis is reversed; that is, there is a preference for hydrolysis of (1–4) linkages. The most rapid hydrolysis occurs in the less dense, less crystalline regions of granules (Van Beynum and Joels, 1985; Wurzburg, 1986c). For acid-treated waxy maize starch, Chabot, Allen, and Hood (1978) suggested two different granular patterns for hydrolysis. In some cases, the hydrolysis front appeared to proceed from the outer granule surface inward, whereas in other instances hydrolysis was thought to progress from the hilum region of the granule outward in radial fashion; the latter hydrolysis pattern is most likely facilitated by channels within granules.

When to stop a conversion is usually determined empirically by monitoring the viscosity of a hot paste of the acid-modified starch. In the starch industry, this is known as determining fluidity. Acid conversion is practiced on both native starches and chemically modified starches. Compared to the parent starch, acid-converted starches disintegrate in hot water more readily upon cooking, and have reduced molecular weights for both linear and branched components, reduced intrinsic viscosity values, increased paste clarity, high solubility in hot water, reduced hot paste viscosity, increased gel strength and opacity, and better film-forming capabilities. Specific characteristics are a function of the parent starch and the conversion conditions.

Conversion by oxidation is based on the principle that a β-alkoxy carbonyl system is formed when a carbonyl unit is introduced at C2, C3, or C6 of a 4-substituted D-glucopyranosyl unit, leading to β elimination and chain cleavage in an alkaline system (BeMiller, 2007). Oxidation is often done by treatment with a solution of sodium hypochlorite. The hypochlorite anion oxidizes the hydroxymethyl group (C6) to an aldehydo group, oxidizes the hydroxyl groups at C2 or C3 to keto groups, and effects cleavage of the C2–C3 bond, resulting in a dialdehyde (aldehydo groups from C2 and C3) (Figure 8.2). Any of the aldehydo groups may subsequently be oxidized to carboxylate groups. Oxidized starch may also be produced using hydrogen peroxide and copper(II) ions, a system that depolymerizes the starch without introducing carbonyl groups. Ammonium persulfate is used to depolymerize starches oxidatively for paper sizing and coating processes. The decision as to when to stop an oxidation is generally determined in the way previously noted for an acid-converted starch (i.e., by monitoring hot paste viscosity or fluidity).

The finding that reaction with periodate ion occurred first in the center of granules led Gallant and Guilbot (1969a,b) to hypothesize that the central area surrounding the hilum is the least organized region of granules. This hypothesis is supported by evidence that digestion of ungelatinized granules with pores and channels begins in the central cavity region of the hilum and proceeds outward (Fannon et al., 2003). It is reasonable that reactions such as hypochlorite oxidation proceed in the same manner.

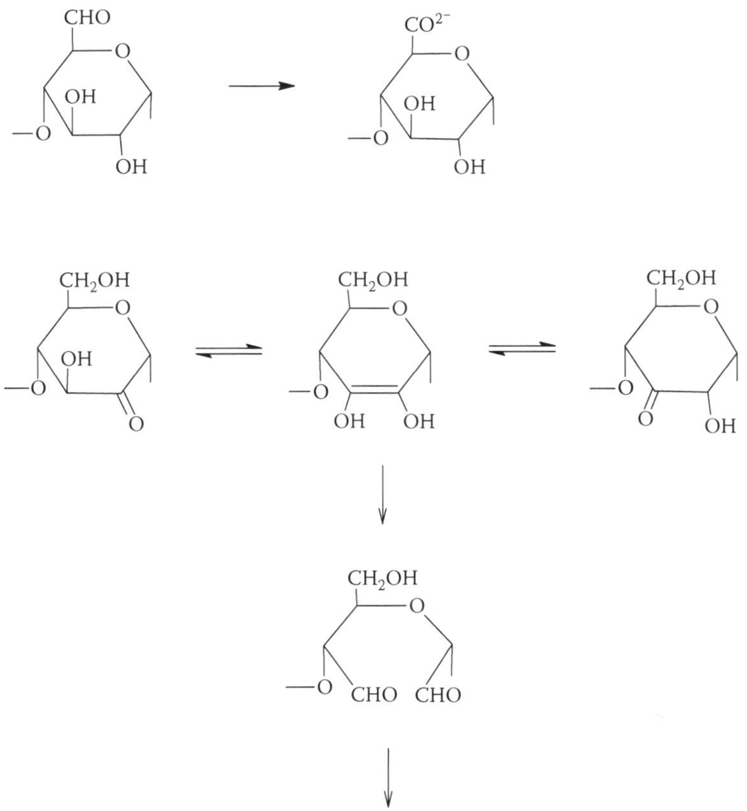

Figure 8.2 Products of oxidation of starch with hypochlorite ions.

The properties of starches converted by oxidative cleavage generally parallel those of starches converted by acid-catalyzed hydrolysis, with two differences. Pastes of oxidized starches generally are clearer and more stable than those made from acid-modified starches, most likely because carboxylate groups are larger than hydroxymethyl groups, providing steric hindrance to retrogradation, and because carboxylate groups have a negative charge. Repulsion of like charges prevents starch chains from approaching each other. As a result, the pastes and gels made from oxidized starches have properties similar to those of acid-thinned, stabilized products (Section 8.3.1.2.4). The other difference is that greater variability in products is possible via choice of the parent starch, oxidizing agent, and conditions of oxidation. For example, starches treated with low levels of hypochlorite at an acidic pH (rather than the usual alkaline pH condition) result in products with a greater (rather than a reduced) hot paste viscosity.

8.3.1.2.2 Transglycosidation Dextrinization is the process of making starch dextrins, which are products with a greater degree of conversion than found in the products termed acid-modified starches (Section 8.3.1.2.1).

Starch dextrins are made by heating a starch with or without addition of an acid (Van Beynum and Joels, 1985; Wurzburg, 1986c). There are three general types of these pyrodextrins: white dextrins, yellow dextrins, and British gums. The variables involved in making dextrins are the base starch used, the amount and type of acid used, the percentage of moisture, the temperature employed, and the time of conversion. In general, white dextrins are made using an acid catalyst, relatively low temperatures, and short times. Yellow dextrins are products of higher temperatures and longer times. British gums are also made using higher temperatures and longer times, but with less or no added acid. Hydrochloric acid is usually used, particularly for food-grade dextrins; other mineral acids may also be used. Starch at 10–22% moisture is either treated with gaseous hydrogen chloride or a solution of hydrochloric, sulfuric, or phosphoric acid is sprayed onto it as an atomized mist. The starch is then pre-dried to a moisture content of 2–5% by heating it at a relatively low temperature, sometimes under reduced pressure, to keep hydrolysis to a minimum. Pre-drying is not as critical in the production of white dextrins, because the goal of the process is limited acid-catalyzed hydrolysis of glycosidic bonds at a temperature low enough that color production is minimal. In contrast, pre-drying is an essential step in the production of yellow dextrins and British gums. During the actual dextrinization step (pyroconversion), when the pre-dried starch is heated at 100–200°C, more water is driven off and the moisture content drops to less than 5%. The product is then cooled and rehumidified. Alternatively, the starch is first dried to a moisture content of 2–5%, and is then treated with either gaseous HCl or hydrochloric acid before being heated.

Three types of reaction take place during dextrinization, the relative amounts of which are a function of the moisture content of the starch and the temperature. (1) Hydrolysis of glycosidic bonds is the predominant reaction when the moisture content is high and is, therefore, the main reaction in the preparation of white dextrins with a low degree of conversion. Because of glycosidic bond cleavage and the resulting reduction in molecular weights, final products have lower solution viscosities. (2) Transglycosidation (transglycosylation) results in transfers of portions of glucan chains to hydroxyl groups on the same or different chains to create a branch point. The new glycosidic bond can be to either O6, O3, or O2 and can be in α or β configuration. The result is more highly branched and more soluble macromolecules. (3) Repolymerization (reversion) occurs with catalytic amounts of acid at high temperatures and low moisture content. Under these conditions, glycosidic bonds are formed from the reducing ends of the malto-oligosaccharides (and any glucose) released during hydrolysis. The result is higher molecular-weight, more branched macromolecules than were present before reversion occurred. When acid-ified starch of a low moisture content is heated, hydrolysis of glycosidic bonds occurs. Some water molecules are consumed in the reaction, and the temperature increases. After the temperature exceeds 100°C, moisture is lost and transglycosidation and reversion begin. White dextrins are the least converted and will form firm pastes and gels. Also, in general, the lower the viscosity and the darker the color of a dextrin solution (at equal concentration), the greater the stability of those cooks, except that solutions of yellow dextrins, which are generally more stable than those of British gums, are usually darker, because they are made with greater amounts of acid (Wurzburg, 1986c).

8.3.1.2.3 Cross-linking Starch cross-linking reactions are employed to strengthen the structure of swollen granules upon gelatinization, enhancing the resistance to viscosity breakdown as a result of mechani-cal shear, acid conditions, or high temperature. Depending on the extent of the reaction and the starch botanical source, cross-linking can have variable effects on specific modified starch characteristics and properties. Very low levels of cross-linking usually stabilize the granule structure to allow the modified starch to attain a higher degree of granule swelling during heating than would be achieved by the unmodified native starch. In such cases, higher swelling powers and paste peak viscosities can be observed (Wurzburg and Symanski, 1970; Howling, 1980; Kasemsuwan and Jane, 1994; Jyothi, Moorthy, and Rajasekharan, 2006; Kaur, Singh, and Singh, 2006).

In contrast, progressively higher levels of cross-linking generally lead to reduced granule swelling (Hoover and Sosulski, 1986; Colas, 1986; Ziegler, Thompson, and Casasnovas, 1993; Zheng, Han, and Bhatty, 1999; Chatakanonda, Varavinit, and Chinachoti, 2000; Hirsch and Kokini, 2002; Van Hung and Morita, 2005a,b; Gunaratne and Corke, 2007b), solubil-ity (Colas, 1986; Chel-Guerrero and Betancur, 1998; Jyothi, Moorthy, and Rajasekharan, 2006; Kaur, Singh, and Singh, 2006), extent of AM leaching (Hoover and Sosulski, 1986; Gunaratne and Corke, 2007b), paste clarity (Waliszewski et al., 2003; Van Hung and Morita, 2005a,b; Jyothi, Moorthy, and Rajasekharan, 2006; Kaur, Singh, and Singh, 2006), and paste peak vis-cosity (Yook, Pek, and Park, 1993; Hirsch and Kokini., 2002; Wongsagonsup et al., 2005). Furthermore, cross-linked starches generally exhibit increased pasting temperatures (Hoover and Sosulski, 1986; Zheng, Han, and Bhatty, 1999; Wongsagonsup et al., 2005), stability to shear (Colas, 1986; Yeh and Yeh, 1993; Kuhn and Schlauch, 1994; Liu, Ramsden, and Corke, 1999; Wongsagonsup et al., 2005; Gunaratne and Corke, 2007b), and tolerance to acid pH conditions (Gunaratne and Corke, 2007b). Mirroring the pattern described previously for swelling and pasting behaviors, gelatinization temperatures and enthalpies decrease with very low levels of cross-linking, but increase with more substantial degrees of cross-linking (Yoneya et al., 2003; Jyothi, Moorthy, and Rajasekharan, 2006; Kaur, Singh, and Singh, 2006). In the majority of investigations, gelatinization temperatures or enthalpies or both were observed to remain unchanged (Yeh and Yeh, 1993; Liu, Ramsden, and Corke, 1999; Choi and Kerr, 2004) or to increase upon cross-linking (Lelievre, 1985; Yeh and Yeh, 1993, 1996; Zheng, Han, and Bhatty, 1999; Chatakanonda, Varavinit, and Chinachoti, 2000; Choi and Kerr, 2004; Gunaratne and Corke, 2007b); an increase in enthalpy was attributed to a reduced molecular mobility within granule amorphous regions due to increasing numbers of molecular cross-links.

Cross-linked starch rheology is ultimately defined by the nature and concentration of swollen granules within a paste (Nayouf, Loisel, and Doublier, 2003), with the firmness of swollen granules increasing with increasing cross-linking levels (Tsai, Li, and Lii, 1997). Retrogradation stability and freeze–thaw stability are not generally achieved by cross-linking alone (Yook, Pek, and Park, 1993; Yeh and Yeh, 1993; Zheng, Han, and Bhatty, 1999; Jyothi, Moorthy, and Rajasekharan, 2006; Kaur, Singh, and Singh, 2006; Tran, Piyachomkwan, and Sriroth, 2007; Gunaratne and Corke, 2007b), but require additional stabilization (Section 8.3.1.2.4) to create dual-modified (cross-linked/stabilized) starches. Susceptibility to enzyme-catalyzed hydrolysis is somewhat reduced with increasing levels of cross-linking (Hoover and Sosulski, 1986; Jyothi, Moorthy, and Rajasekharan, 2006; Gunaratne and Corke, 2007b).

Cross-linking reactions utilize bi- or multifunctional reagents capable of forming both intramolecular or intermolecular cross-links between adjacent starch chains. Common reagents employed for commercial production of cross-linked starches include phosphoryl chloride ($POCl_3$), sodium trimetaphosphate (STMP), sodium tripolyphosphate (STTP), epichlorohydrin (EPI), and adipic–acetic mixed anhydride. Optimal reaction conditions and schemes vary according to reagent type. For reactions with STMP and STPP, starch is generally impregnated with both reagent and catalyzing base within an aqueous granule slurry, after which slurry moisture levels are reduced (<15%) and granules are heated (130°C) in the semi-dry state to drive the reaction (Lim and Seib, 1993). Cross-linking is favored at pH values above 8 and 10 for STPP and STMP, respectively (Lim and Seib, 1993; Liu, Ramsden, and Corke, 1999; Muhammad et al., 2000). In contrast, $POCl_3$, EPI, and adipic–acetic mixed anhydride cross-linking reactions are conducted as aqueous granular suspensions under basic conditions at pH values of 11–12, 8, and 10.5, respectively (Wu and Seib, 1990; Kasemsuwan and Jane, 1994). Addition of a stabilizing salt (e.g., sodium sulfate) is reported to enhance reaction with $POCl_3$ and STMP (Wu and Seib, 1990; Woo and Seib, 1997). An extensive list of possible multifunctional cross-linking agents for starch has been summarized elsewhere (Rutenberg and Solarek, 1984). More recently, azides (Shey et al., 2006), urea (Khalil et al., 2002), hydrocolloids (Lim et al., 2002), and free carboxylic acids (carboxymethyl starch substrates) (Seidel et al., 2001) have been reported as possible cross-linking agents. Reactive extrusion (Section 8.4.2) (Mali and Grossman, 2001; Nabeshima and Grossman, 2001; Seker et al., 2003; Seker, Sadikoglu, and Hanna, 2004; Seker and Hanna, 2005, 2006) and extended exposure to UV or polarized light (Fiedorowicz et al., 1999; Fiedorowicz, Lii, and Tomasik, 2002) have been used to generate cross-linked starches.

Cross-linked starches cannot be differentiated from their native starch counterparts on the basis of granule morphology, inasmuch as reaction does not appear to alter granule appearance (Hoover and Sosulski, 1986; Kaur, Singh, and Singh, 2006). As a highly reactive reagent, $POCl_3$ has been shown to react predominantly at external granule surfaces, including those of channels and cavities (Whistler and Spencer, 1960; Gluck-Hirsch and Kokini, 1997; Huber and BeMiller, 2001; Hirsch and Kokini, 2002; Gray and BeMiller, 2004). In contrast to the surface reaction tendency of $POCl_3$, modification with STMP and EPI (slower reacting reagents) are reported to yield more uniform reaction patterns throughout starch granules (Berghofer and Klaushofer, 1977; Hirsch and Kokini, 2002), based on the hypothesis that the reagents have sufficient time to infiltrate and disperse into granules prior to reacting. Despite the differential granular reaction patterns noted for the various reagents, all cross-linking reactions are thought to occur primarily within granule amorphous regions, based on the lack of difference in x-ray diffraction patterns between the modified and unmodified starch products (Hoover and Sosulski, 1986; Zheng, Han, and Bhatty, 1999). For both granular starches reacted with EPI and $POCl_3$, molecular cross-links occur primarily between AM and AP, with no cross-links detected between AM molecules (Jane et al., 1992; Kasemsuwan and Jane, 1994). As cross-linking reactions may yield both monostarch and distarch derivatives, the degree of cross-linking or proportion of distarch derivatives (cross-linking efficiency) ultimately affects starch properties. For phosphorylated derivatives prepared under standard cross-linking conditions, the distarch form predominates with phosphodiester:monoester ratios ranging from 3:1 to more than 4:1 (Koch, Bommer, and Koppers, 1982; Kasemsuwan and Jane, 1994; Huber and BeMiller, 2001). A similar result was observed for EPI-cross-linked starches, in which the distarch derivative is reported to account for approximately 90% of total incorporated substituent (Hamerstrand, Hofreiter, and Mehltreiter, 1960).

8.3.1.2.4 Stabilization/substitution So-called stabilized starches are produced by reacting starches with monofunctional reagents (Wurzburg, 1986a; Rapaille and Vanhemelrjck, 1997). By converting hydroxyl groups of starch molecules into larger ester or ether groups, interchain associations are blocked, resulting in more stable pastes and gels, that is, pastes and gels with a reduced tendency to undergo retrogradation. Some of the groups introduced via this derivatization have a negative charge, further reducing formation of interchain associations (through like-charge repulsion) to increase paste stability.

Both esters and ethers are made in the same general way. A gelatinization-inhibiting salt (10–30% concentration, most often sodium sulfate, sometimes sodium chloride) is added to a stirred slurry of starch granules (30–45% solids). The pH is adjusted to 8–12, the exact value depending on the reaction to be carried out. An alkaline pH converts some hydroxyl groups into alkoxide ions for participation in nucleophilic substitution reactions. The temperature is often adjusted to about 49°C. Use of a gelatinization-inhibiting salt and a temperature below the gelatinization temperature of both the native starch and the product prevents pasting, and allows the starch to be recovered in granular form. Following reaction with a monofunctional reagent to the desired DS, the stabilized (substituted) starch product is recovered by filtration or centrifugation, washed, and dried.

Unlike cross-linked distarch esters, manufactured monostarch esters include starch acetates, l-octenylsuccinates, phosphates, and succinates. Acetylated starch is prepared using acetic anhydride or, in countries in which acetic anhydride is prohibited, vinyl acetate. Although starch succinates, which are made by reaction with succinic anhydride, are approved as food additives, they are not common. Reaction with the corresponding 1-octenylsuccinic anhydride (OSA) produces the 1-octenylsuccinate ester (so-called OSA starch). Monostarch esters result from reaction with sodium dihydrogen phosphate or sodium tripolyphosphate (STTP). All products of substitution reactions have a reduced tendency to undergo retrogradation. They, therefore, produce more stable pastes and gels, resulting in fewer intermolecular associations, greater clarity, less syneresis, and greater freeze–thaw stability. Some hydrophobicity is also introduced in the case of OSA starch. A large variety of other esters are made by selection of the proper reagent. These other esters include starch and AM triacetate, which are water-insoluble and used to obtain thermoplastic materials. Properties and production of thermoplastic-based starches are discussed in Chapter 6.

Manufactured monostarch ethers include hydroxyethyl, hydroxypropyl, carboxymethyl, and cationic (Section 8.3.1.2.5) starches. Reaction with ethylene oxide produces hydroxyethyl starch, reaction with propylene oxide produces hydroxypropyl derivatives, and reaction with sodium monochloroacetate produces sodium carboxymethyl starch (also known as sodium starch glycolate). The products generally have the same properties as the monosubstituted starch esters, a difference being that ethers are stable to acids and bases whereas esters are not. As with starch esters, a large variety of ethers, including silyl ethers, can be made by selection of the proper reagent (Rutenberg and Solarek, 1984).

Reagents used to produce monostarch esters are also relatively reactive. These reagents are acetic anhydride (used to make acetylated starch), OSA (used to make starch 1-octenylsuccinate), and succinic anhydride (used to make starch succinate). Because of their reactivity, it can be hypothesized that reactions with these anhydrides take place at or near surfaces before the reagent penetrates very far into the granule matrix. On the other hand, the reagents that make monostarch ethers are much less reactive. It was found that reaction with an analogue of propylene oxide occurred rather uniformly throughout the granule matrix, whether or not the granule contained pores and channels (Huber and BeMiller, 2001; Gray and BeMiller, 2004, 2005).

Reactions resulting in stabilization may be performed on starches that have been, or will be, cross-linked (Section 8.3.1.2.3). Average molecular weights of any of the stabilized starch products may be reduced; that is, the products may be "thinned" by any of the processes described in Section 8.3.1.2.1. They may also be pregelatinized (Section 8.3.2.1.1) and are considered to be products of multiple modifications (Section 8.3.3).

8.3.1.2.5 Cationization   As implied, cationization imparts a positive charge to starch chains via derivatization with reagents containing imino, amino, ammonium, sulfonium, or phosphonium moieties (Solarek, 1986). From a commercial perspective, the most industrially significant and commonly utilized reagent is 3-chloro-2-hydroxypropyltrimethylammonium chloride (CHPTMA) (Siau et al., 2004; Xie, Liu, and Cui, 2006). An epoxide version of this reagent, 2,3-epoxypropyltrimethylammonium chloride, is likewise employed in cationization reactions, and reportedly exhibits the higher reaction efficiency of these two similar reactants (Ayoub and Bliard, 2003; Tara et al., 2004). Cationic starch derivatives are typically employed as wet end and surface sizing and coating additives in the paper industry for enhancement of sheet strength and retention of fines, as well as industrial flocculants in sludge dewatering and mineral mining operations. Degree of substitution values for cationic starch derivatives in papermaking applications generally range from 0.02–0.07, while those for flocculent applications are significantly higher (DS ~0.2–0.7) (Xie, Liu, and Cui, 2006).

Cationic starch derivatives may be prepared batchwise via granular slurry, semi-dry, and solubilized paste (nongranular) processes, or by continuous means such as reactive extrusion (Section 8.4.2). For slurry reactions that retain the starch in granular form, both aqueous and aqueous alcohol reaction systems have been investigated. For the aqueous slurry process, a reaction efficiency of 88% has been reported for a 35% (w/v) starch slurry reacted with CHPTMA (0.05 moles reagent/mole starch) at 50°C under alkaline conditions (NaOH:reagent mole ratio = 2.8:1) with inclusion of Na_2SO_4 as a gelatinization inhibitor (Carr and Bagby, 1981). In contrast, optimized conditions for the aqueous alcohol system were achieved using a 65% (v/v) aqueous ethanol reaction medium with a 1:1 (w/w) water:starch ratio reacted at 50°C in the presence of NaOH as a catalyst (Kweon, Bhirud, and Sosulski, 1996; Kweon, Sosulski, and Bhirud, 1997). The aqueous ethanol medium exhibited a reaction efficiency approaching 80%, and eliminated the need for a gelatinization-inhibiting salt. The semi-dry cationization process entails heating dry granular starch (<120°C) in the presence of a catalyst (NaOH) and reagent. Although the semi-dry process maintains starch in the granular form, some localized damage at granule surfaces has been observed microscopically (Radosta et al., 2004; Tuting, Wegemann, and Mischnick, 2004). For cationization via the paste process, a starch slurry is adjusted to an alkaline pH ($\approx$11.3) and gelatinized via steam injection followed by reaction at 50°C (Radosta et al., 2004). For a reactive extrusion process (Section 8.4.2.), reaction efficiencies in the range of 90% have been achieved for starch modified under alkaline conditions in the molten state (Carr, 1994a; Gimmler and Meuser, 1995; Berzin et al., 2007), although some degree of thermomechanical degradation of starch molecules may occur depending on processing parameters (Ayoub, Gruyer, and Bliard, 2003; Ayoub et al., 2004). In contrast to the slurry and semi-dry reaction schemes, both paste and reactive extrusion cationization processes result in loss of starch granular order.

For cationic starches modified in the granule form, relative granule crystallinity is reported to decrease with increasing DS levels (Siau et al., 2004). Similarly, starch gelatinization temperatures and enthalpies also decrease upon cationization (Craig, Sieb, and Jane, 1987; Yook, Sosulski, and Bhirud, 1994; Kweon, Sosulski, and Han, 1997; Radosta et al., 2004; Siau et al., 2004). Craig, Sieb, and Jane (1987) reported a reduced melting temperature for AM–lipid complexes following wheat starch cationization. Relative to native starches, cationic starch derivatives exhibit increased swelling powers (Kweon, Sosulski, and Han, 1997b), water sorption properties (Ayoub and Bliard, 2003), stability to syneresis (Yook, Sosulski, and Bhirud, 1994; Kweon, Sosulski, and Han, 1997b; Siau et al., 2004), and susceptibility to amylolysis (Kweon, Sosulski, and Han, 1997). Cationic groups also enhance the fat-binding capacity and emulsion stability of starch (Vermeire et al., 1999; Siau et al., 2004). In regard to pasting behavior, cationic starch derivatives exhibit reduced pasting temperatures and increased peak and breakdown viscosities compared to their respective unmodified starch controls (Yook, Sosulski, and Bhirud, 1994; Kweon, Sosulski, and Han, 1997; Han and Sosulski, 1998; Siau et al., 2004). From a sequestering standpoint, cationic starches are very efficient flocculants for removal of Hg^{2+}, Cu^{2+}, and Zn^{2+} (Khalil and Farag, 1994, 1998), with those cationic derivatives possessing primary amino groups exhibiting the highest sorption efficiency (Khalil and Farag, 1994). At the molecular level, solution behaviors of cationic starch derivatives vary according to ionic strength. At low ionic strength, cationic starch molecules exist as flexible chains (Modig, Nilsson, and Wahlund, 2006), possessing larger hydrodynamic volumes than those of the respective unmodified starch due to repulsive forces between the modified polymer chains (Manelius, Nurmi, and Bertoft, 2005). However, with increasing ionic strength, cationic starch polymers take on a less extended and more compact three-dimensional shape, and are subject to increased molecular aggregation (Larsson, 1998; Modig, Nilsson, and Wahlund, 2006).

The granular pattern of reaction and distribution of cationic substituents on starch chains varies according to the method of preparation. In contrasting granular reaction patterns of aqueous slurry and semi-dry processed starch cationic derivatives, Hamunen (1995) detected no differences in the nitrogen distributions for the two modification schemes. However, the greater body of evidence would suggest that the aqueous slurry reaction results in a reasonably homogeneous distribution of cationic groups throughout the granule, whereas the majority of reaction sites for the semi-dry process occur predominantly at or near granule surfaces (Vihervaara et al., 1990; Manelius et al., 2000, 2002; Tuting, Wegemann, and Mischnick, 2004). Even presuming a relatively homogeneous reaction pattern at the rudimentary granular level, it should be further noted that starch cationization occurs preferentially in the amorphous regions of the granule (Manelius, Nurmi, and Bertoft, 2000). However, it has also been demonstrated that the level of granule crystallinity itself may become altered or reduced due to increasing levels of cationization (Siau, et al., 2004).

At the molecular level, substitution patterns on starch molecules differ for granular (e.g., slurry, semi-dry) and agranular (paste/solution, reactive extrusion) reaction methods. For agranular reaction schemes, substituent distribution along and between starch chains is reasonably homogeneous, whereas that for granular reactions is not simply random (Richardson et al., 2003; Manelius, Nurmi, and Bertoft, 2005). For granular reactions, molecular derivatization sites occur predominantly within AP branch point regions and at nonreducing ends of starch chains (Manelius et al., 2000, 2002; Tuting, Wegemann, and Mischnick, 2004). Furthermore, long chains (including B-chains of AP and those of AM) are favored for cationization over short chains (Manelius, Nurmi, and Bertoft, 2005). For cationization reactions involving solubilized starch, neither AM nor AP was preferentially favored for reaction. For granular starch reactions, AM is thought to be preferentially reacted (Steeneken, 1984; Manelius, Nurmi, and Bertoft, 2005), although a contrasting report of AP being more reactive to cationic reagents than AM also has been published (Salomonsson, Fransson, and Theander, 1991). No matter the preparation method, O2 is the favored site of reaction on the starch glucosyl unit irrespective of the reaction type (Shiroza et al., 1982; Wilke and Mischnick, 1997; Heinze, Haack, and Rensing, 2004; Tuting, Wegemann, and Mischnick, 2004).

8.3.1.2.6 Graft copolymerization Graft copolymers are chemical derivatives (considered separately from those described in Section 8.3.1.2.4, although ether linkages are present) in which synthetic polymers are covalently bonded to starch molecules. Most often, these products are obtained by generating free radicals from hydroxyl groups of native or hydroxyalkylated starch molecules, followed by reaction with an unsaturated monomer. A wide range of starch graft copolymers has been made using different methods of producing starch free radicals, different vinyl and acrylic monomers, mixtures of monomers, and different starches and modified starch products (Willett, 2008). As an example, one commercial product is a graft copolymer of hydroxyethyl starch and styrene–butadiene latex (Nguyen, Martin, and Pauley, 1990). Other examples of products that have been made are starch graft-polyacrylamide, -poly(acrylic acid), -polyacrylonitrile, -poly(butyl acrylate), -poly(methyl acrylate), -poly(methyl methacrylate), -polystyrene, and -poly(vinyl acetate) (Willett, 2008). Synthesis of starch graft copolymers is sometimes performed by reactive extrusion (Section 8.4.2). Uses and potential uses of some of the starch graft copolymers have been reviewed by Fanta and Doane (1986).

8.3.2   Physically modified starches

8.3.2.1   Types of physical modifications
Most starch physical modifications are accomplished by heating and by mechanical shearing. The properties achieved through these types of modifications vary according to the treatment temperature, heating time, and botanical origin of the starch, but are especially impacted by the starch moisture content during treatment. Details about the influence of water on macromolecular and rheological changes to starch during heating are also discussed in Chapter 5.

8.3.2.1.1 Pregelatinization This modification involves application of heat sufficient to bring about gelatinization of starch, followed by drying and grinding. The objective is to generate starch ingredients with instantaneous cold-water solubility and thickening/gelling capabilities. In an industrial setting, pregelatinized starches are traditionally prepared by drum drying or extrusion cooking (Colonna et al., 1984; Doublier, Colonna, and Mercier, 1986; Alves, Grossmann, and Silva, 1999; Shim and Mulvaney, 1999; Bindzus et al., 2002), although other processes such as ohmic heating (Martinez-Bustos et al., 2005; An and King, 2007) and high-pressure thermal processing (Loisel et al., 2006) continue to be investigated. The starch slurry for drum drying can have a solids content up to 42%. However, a slurry moisture content greater than 70% and a processing temperature of 90–95°C are recommended for maximal granule swelling and optimal starch rheological characteristics (Takahash and Ojima, 1969; Doublier, Colonna, and Mercier, 1986).

Pregelatinized starch particles exhibit a complete lack of birefringence, and generally retain very little, if any, of the original native granule structure (Colonna et al., 1984; Anastasiades et al., 2002), although there are almost always some ungelatinized granules in any commercial preparation made on a hot roll, i.e., via drum drying. Drum-dried starches do exhibit some paste rheology attributes typical of swollen particles (Doublier, Colonna, and Mercier, 1986); however, this phenomenon is more likely a function of the particle size of the milled material. During preparation, drum-dried pregelatinized starches undergo some depolymerization as a result of the heat/shear imposed by the process; for wheat starch, AM and AP molecular weights were reported to decrease by factors of 1.1 and 2.6, respectively (Colonna et al., 1984; Doublier, Colonna, and Mercier, 1986). For extruded pregelatinized starches, decreases in molecular weight of AM and AP were considerably larger (by factors of 1.5 and 15, respectively.) Consequently, even drum-dried products generally exhibit slightly reduced intrinsic viscosities relative to the respective native starch (Colonna et al., 1984; Anastasiades et al., 2002). Pastes are often observed to possess a dull or grainy appearance. Swelling powers at ambient temperature for drum-dried wheat starch were reported to range from 15.6–16.3 (Doublier, Colonna, and Mercier, 1986). Generally, greater viscosities can be achieved by heating a dispersion of pregelatinized starch. Solutions/pastes of pregelatinized native or modified starches exhibit the same characteristics as those observed for their parent starches (upon cook-up). Properties of pregelatinized starches may be further enhanced by utilizing a chemically modified starch as the substrate for the pregelatinization process.

8.3.2.1.2 Granular cold-water soluble (GCWS) starch Although the rationale for generating granular cold-water soluble (GCWS) and standard pregelatinized starches is similar (solubility/swelling without cooking), GCWS starches provide the added advantage of a retained granule structure, resulting in superior starch properties (Jane, 1995). Thus, the strategy associated with creation of GCWS starches involves disruption of the native granule crystallinity, while maintaining the starch in some degree of granular form. Industrially, GCWS starch is made by spray-cooking (Pitchon, O'Rourke, and Joseph, 1981) or heating in an aqueous alcohol slurry (Eastman and Moore, 1984; Rajagopalan and Seib, 1992). Other processes for generation of GCWS starches include treatment with liquid ammonia (Jackowski, Czuchajowska, and Baik, 2002; Baik and Jackowski, 2004) or suspension in alcoholic–alkaline solution (Chen and Jane, 1994a,b; Singh and Singh, 2003). By definition, granules remain intact, however, all processes used to generate GCWS starches report some degree of granular distortion, ranging from cracks, fissures, and indentations at granule surfaces to slight increases in granule size.

In general, GCWS starches exhibit a greatly reduced or a complete lack of both a gelatinization endotherm and native granule crystallinity, and no longer exhibit the characteristic polarization cross under plane-polarized light. For GCWS starches prepared by heating aqueous alcohol slurries, double helices undergo transformation from the native crystalline packing arrangement to a V single-helix form, providing an explanation for the enhanced cold-water solubility of the modified products (Jane et al., 1986). Cold-water solubilities greater than 90% are reported for various GCWS starches (Jane et al., 1986; Chen and Jane, 1994a; Singh and Singh, 2003), and swelling powers as high as 15.3% have been observed, depending on the preparation process and the starch botanical source (Baik and Jackowski, 2004). Similar final viscosities are observed for GCWS and native corn starches subjected to pasting at 25°C and 95°C, respectively (Jane et al., 1986). Like conventional pregelatinized starches, pastes of GCWS starches have properties similar to those of their parent starches. Furthermore, properties of GCWS starches may be extended through various types of chemical modification.

8.3.2.1.3 Ball milling   Pulverizing mixer or ball mills can be utilized to alter the granule morphology, crystallinity, solubility, and swelling behavior of a starch by subjecting it to high-energy impact. At a low degree of ball-milling, visual damage is detected primarily at granule surfaces, which take on a rough appearance (Tamaki et al., 1997, 1998; Chen, Lii, and Lu, 2003), and exhibit some localized loss of surface material and partial loss of the characteristic polarization cross within damaged granular regions. A comprehensive categorization of the visual types of granule damage incurred during ball milling have previously been detailed (Adler, Baldwin, and Melia, 1995; Baldwin et al., 1995). With high levels of ball milling, starch granules become fractured into smaller pieces. Over the course of ball milling, there is a gradual loss of granule crystallinity and starch double-helical order, both of which can be reduced to undetectable levels with extensive treatment (Morrison, Tester, and Gidley, 1994; Tamaki et al., 1997; Chen, Lii, and Lu, 2003), resulting in an entirely amorphous starch material that no longer exhibits a discernable melting endotherm (Morrison, Tester, and Gidley, 1994; Huang et al., 2007). With extended periods of ball milling, amorphous granule fragments may also agglomerate to form composite particles exceeding the size of the original granular starch prior to treatment (Chen, Lii, and Lu, 2003; Huang et al., 2007). It was speculated that the potential for hydrogen bonding increases as the native granule structure is destroyed during ball milling, allowing formation of new hydrogen bonds between amorphous granule fragments. At the molecular level, ball milling induces starch depolymerization, which occurs predominantly at glycosidic linkages (Tamaki et al., 1997, 1998). Amylopectin is more susceptible to depolymerization than AM and the primary points of molecular cleavage on AP seem to occur on longer branch chains (B2–B4) between adjoining clusters (Morrison and Tester, 1994; Tamaki et al., 1997, 1998).

Parameters reported to influence ball milling of starch granules include moisture content, pretreatment with acid, and starch botanical source. Incorporation of some moisture into otherwise dry starch facilitates greater damage during ball milling and enhances formation of starch agglomerates (Huang et al., 2007; Martinez-Bustos et al., 2007), whereas acid hydrolysis pretreatment promotes digestion of granule amorphous regions and increases granule susceptibility to breakage and size reduction (Jane et al., 1992; Sanguanpong et al., 2003). Waxy starches are generally more prone to damage than their normal counterparts (Han et al., 2002), whereas for wheat starch, the degree of damage differs according to wheat class (durum > hard > soft) (Mok and Dick, 1991).

With respect to properties, ball-milled granular starches exhibit increased water absorption, swelling capacity, and solubility (Morrison and Tester, 1994; Morrison, Tester, and Gidley, 1994; Tester and Morrison, 1994; Sangaunpong et al., 2003; Martinez-Bustos et al., 2007; Huang et al., 2007), which correspond to the relative level of incurred starch damage (both granular and molecular). Starch pasting temperatures and times are reduced proportionally to the extent of ball milling, although both the viscosity of the resulting starch pastes and the elastic modulus values of subsequent starch gels decrease as treatment length is increased (Han et al., 2002) as a result of depolymerization. Ball-milled starches also exhibit increased susceptibility to hydrolysis by amylolytic enzymes (Baldwin et al., 1995; Tamaki et al., 1997, 1998).

8.3.2.1.4   Annealing   By definition, annealing involves heating of granular starch in water (>40% w/w) at a temperature above the glass transition temperature (T_g), but below that of the gelatinization onset temperature (Jacobs and Delcour, 1998; Tester and Debon, 2000). The process appears to induce little, if any, visible external change to starch granules (Hoover and Vasanthan, 1994), although there are a few reports of increased granule porosity, increased defects, and alteration of granule size (Gough and Pybus, 1971; Nakazawa and Wang, 2003). The most common and consistent effect of annealing is observed in starch gelatinization characteristics, with annealed starches exhibiting increased onset (T_o) and peak (T_p) gelatinization temperatures and a narrowing of the gelatinization temperature range (Tester and Debon, 2000). Gelatinization enthalpy (ΔH) remains either unchanged (Wang, Powell, and Oaks, 1997a; Vermeylen, Goderis, and Delcour, 2006) or is slightly increased with annealing (Hoover and Vansanthan, 1994; Nakazawa and Wang, 2003; Genkina et al., 2004). The majority of published studies have not detected any change in the native starch crystalline packing arrangement upon annealing (Tester and Debon, 2000); however, more recent reports have observed a partial C (mixed) to A polymorph transition upon extended annealing (Genkina et al., 2004; Gomes et al., 2004; Waduge et al., 2006).

At the molecular level, annealing does not appear to introduce any new double helices within normal or waxy starch (Tester, Debon, and Karkalas, 1998), but instead increases the ordering and improves the registration of existing AP double helices in granule crystalline regions (Hoover, 2000; Tester and Debon, 2000; Kiseleva et al., 2004). However, for high-AM maize starches, there is some evidence that annealing does result in formation of some new double helices (including AM–AM, AM–AP, and AP–AP structures), possibly leading to some compartmentalization of like structures within the granule (Knutson,1990; Tester, Debon, and Sommerville, 2000). Upon annealing, monophosphate moieties on AP molecules tend to retard crystallite perfection (based on a lesser ΔT_p for high P starches), although starches with high native monophosphate levels have greater potential for an increase in ΔH (relative to native starch) after annealing (Muhrbeck and Svensson, 1996; Muhrbeck and Wischmann, 1998). Annealing is also thought to increase the T_g and, thus, the rigidity of granule amorphous regions (Seow and Teo, 1993; Tester and Debon, 2000). In short, changes occurring within both the crystalline and amorphous regions during annealing would appear to explain, in part, the observed changes in starch gelatinization behavior.

As a consequence of the induced molecular interactions within starch granules, annealed starches generally exhibit reduced solubility, swelling power, and AM leaching (Hoover and Vasanthan, 1994; Jacobs et al., 1995, 1998; Tester and Debon, 2000; Tester, Debon, and Sommerville, 2000; Chung, Moon, and Chun, 2000; Gomes et al., 2004; Waduge et al., 2006). However, existing AM–lipid complexes are anticipated to be influenced very little, if at all, by starch granule annealing, because the transition temperatures for these complexes far exceed those of starch gelatinization and, thus, those employed by traditional annealing (Tester and Debon, 2000). Although gels of annealed starches tend to exhibit larger elastic modulus values relative to those of native starches (Jacobs et al., 1998), the precise effect of annealing on starch pasting properties is difficult to reconcile from the varied literature reports, as observed properties tend to differ with both the starch botanical source and the rheological instrument employed (Jacobs et al., 1995, 1998; Tester and Debon, 2000). A similar starch-specific effect is observed for the susceptibility of annealed starches to both acid- and enzyme-catalyzed hydrolysis (Jacobs et al., 1998; Tester and Debon, 2000).

Annealing can theoretically occur with starch moisture contents as low as 20% (w/w) and at temperatures as low as 15°C below T_o, but the process is markedly enhanced at excess moisture levels (60% w/w) and at temperatures approaching T_o (Tester, Debon, and Karkalas, 1998). Under optimum conditions, the greatest effect is generally achieved in the first six hours of annealing (Larsson and Eliasson, 1991; Chung, Moon, and Chun, 2000), although crystalline polymorph transitions require more extended periods of time (Genkina et al., 2004). Annealing can be conducted as a single- or multistep process, with higher gelatinization temperatures typically achieved with temperature cycling (Jacobs et al., 1998).

8.3.2.1.5   Heat-moisture treatment (HMT)   This modification also involves heating of starch granules above the glass transition temperature for varied lengths of time, but sets itself apart from annealing processes in that it is generally conducted at reduced levels of moisture (<35%) at relatively higher processing temperatures (80–140°C). An in-depth review of this topic is provided by Jacobs and Delcour (1998). Similar to annealing, traditional HMT starches retain their granular form and characteristic polarization cross (Stute, 1992; Abraham, 1993; Hoover and Vasanthan, 1994; Vasanthan, Sosulski, and Hoover, 1995; Gunaratne and Hoover, 2002; Adebowale and Lawal, 2003), although some subtle increases in granule size, surface cracking, and hollowing at granule centers have been reported (Kawabata et al., 1994; Vasanthan, Sosulski, and Hoover, 1995; Vermeylen, Goderis, and Delcour, 2006). Changes within both the crystalline and amorphous regions are thought to occur during HMT (Kawabata et al., 1994; Hoover and Manuel, 1996a, 1996b; Jacobs and Delcour, 1998; Miyoshi, 2002).

In regard to granule crystalline regions, HMT starches consistently exhibit both an upward shift in gelatinization temperature (T_o, T_p, and often T_c) and a broadened gelatinization temperature range (Jacobs et al., 1998; Adebowale and Lawal, 2003; Shin et al., 2005; Vermeylen, Goderis, and Delcour, 2006), whereas gelatinization enthalpy is reported to either decrease (Vasanthan, Sosulski, and Hoover, 1995; Perera, Hoover, and Martin, 1997; Shin et al., 2005; Vermeylen, Goderis, and Delcour, 2006) or to remain unchanged (Hoover, Swamidas, and Vasanthan, 1993; Hoover and Manuel, 1996a,b). For potato starch, Vermeylen, Goderis, and Delcour (2006) observed a decrease in both gelatinization enthalpy and total granule crystallinity for HMT temperatures up to 120°C, above which temperature (130°C) total crystallinity increased.

It is well established that native starches exhibiting a B- or C-type (mixed) packing arrangement can undergo a gradual transition to the A polymorphic form upon HMT (Stute, 1992; Perera, Hoover, and Martin,

1997; Gunaratne and Hoover, 2002). This transition is favored by a high temperature during HMT and a gradual cooling thereafter (Vermeylen, Goderis, and Delcour, 2006). For potato starch, the gradual disappearance of the B polymorph with increasing treatment temperatures (90–120°C) coincided with an observed decrease in total crystallinity, the decrease of which was offset and eventually overcome by simultaneous formation of the A polymorph (at a treatment temperature of 130°C), accounting for the final increase in total crystallinity (Vermeylen, Goderis, and Delcour, 2006). This polymorphic transition was hypothesized to be enhanced by a release of linear segments of AP branch chains arising from thermal degradation (Vermeylen, Goderis, and Delcour, 2006), which has been previously reported to occur during HMT (Lu, Chen, and Lii, 1996). Thus, relative crystallinity can either increase or decrease with HMT, depending on the starch source and conditions employed (Jacobs et al., 1998). Possible changes within granule crystalline regions associated with HMT have been attributed to disruption of low-level crystallites and strengthening of higher melting crystallites (Luo et al., 2006), alteration of the starch packing arrangement (Stute, 1992; Hoover, Swamidas, and Vasanthan, 1993; Hoover and Vasanthan, 1994; Kawabata et al., 1994; Jacobs et al., 1998), growth or perfection of existing crystallites (Hoover and Vasanthan, 1994; Jacobs et al., 1998), disruption of stacked lamellae (Hoover and Vasanthan, 1994; Vermeylen, Goderis, and Delcour, 2006), and recrystallization of starch chains (Vermeylen, Goderis, and Delcour, 2006).

Aside from the noted impact on crystalline regions, the greatest changes associated with HMT have been proposed to occur within granule amorphous regions (Hoover and Manuel, 1996a; Miyoshi, 2002). Reported alterations within amorphous regions include development of new order and/or crystallites (AM–AM, AM–AP interactions) (Hoover and Vasanthan, 1994; Hoover and Manuel, 1996a), formation or enhancement of AM–lipid complexes (Hoover and Vasanthan, 1994; Kawabata et al., 1994; Hoover and Manuel, 1996a; Miyoshi, 2002), and conversion of AM from an unordered to a partial helical form (Lorenz and Kulp, 1982; Hoover and Vasanthan, 1994). Alterations within granule amorphous regions affect both the gelatinization (melting) and physical properties of starch.

Due to the diverse combinations of experimental conditions employed for HMT (moisture content, temperature, length of treatment, heat source, starch source, etc.), the characteristics and properties of HMT starches vary and are not easily defined in a consistent and unified fashion. In almost all cases, HMT results in decreased levels of starch solubility, swelling power, and AM leaching (Abraham, 1993; Hoover, Swamidas, and Vasanthan, 1993; Hoover and Vasanthan, 1994; Kawabata et al., 1994; Hoover and Manuel, 1996b; Kurakake et al., 1997; Lewandowicz, Jankowski, and Fornal, 1997; Perera, Hoover, and Martin, 1997; Gunaratne and Hoover, 2002; Lawal and

Adebowale, 2005; Luo et al., 2006). In addition, HMT starches generally display a decreased peak viscosity and an increased pasting temperature (Hoover, Swamidas, and Vasanthan, 1993; Hoover and Vansanthan, 1994; Hoover and Manuel, 1996a; Eerlingen et al., 1997; Kurakake et al., 1997; Jacobs and Delcour; 1998; Adebowale and Lawal, 2003; Singh et al., 2005; Stevenson, Biswas, and Inglett, 2005; Luo et al., 2006), although effects for other pasting attributes are not always consistent among literature reports. Concentrated HMT starch gels exhibit increased gel hardness relative to those of native starches (Eerlingen et al., 1997; Gunaratne and Corke, 2007a), whereas the opposite effect is observed for dilute HMT starch gels. Reports on the susceptibility of HMT starches to retrogradation (Takaya, Sano, and Nishinari, 2000; Miyoshi, 2002; Gunaratne and Hoover, 2002; Lawal and Adebowale, 2005), acid/enzyme-catalyzed hydrolysis (Jacobs and Delcour, 1998), and chemical modification (Vasanthan, Sosulski, and Hoover, 1995; Perera, Hoover, and Martin, 1997; Liu, Corke, and Ramsden, 2000; Gunaratne and Corke, 2007a) are mixed, and vary according to botanical source and treatment parameters.

To facilitate more rapid processing of HMT starches, alternative methods utilizing reduced pressure (Maruta et al., 1994) and microwave heating (Stevenson, Biswas, and Inglett, 2005; Anderson and Guraya, 2006; Luo et al., 2006) have been reported. HMT is commonly employed in conjunction with treatment with an acid (Brumovsky and Thompson, 2001; Shin et al., 2004a), miscellaneous processing techniques (Kurakake et al., 1997; Pukkahuta, Shobsngob, and Varavinit, 2007), and chemical modification (Vasanthan, Sosulski, and Hoover, 1995; Perera, Hoover, and Martin, 1997; Perera and Hoover, 1998; Liu, Corke, and Ramsden, 2000; Sang and Seib, 2006; Gunaratne and Corke, 2007a) to create starch products with novel properties.

8.3.2.1.6 Dry heating of starch This process includes heating of starch under low (<15%) moisture conditions at a temperature below that which thermal degradation changes the physical properties of the starch (Goto, 1972; Seguchi, 1984; Seguchi and Yamada, 1988). Starches with acid-, shear-, and temperature-tolerance profiles similar to those of chemically cross-linked starches can be prepared by heating dry starch at temperatures >100°C for several hours (Chiu et al., 1998, 1999b). Drying to <1% moisture before heating and alkalinity (pH 8.0–9.5) facilitate the property change.

8.3.3 Multiple modifications

In many cases, no single type of modification is sufficient to impart all necessary properties to a starch intended for a specific application. Thus, starches are often subjected to a combination of physical or chemical modification schemes to tailor the collective starch properties for a particular end use. Dual modification traditionally refers to a starch that has been both stabilized (Section 8.3.1.2.4) and cross-linked (Section 8.3.1.2.3), and is commonly employed to enhance stability to retrogradation (stabilization) and to impart tolerance to acid, high shear, and prolonged heating conditions (cross-linking). Many commercial starch products have been made with more than two types of modifications. Acid thinning or oxidation (Section 8.3.1.2.1) is often conducted in conjunction with other chemical derivatization schemes to fine-tune paste and gel characteristics. Chemical modification may also be used to modify starch hydroxyl groups in preparation for further derivatization (e.g., graft copolymerization; Section 8.3.1.2.6). Chemical starch derivatives can be converted to cold-water soluble or swellable forms (Sections 8.3.2.1.1 and 8.3.2.1.2), and HMT (Section 8.3.2.1.5) can be used to alter starch reactivity to chemical modifying agents. Because the possibilities for multiple modifications are numerous, and can be applied to a variety of base starches, including genetically modified native starches (Section 8.4.2), there is an excellent potential for realization of starches with novel properties.

8.4 Emerging trends involving chemical or physical modifications

8.4.1 Resistant starch (RS) and slowly digestible starch (SDS)

Most starches, native or modified, contain varying proportions of rapidly digesting, slowly digesting, and resistant starch (Englyst and Hudson, 1996), depending on the source of the starch and any treatments it has received. Resistant starch (RS) is defined as starch material that escapes digestion by human enzymes within the small intestine and passes into the colon, where it is metabolized into secondary products by colon microflora. Conversely, slowly digestible starch (SDS) refers to starch material that is hydrolyzed to glucose, but at a moderated or reduced rate, during its transit through the human small intestine. Although it is eventually absorbed as glucose within the small intestine, hydrolysis and absorption of SDS occur over a more extended period (relative to rapidly digestible starch), thus providing a moderating effect on blood sugar levels. Both RS and SDS are thought to provide physiological benefits, including moderation of the postprandial glycemic response and generation of secondary metabolites (e.g., short-chain fatty acids) that contribute to colonic health (Englyst and Englyst, 2005). Changes to the chemical structures of starch molecules or changes that make the molecules or granules more crystalline tend to reduce the rate of starch digestion. A number of reviews have covered their production (Sajilata, Singhal, and Kulkarni, 2006; Thompson, 2007; Hamaker et al., 2007;

Hamaker, Zhang, and Venkatachalam, 2007; Lehmann and Robin, 2007) and nutritional/physiological effects (Sajilata, Singhal, and Kulkarni, 2006; Björck, 2006; Thompson, 2007; Birkett and Brown, 2007; Lehmann and Robin, 2007). Most of the research to date has centered on RS.

Resistant starch has been subclassified into four primary categories or types (RS1–RS4). Of relevance to this chapter are three types (RS2, RS3, and RS4), which are generated or enhanced via physical or chemical modification schemes. RS2 is ungelatinized (raw) starch or starch granules whose degree of crystallinity has been increased. RS3 is gelatinized/retrograded starch, and RS4 results from chemical derivatization of starch. In general, the greater the AM content of a starch the slower the digestion (Goddard, Young, and Marcus, 1984; Granfeldt, Drews, and Björck, 1995), so increasing the apparent AM content of a starch by genetic means can increase the RS content of otherwise unmodified starches (Yang et al., 2006; Leeman et al., 2006).

The RS2 content of granular starch may be increased by:

- Traditional annealing (Section 8.3.2.1.4) and heat–moisture treatment (Section 8.3.2.1.5) (Jacobasch et al., 2006)
- Heating a slurry of high-AM starch granules in the presence of a swelling-inhibiting salt (Chiu, Shi, and Sedam, 1999a)
- Heating high-AM starch granules in a limited amount of water (Shi and Trzasko, 1996)
- Temperature cycling a slurry of high-AM starch granules in an aqueous alcohol medium (Stanley et al., 2006)
- Heating of starch granules in an acidic aqueous alcohol medium (Binder and McClain, 2006)

The RS3 content of starch may be facilitated by:

- Subjection of pastes and gels to thermal treatments (Wasserman et al., 2007)
- Cooling or freezing a paste of partially or fully gelatinized granules of normal or high-AM starches (Zhao et al., 2005; Chung, Lim, and Lim, 2006; Yamamori et al., 2006)
- Pasting a starch in the presence of an acid, followed by cooling or temperature cycling of the paste (Aparicio-Saguilan et al., 2005; Onyango et al., 2006)
- Pasting a starch or a lintnerized starch, followed by cooling and cross-linking (Shin, Woo, and Seib, 2003)
- Application of high hydrostatic pressure to effect gelatinization, especially with acid-treated starch (Bauer, Wiehle, and Knorr, 2005)
- Extrusion of starch alone (Agustiniano-Osornio et al., 2005; Nehmer, Nobes, and Yackel, 2006; Gonzalez-Soto et al., 2007b) or in the

presence of guar gum and citric acid (Wang, Jin, and Yuan, 2006). However, under certain conditions, extrusion may also increase starch digestibility (Sun et al., 2006)

Just as a slight degree of acid-catalyzed hydrolysis of starch may enhance retrogradation, treatment of starch with various amylases also augments formation of slowly digesting and resistant starch (Ao et al., 2007). These enzymic processes commonly employ the use of a debranching enzyme (Berry, 1986; King and Tan, 2005; Gonzalez-Soto et al., 2007a; Leong, Karim, and Norziah, 2007), followed by various pasting and isothermal holding or temperature cycling processes. The debranching may occur after treatment with α-amylase (Gao, Luo, and Yang, 2007), after successive treatment with α-amylase and glucoamylase (Chen, 2004), or after gelatinization followed by addition of citric acid and freeze–thaw cycling (Gu and Huang, 2007).

The RS4 content of a starch may be increased by:

- Acetylation (Li, Chen, and Li, 2007)
- Phosphorylation of an acid-modified starch (Carmosino, 2005)
- A combination of acid modification, annealing, and cross-linking (Lim et al., 2004)
- Simultaneous cross-linking, phosphorylation, and heat–moisture treatment (Sang and Seib, 2006)
- Oxidation (Woo, Bassi, and Maningat, 2005)
- Pyrodextrinization (Campechano-Carrera et al., 2006)
- Pasting a starch or a lintnerized starch, followed by cooling and cross-linking (Shin, Woo, and Seib, 2003)

It should be noted that there is frequent overlap of RS categories or types within a single RS product, as multiple approaches are often utilized simultaneously to achieve the overall resistance to digestion. Similar treatments of flours, whole grains, and crystalline malto-oligosaccharides are beyond the scope of this chapter and are not addressed.

There is additional overlap in the literature regarding RS and SDS, as the two starch forms differ from a practical standpoint only in their extent of digestion by α-amylase as a function of time. RS and SDS often occur together in a starch preparation, and most likely possess a common or overlapping molecular basis for their moderated digestibilities. The structural basis for the slow digestion of native cereal starches is related to their supramolecular A-type crystal structures (Zhang, Venkatachalam, and Hamaker, 2006). A study of AP from rice revealed that SDS was positively correlated with long- and intermediate/short-chain content and negatively correlated with very short chain content (Benmoussa, Moldenhauer, and Hamaker, 2007). Approaches to production and benefits of SDS have been reviewed (Hamaker, Zhang, and Venkatachalam, 2007b; Lehmann and Robin, 2007). Although a variety of approaches for producing SDS have been investigated, processes that increase the content of RS generally increase the content of SDS and vice versa. The ensuing discussion is limited to those processes centered solely on the generation of SDS.

Approaches to making SDS have included:

- Treatment with various amylases to increase branch density (Ao et al., 2007)
- Controlled partial hydrolysis with α-amylase (Hamaker and Han, 2004; Han et al., 2006b)
- Debranching (Shi, Cui, and Birkett, 2003a,b)
- Debranching followed by isothermal holding of the paste (Guraya, James, and Champagne 2001; Shin et al., 2004b)
- Heat-moisture treatment (Section 8.3.2.1.5) (Anderson et al., 2002; Anderson and Guraya, 2004; Shin et al., 2005)
- Cross-linking (Yeo, Woo, and Seib, 2002; Woo and Seib, 2002)
- Increasing the AM content (Kitahara et al., 2007)
- Entrapping starch in a cross-linked nonstarch matrix (Hamaker et al., 2007)
- Octenylsuccinylation followed by dry heating of the granular starch (Han and BeMiller, 2007)

Han and BeMiller (2007) found that esterification with OSA was highly effective in increasing the content of both SDS and RS in waxy maize starch. Dry heating of the resulting OSA waxy maize starch increased the SDS content and decreased the RS content. Using a combination of esterification with OSA followed by dry heating, the content of SDS was raised to 47% in waxy maize starch, 38% in normal corn starch, 46% in cassava starch, and 33% in potato starch.

8.4.2   Reactive extrusion

Reactive extrusion is the effecting of a chemical reaction in an extruder under the combined influence of temperature, shear, and pressure. Reactions of starch in extruders have been investigated primarily as a means to develop continuous modification processes. Reactive extrusion product characteristics are a function of process and system parameters, which can be widely varied. Process variables (other than the nature and concentration of the reactant and catalyst, and the moisture content) include temperature, shear rate, and thermomechanical parameters, which together define the extent and time of mixing, residence or reaction time, and pressure. Reactive extrusion of starch involves a combination of physical and chemical modification, in which granular starch simultaneously loses granular/molecular order via high mechanical input, and is derivatized in the molten state. This process should not be confused with using an extruder to prepare pregelatinized starch products or to blend starches with other substances. In addition to gelatinization, starches undergo depolymerization to an extent that is dependent on the moisture, heat, and shear conditions of the extrusion process (Brncic et al., 2000; Qu and Wang, 2002), and may also be intentionally converted into low-molecular-weight products by introduction of a thermostable α-amylase (Komolprasert, 1990; Govindasamy, Campanella, and Oates, 1997a,b).

Some advantages of reactive extrusion of starch are the continuous nature of the process, the wide range of possible processing conditions, enhanced contact between the reactants, a lack of effect of granule structure on molecular reaction patterns, more complete reactions (high yields), relatively short reaction times, simultaneous depolymerization of starch polymers, and the lack of process effluents. Primary challenges encountered with reactive extrusion include reaction control under conditions of high viscosity, elevated temperature, and short residence time (Vergnes and Berzin, 2004). Because reactions must be fairly rapid, more esters (using acid anhydrides) and graft co-polymers (made with free-radical reactions) than ethers have been made. Other possible disadvantages of reactive extrusion include the difficulty associated with removal of catalysts and unreacted reagents (due to water solubility of the product), the limitation that only relatively fast reactions are possible, depolymerization of starch polymers, challenges in scale-up, melt viscosity limitations, and potential damage to the screws and barrel by chemicals. The principles, technologies, and challenges related to starch modification by reactive extrusion have been reviewed (Meuser Gimmler, and Oeding, 1990; Xie et al., 2006; Kalambur and Rizvi, 2006). In short, esterification, etherification, cross-linking, oxidation, and hydrolysis have all been perormed via reactive extrusion (Kesselmans, Venema, and Hadderingh, 2001).

Various esters have been made using various carboxylic acid anhydrides (Tomasik, Wang, and Jane, 1995; Wang, Shogren, and Willett, 1997b,c; Miladinov and Hanna, 2000; Bloembergen, Kappan, and VanLeeuwen, 2002; Rudnik and Zukowska, 2004; Narayan et al., 2006), lactones (Tomka, 1998), free dicarboxylic acids (Narayan et al., 2006), citric acid (Narkrugsa, Berghofer, and Camargo, 1992), and sodium tripolyphosphate (San Martin-Martinez et al., 2004). Starch has been acylated by transesterification via reactive extrusion (DeGraff, Broekroelofs, and Janssen, 1998; Vaca-Garcia, Borredon, and Gaset, 2000). Reaction efficiencies may be influenced through the use of different base catalysts (Wang, Shogren, and Willett, 1997a; Rudnik and Zukowska, 2004).

Molecular cross-linking occurs when sodium trimetaphosphate and sodium hydroxide (Nabeshima and Grossman, 2001; Seker et al., 2003;

Seker and Hanna, 2005, 2006) are added. The products differ from conventional cross-linked starches (Section 8.3.1.2.3), which are granular products, in that the shear forces and temperature in the extruder destroy the granular structure and depolymerize the starch polymers, resulting in "thinned," cross-linked, pregelatinized products. Extruded starch molecules have also been cross-linked with acetic–adipic mixed anhydride (Mali and Grossman, 2001), with aldehydes (Bloembergen, Kappan, and VanLeeuwen, 2002; Mackey et al., 2003), and with polyesters in the production of biodegradable starch–polyester composites (Kalambur and Rizvi, 2005; Rizvi and Kalambur, 2005). To make starch compatible with biodegradable polyesters, maleic anhydride and a radical initiator are often used (Maliger et al., 2006).

Starches have been etherified to produce cationic derivatives using both 2,3-epoxypropyl and 3-chloro-2-hydroxypropyl quaternary ammonium reagents (Della Valle, Colonna, and Tayeb, 1991; Narkrugsa, Berghofer, and Camargo, 1992; Carr, 1994a; Gimmler and Meuser, 1995; Esan, Bruemmer, and Meuser, 1996; Tara et al., 2004; Xie et al., 2006). Reactive extrusion has also been used to carboxymethylate starch (Gimmler and Meuser, 1995), but it has been most explored for formation of graft copolymers with starch or a starch derivative (often hydroxyethyl starch) extruded in the presence of a vinyl monomer and a radical initiator. Products that have been claimed are starch-g-polyacrylonitrile (Carr et al., 1992; Yoon, Carr, and Bagley, 1992; Carr, 1994b), starch-g-polymethacrylate (Carr et al., 1992), starch-g-poly(methylacrylate) (Trimnell et al., 1993), starch-g-polyacrylamide (Carr et al., 1992; Carr, 1994b; Willett and Finkenstadt, 2003, 2004, 2006a,b; Finkenstadt and Willett, 2005), and starch-g-polystyrene (DeGraaf and Janssen, 2000). Also claimed is starch chemically bonded to polyethylene (Yoo, Kim, and Cho, 1994) and starch-g vegetable oil polymers (Narayan, Balakrishnan, and Shin, 2006b). Bifunctional fatty acid oxazoline derivatives have also been reacted with starch via a radical process in an extruder (Kosan et al., 2006).

Transglycosylation reactions have been carried out in extruders. The alcohols used in preparing glucosides have been nonvolatile, with two or more hydroxyl groups, such as ethylene glycol and glycerol (Carr, 1991; Carr and Cunningham, 1992; Subramanian and Hanna, 1996). Addition of hydrogen peroxide and an iron(II)–copper(II) sulfate catalyst to the feed of an extruder results in oxidized starches (Wing and Willett, 1997).

8.4.3   Blends and starch–hydrocolloid reactions and interactions

Starch blends may be employed to attain starch characteristics and properties and, in some cases, achieve physical properties akin to those of chemically modified starches (Stute and Kern, 1994; Obanni and BeMiller, 1997; Ortega-Ojeda and Eliasson, 2001; Karam et al., 2005; Novello-Cen and

Betancur-Ancona, 2005). Beneficial characteristics and properties most commonly reported with the use of starch blends include increased hot paste stability and decreased tendency for retrogradation and syneresis; it is also possible to alter paste clarity and gel strength using starch blends. Blends may consist of starches from different botanical origins (Stute and Kern, 1994; Obanni and BeMiller, 1997; Yao, Zhang, and Ding, 2003; Karam et al., 2005; Novello-Cen and Betancur-Ancona, 2005; Karam et al., 2006; Gunaratne and Corke, 2007c), different varieties or genotypes of the same botanical origin (Chen, Lai, and Lii, 2003, 2004; Hagenimana and Ding, 2005; Hagenimana, Pu, and Ding, 2005), or differing granule types or populations within a botanical source (Shinde, Nelson, and Huber, 2003; Kaur et al., 2007). Some starch blends may also include chemically or physically modified starches (Obanni and BeMiller, 1997; Yao, Zhang, and Ding, 2003; Gunaratne and Corke, 2007c).

Starches, including modified food starches and hydrocolloids are often used in combination in processed foods. Most often, they are used together to take advantage of the bulk and body imparted by the starch product and the low cost and improvements in stability (e.g., freeze–thaw stability, reduction in retrogradation) and texture imparted by the hydro-colloid. Particularly, the attributes imparted by the hydrocolloid or a combination of hydrocolloids may allow the use of a native, nonchemically modified starch, so that the food product may be labeled as "all natural". Thus, addition of a hydrocolloid to a starch can be considered another way of modifying the properties of a starch. As a result of the extensive use of starch–hydrocolloid combinations in foods, investigations in this area have been numerous.

The literature describing the effects of hydrocolloids on starches leads to one principal conclusion: namely, that both the nature and the extent of the effects of hydrocolloids on the pasting and paste properties of starches are specific to a particular starch–hydrocolloid combination (Gudmundsson et al., 1991; Shi and BeMiller, 2002). The use of hydrocolloids in combination with starch may be used to alter pasting temperature (Shi and BeMiller, 2002), peak and final viscosities, breakdown, and setback; reduce syneresis and improve paste freeze–thaw stability; alter the melting enthalpy of starch crystallites; and increase the storage and loss moduli and dynamic viscosity of starch pastes. A specific gel-forming interaction between wheat starch and Hsian-tsao leaf gum has been reported (Lii and Chen, 1980; Lai and Liao, 2002).

Two general mechanisms have been proposed to explain these results. One involves formation of starch–hydrocolloid associations (Christianson et al., 1981; Sajjan and Rao, 1987; Shi and BeMiller, 2002), with the mechanism of Shi and BeMiller (2002) being restricted to the apparent decrease in pasting temperature observed upon addition of some hydrocolloids to some starches. The other likely explanation for changes in paste properties is formation of a phase-separated microstructure in which AM- and AP-rich domains are dispersed in a hydrocolloid-rich continuous phase (Alloncle and Doublier, 1991; Annable et al., 1994; Yoshimura, Takaya, and Nishinari, 1996; Biliaderis et al., 1997; Sasaki, Yasui, and Matsuki, 2000).

Lim et al. (2003) reported that dry heat treatment (Section 8.3.2.1.6) in the presence of an anionic polysaccharide produced changes in the pasting and paste properties of starch, with the properties of the treated starch dependent on the specific combination of starch and polysaccharide. Use of xanthan in the treatment gave products with more restricted granular swelling and increased paste shear stability, as compared with the use of carboxymethylcellulose or sodium alginate (Lim, BeMiller, and Lim, 2003). The properties of the products were a function of the pH of the starch–hydrocolloid mixtures. By dry heating the starch with anionic hydrocolloid mixtures, such as xanthan–alginate and xanthan–CMC, both a viscosity increase and good shear stability were achieved.

8.4.4   Genetic modification

There is interest in in situ alteration of starch characteristics within plants as a means of modifiying starch properties for specific purposes or novel applications. A classic example is waxy maize, a mutant of corn which produces a starch devoid of AM. Plants that produce starches with altered AM and/or PM proportions, or fine structures are also possible (Preiss, 2008). Such starches offer the possibility of novel properties both in their native state and after modification. Detailed discussions have been provided by Blennow (2004) and White and Tzoitis (2004). Rice plants that produce starches with a range of AM contents, from less than 1% to at least 33%, are found among commercial cultivars (Bao et al., 2006; Chen and Bergman, 2007). Using various genetic and molecular biological techniques, all-AP and reduced AM starches from wheat, rice, and potato and increased AM starches from corn, wheat, rice, and potato have been developed (Blennow, 2004; White and Tzoitis, 2004; Grommers and van der Krogt, 2008; Maningat et al., 2008; Watson and Eckhoff, 2008). Starches from maize plants that biosynthesize starch with high (~55% and ~70%) contents of AM are available commercially.

8.5   Final remarks

More demands are being placed upon starches, requiring that they exhibit a more diverse range of properties. For starch derivatives intended for use in foods or in materials comprising contact surfaces (edible films, coatings, etc.), it is unlikely that either new modification reagents or reagent levels will be approved. Thus, for food applications, any new modified starch products offering novel properties or improved functionalities will likely need to be created within the constraints of current regulations. Under this constraint, advances in food starch chemical modification practices are most likely to be realized through a greater understanding of starch granule reactivity, enhancing the potential to manipulate or direct starch granular and molecular reaction patterns for development of novel products. Significant progress has been achieved in understanding where reactions take place within starch granules and on starch molecules, as well as an understanding of factors that influence chemical reactivity of starch modified in the conventional manner (Section 8.3.1.1).

Because of a negative perception regarding the use of chemically modified starch products in food applications, much attention has been focused on genetic, and more recently, physical modifications. Genetic modification, through means of traditional breeding or molecular biology techniques, will likely continue to expand the repertoire of starch materials and properties available in the future (Section 8.4.4). Certain physical treatments may be altered and complemented through blending of starches or by the addition of other hydrocolloids (Section 8.4.3), providing a basis for premixed starch and starch–hydrocolloid blends. Although all these developments have extended the properties of starch products within food systems, no one of these alternatives to date has proven entirely capable of eliminating the need for chemically modified starches in their multitude of applications. Further diversification of starch properties is likely through a combination of genetic, physical, and chemical modification strategies (Section 8.3.3).

There is also strong interest in developing starches with modified digestion characteristics (Section 8.4.1). Starches with higher percentages of RS and SDS have been made by physical treatments; chemical derivatizations, particularly via the octenylsuccinate ester; increases in starch amylose content; treatment with enzymes; other miscellaneous means; and by various combinations of these treatments. These modified products have the potential to benefit human health through moderation of the glycemic response and as a source of prebiotic carbohydrate.

For industrial applications, there is significant interest in replacement or substitution of synthetic polymers with starch and other biopolymers, due to their biodegradable and sustainable natures. As some form of chemical modification is often necessary to enhance the properties of starch for optimal performance in these applications, reactive extrusion, a recent development in the modification process (Section 8.4.2), should prove to be useful for generating modified starch materials with targeted properties for industrial purposes. Modified starches produced for industrial purposes do not face the same regulatory constraints previously noted for chemically modified food starches; thus, a broader range of reagents and reagent levels is potentially available for the generation of industrial starch products.

References

Abraham, T.E. 1993. Stabilization of paste viscosity of cassava starch by heat moisture treatment. *Starch/Stärke.* 45:131–135.

Adebowale, K.O. and Lawal, O.S. 2003. Microstructure, physicochemical properties and retrogradation behaviour of mucuna bean (*Mucuna pruriens*) starch on heat moisture treatments. *Food Hydrocoll.* 17:265–272.

Adler, J., Baldwin, P.M., and Melia, C.D. 1995. Starch damage. 2. Types of damage in ball-milled potato starch, upon hydration observed by confocal microscopy. *Starch/Stärke.* 47:252–256.

Agustiniano-Osornio, J.C., Gonzalez-Soto, R.A., Flores-Huicochea, E., Manrique-Quevedo, N., Sanchez-Hernandez, L., and Bello-Perez, L.A. 2005. Resistant starch production from mango starch using a single-screw extruder. *J. Sci. Food Agric.* 85:2105–2100.

Allen, J.E., Hood, L.F., and Parthasarathy, M.V. 1976. The ultrastructure of unmodified and chemically-modified tapioca starch granules as revealed by the freeze-etching technique. *J. Food Technol.* 11:537–541.

Alloncle, M. and Doublier, J.L. 1991. Viscoelastic properties of maize starch/hydrocolloid pastes and gels. *Food Hydrocoll.* 5:455–467.

Alves, R.M.L., Grossmann, M.V.E., and Silva, R.S.S.F. 1999. Gelling properties of extruded yam (*Dioscorea alata*) starch. *Food Chem.* 67:123–127.

An, H.J. and King, J.M. 2007. Thermal characteristics of ohmically heated rice starch and rice flours. *J. Food Sci.* 72:C84–C88.

Anastasiades A., Thanou, S., Loulis, D., Stapatoris, A., and Karapantsios, T.D. 2002. Rheological and physical characterization of pregelatinized maize starches. *J. Food Eng.* 52:57–66.

Anderson, A.K. and Guraya, H. 2004. Microwave heat applications to produce slowly-digestible starches. *Abstr. Papers, Am. Chem. Soc. 227th Annual Meeting,* AGFD-050.

Anderson, A.K. and Guraya, H.S. 2006. Effects of microwave heat-moisture treatment on properties of waxy and non-waxy rice starches. *Food Chem.* 97:318–323.

Anderson, A.K., Guraya, H.S., James, C., and Salvaggio, L. 2002. Digestibility and pasting properties of rice starch heat-moisture treated at the melting temperature (Tm). *Starch/Stärke.* 54:401–409.

Annable, P., Fitton, M.G., Harris, B., Phillips, G.O., and Williams, P.A. 1994. Phase behavior and rheology of mixed polymer systems containing starch. *Food Hydrocoll.* 8:351–359.

Ao, Z., Simsek, S., Zhang, G., Venkatachalam, M., Reuhs, B.L., and Hamaker, B.R. 2007. Starch with a slow digestion property produced by altering its chain length, branch density, and crystalline structure. *J. Agric. Food Chem.* 55:4540–4547.

Aparicio-Saguilan, A., Flores-Huicochea, E., Tovar, J., Garcia-Suarez, F., Gutierrez-Meraz, F., and Bello-Perez, L.A. 2005. Resistant starch-rich powders prepared by autoclaving of native and lintnerized banana starch: Partial characterization. *Starch/Stärke.* 57:405–412.

Ayoub, A. and Bliard, C. 2003. Cationisation of glycerol plasticised wheat starch under microhydric molten conditions. *Starch/Stärke.* 55:297–303.

Ayoub, A., Berzin, F., Tighzert, L., and Bliard, C. 2004. Study of the thermoplastic wheat starch cationisation reaction under molten condition. *Starch/Stärke.* 56:513–519.

Ayoub, A., Gruyer S., and Bliard, C. 2003. Enzymatic degradation of hydroxypropyltrimethylammonium wheat starches. *Int. J. Biol. Macromol.* 32:209–216.

Azemi, B.M.N.M. and Wootton, M. 1995. Distribution of partial digestion products of hydroxypropyl derivatives of maize (NM), waxy maize (WM) and high amylose (HA) maize starches. *Starch/Stärke.* 47:465–469.

Baik, B.-K. and Jackowski, R. 2004. Characteristics of granular cold-water gelling starches of cereal grains and legumes prepared using liquid ammonia and ethanol. *Cereal Chem.* 81:538–543.

Baldwin, P.M., Adler, J., Davies, M.C., and Melia, C.D. 1995. Starch damage. Part 1: Characterisation of starch damage in ball-milled potato starch study by SEM. *Starch/Stärke.* 46:247–251.

Bao, J., Shen, S., Sun, M., and Corke, H. 2006. Analysis of genotypic diversity in the starch physicochemical properties of nonwaxy rice: Apparent amylose content, pasting viscosity and gel texture. *Starch/Stärke.* 58:259–267.

Bauer, B.A., Wiehle, T., and Knorr, D. 2005. Impact of high hydrostatic pressure treatment on the resistant starch content of wheat starch. *Starch/Stärke.* 57:124–133.

BeMiller, J.N. 1997. Structure of the starch granule. *J. Appl. Glycosci.* 44:43–49.

BeMiller, J.N. 2007. *Carbohydrate Chemistry for Food Scientists.* St. Paul: AACC International, Inc.

BeMiller, J.N. and Pratt, G.W. 1981. Sorption of water, sodium sulfate, and water-soluble alcohols by starch granules in aqueous suspension. *Cereal Chem.* 58:517–520.

Benmoussa, M., Moldenhauer, K.A.K., and Hamaker, B.R. 2007. Rice amylopectin fine structure variability affects starch digestion properties. *J. Agric. Food Chem.* 55:1475–1479.

Berghofer, V.E. and Klaushofer, H. 1977. Bestimmung der verteilung von substituenten in den koernern von staerkederivaten. *Starch/Stärke.* 29:296–298.

Berry, C.S. 1986. Resistant starch: Formation and measurement of starch that survives exhaustive digestion with amylolytic enzymes during the determination of dietary fiber. *J. Cereal Sci.* 4:301–314.

Bertolini, A.C., Souza, E., Nelson, J.E., and Huber, K.C. 2003. Composition and reactivity of A- and B-type starch granules of normal, partial waxy, and waxy wheat. *Cereal Chem.* 80:544–549.

Berzin, F., Tara, A., Tighzert, L., and Vergnus, B. 2007. Computation of starch cationization performances by twin-screw extrusion. *Polym. Eng. Sci.* 47:112–119.

Bhattacharyya, D., Singhal, R.S., and Kulkarni, P.R. 1995. A comparative account of conditions for synthesis of sodium carboxymethyl starch from corn and amaranth starch. *Carbohydrate Polym.* 27:247–253.

Biliaderis, C.G., Arvanitoyannis, I., Izydorczyk, M.S., and Prokopowich, D.J. 1997. Effect of hydrocolloids on gelatinization and structure formation in concentrated waxy maize and wheat starch gels. *Starch/Stärke.* 49:278–283.

Binder, T.P. and McClain, J.A. 2006. Methods of producing resistant starch and products therefrom. U.S. Patent Appl. Publ. 2006073263; *Chem. Abstr.* 144:349550.

Bindzus, W., Fayard, G., van Lengerich, B., and Meuser, F. 2002. Application of an in-line viscometer to determine the shear stress of plasticised wheat starch. *Starch/Stärke*. 54:243–251.

Birkett, A.M. and Brown, I.L. 2007. Resistant starch. In: *Novel Food Ingredients for Weight Control*, C.J.K. Henry (Ed.). Cambridge, UK: Woodhead, pp.174–197.

Björck, I. 2006. Starch: Nutritional aspects. In: *Carbohydrates in Food*, A.C. Eliasson (Ed.). Boca Raton, FL: CRC Press, pp. 471–521.

Blennow, A. 2004. Starch bioengineering. In: *Starch in Food*, A.C. Eliasson (Ed.). Cambridge, UK: Woodhead, pp. 97–127.

Bloembergen, S., Kappan, F., and VanLeeuwen, M. 2002. Biopolymer latex adhesive for the manufacture of corrugated paperboard. Eur. Patent Appl. 1,254,939; *Chem. Abstr.* 137:33924.

Boruch, M. 1985. Transformations of potato starch during oxidation with hypochlorite. *Starch/Stärke*. 37:91–98.

Brncic, M., Mrkic, D., Jezek, D., and Tripalo, B. 2000. Reaction models during wheat starch extrusion. *Kemija u Industríjí* 49:101–110; *Chem. Abstr.* 133:42300.

Brumovsky, J.O. and Thompson, D.B. 2001. Production of boiling-stable granular resistant starch by partial acid hydrolysis and hydrothermal treatments of high-amylose maize starch. *Cereal Chem.* 78:680–389.

Campechano-Carrera, E., Corona-Cruz, A., Chel-Guerrero, L., and Betancur-Ancona, D. 2006 Effect of pyrodextrinization on available starch content of lima bean (*Phaseolus lunatus*) and cowpea (*Vigna unguiculata*) starches. *Food Hydrocoll.* 21:472–479.

Carmonsino, J.K. 2005. A resistant starch and process for the production thereof. U.S. Patent Appl. Publ. 2005271793; *Chem. Abstr.* 144:5891.

Carr, M.E. 1991. Starch-derived glycol and glycerol glucosides prepared by reactive extrusion processing. *J. Appl. Polym. Sci.* 42:45–53.

Carr, M.E. 1994a. Preparation of cationic starch containing quaternary ammonium substituents by reactive twin-screw extrusion processing. *J. Appl. Polym. Sci.* 54:1855–1861.

Carr, M.E. 1994b. Preparation of starch derivatives by reactive extrusion. *52nd Ann. Tech. Conf. Soc. Plastic Eng.* 1444–1448.

Carr, M.E. and Bagby, M.O. 1981. Preparation of cationic starch ether: A reaction efficiency study. *Starch/Stärke*. 33:310–312.

Carr, M.E. and Cunningham, R.L. 1992. Glycol glucoside extrudate from cornstarch as the polymer polyol for polyurethane foam preparation. *Polym. Preprints*. 33:946–947.

Carr, M.E., Kim, S., Yoon, K.J., and Stanley, K.D. 1992. Graft polymerization of cationic methacrylate, acrylamide, and acrylonitrile monomers onto starch by reactive extrusion. *Cereal Chem.* 69:70–75.

Chabot, J.F., Allen, J.E., and Hood, L.F. 1978. Freeze-etch ultrastructure of waxy maize and acid hydrolyzed waxy maize starch granules. *J. Food Sci.* 43:727–730, 734.

Chatakanonda, P., Varavinit, S., and Chinachoti, P. 2000. Effect of crosslinking on thermal and microscopic transitions of rice starch. *Lebensm. Wiss. Technol.* 33:276–284.

Chel-Guerrero, L. and Betancur, A.D. 1998. Cross-linkage of *Canavalia ensiformis* starch with adipic acid: Chemical and functional properties. *J. Agric. Food Chem.* 46:2087–2091.

Chen, J. and Jane, J.-L. 1994a. Preparation of granular cold-water-soluble starches by alcoholic-alkaline treatment. *Cereal Chem.* 71:618–622.

Chen, J. and Jane, J.-L. 1994b. Properties of granular cold-water-soluble starches prepared by alcoholic-alkaline treatment. *Cereal Chem.* 71:623–627.

Chen, J., Lai, V.M.F., and Lii, C. 2003. Effects of compositional and granular properties on the pasting viscosity of rice starch blends. *Starch/Stärke.* 55:203–212.

Chen, J.-J., Lai, V.M.F., and Lii, C-Y. 2004. Composition dependencies of the rheological properties of rice starch blends. *Cereal Chem.* 81:267–274.

Chen, J.-J., Lii, C-Y., and Lu, S. 2003. Physicochemical and morphological analysis on damaged rice starches. *J. Food Drug Anal.* 11:283–289.

Chen, L. 2004. Manufacture and application of resistant starch. CN Patent 1546678; *Chem. Abstr.* 143:345500.

Chen, M.-H. and Bergman, C.J. 2007. Method for determining the amylose content, molecular weights, and weight- and molar-based distributions of degree of polymerization of amylose and fine-structure of amylopectin. *Carbohydrate Polym.* 69:562–578.

Chen, Z., Schols, H.A., and Voragen, A.G.J. 2004. Differentially sized granules from acetylated potato and sweet potato starches differ in the acetyl substitution pattern of their amylose populations. *Carbohydrate Polym.* 56:219–226.

Chiu, C.-W., Schiermeyer, E., Thomas, D.J., and Shah, M.B. 1998. Thermally inhibited starches and flours and process for their production. U.S. Patent 5,725,676; *Chem. Abstr.* 128:193928.

Chiu, C.-W., Schiermeyer, E., Thomas, D.J., and Shah, M.B. 1999. Thermally inhibited non-pregelatinized granular starches and flours and preparation thereof. U.S. Patent 5,932,017; *Chem. Abstr.* 131:131452.

Chiu, C.-W., Shi, Y.-C., and Sedam, M. 1999. Process for producing amylase resistant starch. U.S. Patent 5,902,410; *Chem. Abstr.* 130:311002.

Choi, S.-G. and Kerr, W.L. 2004. Swelling characteristics of native and chemically modified wheat starches as a function of heating temperature and time. *Starch/Stärke.* 56:181–189.

Christianson, D.D., Hodge, J.E., Osborne, D., and Detroy, R.W. 1981. Gelatinization of wheat starch as modified by xanthan gum, guar gum, and cellulose gum. *Cereal Chem.* 58:513–517.

Chung, H.-J., Lim, H.S., and Lim, S.-T. 2006. Effect of partial gelatinization and retrogradation on the enzymatic digestion of waxy rice starch. *J. Cereal Sci.* 43:353–359.

Chung, K.M., Moon, T.W., and Chun, J.K. 2000. Influence of annealing on gel properties of mung bean starch. *Cereal Chem.* 77:567–571.

Colas, B. 1986. Flow behavior of crosslinked corn starches. *Lebensm. Wiss. Technol.* 19:308–311.

Colonna, P., Doublier, J.L., Melcion, J.P., Demonredon, F., and Mercier, C. 1984. Extrusion cooking and drum drying of wheat-starch. 1. Physical and macromolecular modifications. *Cereal Chem.* 61:538–543.

Craig, S.A.S., Sieb, P.A., and Jane, J.-L. 1987. Differential scanning calorimetry properties and paper-strength improvement of cationic wheat starch. *Starch/Stärke.* 39:167–170.

DeGraaf, R.A. and Janssen, L.P.B.M. 2000. The production of a new partially biodegradable starch plastic by reactive extrusion. *Polym. Eng. Sci.* 40:2086–2094.

DeGraaf, R.A., Broekroelofs, A., and Janssen, L.P.B.M. 1998. The acetylation of starch by reactive extrusion. *Starch/Stärke.* 50:198–205.

Della Valle, G., Colonna, P., and Tayeb, J. 1991. Use of a twin-screw extruder as a chemical reactor for starch cationization. *Starch/Stärke.* 43:300–307.

Donovan, J.W. 1979. Phase-transitions of the starch-water system. *Biopolymers.* 18:263–275.

Doublier, J.L., Colonna, P., and Mercier, C. 1986. Extrusion cooking and drum drying of wheat-starch. 2. Rheological characterization of starch pastes. *Cereal Chem.* 63:240–246.

Eastman, J.E. and Moore, C.O. 1984. Cold water-soluble granular starch. U.S. Patent 4,465,702; *Chem. Abstr.* 101:74688.

Eerlingen, R.C., Jacobs, H., Block, K., and Delcour, J.A. 1997. Effects of hydrothermal treatments on the rheological properties of potato starch. *Carbohydrate Res.* 297:347–356.

Englyst, H.N. and Hudson, G.J. 1996. The classification and measurement of dietary carbohydrates. *Food Chem.* 57:15–21.

Englyst, K.N. and Englyst, H.N. 2005. Carbohydrate bioavailability. *Brit. J. Nutr.* 94:1–11.

Esan, M., Bruemmer, T.M., and Meuser, F. 1996. Chemische und verfahrenstechnische Gesichtspunkte zur Herstellung von kationischer Kartoffelstärke durch Kochextrusion. *Starch/Stärke.* 48:131–136.

Fannon, J.E., Gray, J.A., Gunawan, N., Huber, K.C., and BeMiller, J.N. 2003. The channels of starch granules. *Food Sci. Biotechnol.* 12:700–704.

Fannon, J.E., Gray, J.A., Gunawan, N., Huber, K.C., and BeMiller, J.N. 2004. Heterogeneity of starch granules and the effect of granule channelization on starch modification. *Cellulose.* 11:247–254.

Fannon, J.E., Hauber, R.J., and BeMiller, J.N. 1992. Surface pores of starch granules. *Cereal Chem.* 69:284–288.

Fannon, J.E., Schull, J.M., and BeMiller, J.N. 1993. Interior channels of starch granules. *Cereal Chem.* 70:611–613.

Fanta, G.F. and Doane, W.M. 1986. Grafted starches. In: *Modified Starches: Properties and Uses*, O.B. Wurzburg (Ed.). Boca Raton, FL: CRC Press, pp. 149–178.

Fiedorowicz, M., Lii, C.-I., and Tomasik, P. 2002. Physicochemical properties of potato starch illuminated with visible polarised light. *Carbohydrate Polym.* 50:57–62.

Fiedorowicz, M., Tomasik, P., You, S., and Lim, S.-T. 1999. Molecular distribution and pasting properties of UV-irradiated corn starches. *Starch/Stärke.* 51:126–131.

Finkenstadt, V.L. and Willett, J.L. 2005. Reactive extrusion of starch-polyacrylamide graft copolymers: Effects of monomer/starch ratio and moisture content. *Macromol. Chem. Phys.* 206:1648–1652.

Gallant, D.J. and Guilbot, A. 1969a. Etude de l'ultrastructure du grain d'amidon a l'aide de nouvelles methods de preparations en microscopie electronique. *Starch/Stärke.* 21:156–163.

Gallant, D.J. and Guilbot, A. 1969b. Application de l'oxydation periodique a l'etude de l'ultrastructure de l'amidon de pomme de terre. *J. Microscopie* 8:549–568.

Gao, Q., Luo, Z., and Yang, L. 2007. Method for manufacturing resistant starch with amylase combined with heat-moisture treatment. CN Patent 1973688; *Chem. Abstr.* 147:94688.

Genkina, N.K., Wasserman, L.A., Noda, T., Tester, R.F., and Yuryev, V.P. 2004. Effects of annealing on the polymorphic structure of starches from sweet potatoes (Ayamurasaki and Sunnyred cultivars) grown at various soil temperatures. *Carbohydrate Res.* 339:1093–1098.

Gimmler, N. and Meuser, F. 1995. Influence of extrusion cooking conditions on the efficiency of the cationization and carboxymethylation of potato starch granules. 1995. *Starch/Stärke*. 46:268–276.

Gluck-Hirsch, J.B. and Kokini, J.L. 1997. Determination of the molecular weights between crosslinks of waxy maize starches using the theory of rubber elasticity. *J. Rheol.* 41:129–139.

Goddard, M.S., Young, G., and Marcus, R. 1984. The effect of amylose content on insulin and glucose responses to ingested rice. *Am. J. Clin. Nutr.* 39:388–392.

Gomes, A.M.M., Mendes da Silva, C.E., Ricardo, N.M.P.S., Sasaki, J.M., and Germani, R. 2004. Impact of annealing on the physicochemical properties of unfermented cassava starch (*Polvilho Doce*). *Starch/Stärke*. 56:419–423.

Gonzalez-Soto, R.A., Mora-Escobedo, R., Hernandez-Sanchez, H., Sanchez-Rivera, M., and Bello-Perez, L.A. 2007a. The influence of time and storage temperature on resistant starch formation from autoclaved debranched banana starch. *Food Res. Int.* 40: 304–310.

Gonzalez-Soto, R.A., Mora-Escobedo, R., Hernandez-Sanchez, H., Sanchez-Rivera, M., and Bello-Perez, L.A. 2007b. Extrusion of banana starch: Characterization of the extrudates. *J. Sci. Food Agric.* 87:348–356.

Goto, F. 1972. Determination of gelatinization property of highly concentrated starch suspensions by Brabender plastograph. IV. Plastograms of heat-treated starches. *J. Jpn. Soc. Starch Sci.* 19:90–99.

Gough, B.M. and Pybus, J.N. 1971. Effect on gelatinization temperature of wheat starch granules of prolonged treatment with water at 50°C. *Starch/Stärke*. 23:210–212.

Govindasamy, S., Campanella, O.H., and Oates, C.G. 1997a. High moisture twin screw extrusion of sago starch. II. Saccharification as influenced by thermomechanical history. *Carbohydrate Polym.* 32:267–274.

Govindasamy, S., Campanella, O.H., and Oates, C.G. 1997b. The single screw extruder as a bioreactor for sago starch hydrolysis. *Food Chem.* 60:1–11.

Granfeldt, Y., Drews, A., and Björck, I. 1995. Arepas made from high amylose corn flour produce favorably low glucose and insulin responses in healthy humans. *J. Nutr.* 125:459–465.

Gray, J.A. and BeMiller, J.N. 2004. Development and utilization of reflectance confocal laser scanning microscopy to locate reaction sites in modified starch granules. *Cereal Chem.* 81:278–286.

Gray, J.A. and BeMiller, J.N. 2005. Influence of reaction conditions on the location of reactions in waxy maize starch granules reacted with a propylene oxide analog at low substitution levels. *Carbohydrate Polym.* 60:147–162.

Grommers, H.E. and van der Krogt, D.A. 2008. Potato starch: Production, modification and uses. In: *Starch: Chemistry and Technology,* 3rd ed., J.N. BeMiller and R.L. Whistler (Eds.). Orlando, FL: Elsevier, Chapter 11.

Gu, Z. and Huang, G. 2007. Method for producing starch with high content of non-absorbable resistant starch. CN Patent 1995067; *Chem. Abstr.* 147:210895.

Gudmundsson, M., Eliasson, A.-C., Bengtsson, S., and Aman, P. 1991. The effects of water soluble arabinoxylan on gelatinization and retrogradation of starch. *Starch/Stärke*. 43:5–10.

Gunaratne, A. and Corke, H. 2007a. Effect of hydroxypropylation and alkaline treatment in hydroxypropylation on some structural and physicochemical properties of heat-moisture treated wheat, potato and waxy maize starches. *Carbohydrate Polym.* 68:305–313.

Gunaratne, A. and Corke, H. 2007b. Functional properties of hydroxypropylated, cross-linked, and hydroxypropylated cross-linked tuber and root starches. *Cereal Chem.* 84:30–37.

Gunaratne, A. and Corke, H. 2007c. Gelatinizing, pasting, and gelling properties of potato and amaranth starch mixtures. *Cereal Chem.* 84:22–29.

Gunaratne, A. and Hoover, R. 2002. Effect of heat-moisture treatment on the structure and physicochemical properties of tuber and root starches. *Carbohydate Polym.* 49:425–437.

Guraya, H.S., James, C., and Champagne, E.T. 2001. Effect of enzyme concentration and storage temperature on the formation of slowly digestible starch from cooked debranched rice starch. *Starch/Stärke.* 53:131–139.

Hagenimana, A. and Ding, X.L. 2005. A comparative study on pasting and hydration properties of native rice starches and their mixtures. *Cereal Chem.* 82:70–76.

Hagenimana, A., Pu, P.P., and Ding, X.L. 2005. Study on thermal and rheological properties of native rice starches and their corresponding mixtures. *Food Res. Int.* 38:257–266.

Hall, D.M. and Sayre, J.G. 1970. A scanning electron-microscope study of starches. Part II: Cereal starches. *Text. Res. J.* 40:147–157.

Hamaker, B.R. and Han, X.-Z. 2004. Slowly digestible starch. PCT Int. Patent Appl. WO 2004066955; *Chem. Abstr.* 141:173389.

Hamaker, B.R., Venkatachalam, M., Zhang, G., Keshavarzian, A., and Rose, D.J. 2007. Slowly digestible starch and fermentable fiber. U.S. Patent Appl. Publ. 2007196437.

Hamaker, B.R., Zhang, G., and Venkatachalam, M. 2007. Modified carbohydrates for lower glycemic index. In: *Novel Food Ingredients for Weight Control*, C.J.K. Henry (Ed.). Cambridge, UK: Woodhead, pp. 198–217.

Hamerstrand, G.E., Hofreiter, B.T., and Mehltreiter, C.L. 1960. Determination of the extent of reaction between epichlorohydrin and starch. *Cereal Chem.* 37:519–524.

Hamunen, R. 1995. Distribution of nitrogen in the cationized potato starch granule. *Starch/Stärke.* 47:215–219.

Han, H.L. and Sosulski, F.W. 1998. Cationization of potato and tapioca starches using an aqueous alcoholic-alkaline process. *Starch/Stärke.* 50:487–492.

Han, J.-A. and BeMiller, J.N. 2005. Rate of hydroxypropylation of starches as a function of reaction time. *Starch/Stärke.* 57:395–404.

Han, J.-A. and BeMiller, J.N. 2006. Influence of reaction conditions on MS values and physical properties of waxy maize starch derivatized by reaction with propylene oxide. *Carbohydrate Polym.* 64:158–162.

Han, J.-A. and BeMiller, J.N. 2007. Preparation and physical characteristics of slowly digesting modified food starches. *Carbohydrate Polym.* 67:366–374.

Han, J.-A., Gray, J.A., Huber, K.C., and BeMiller, J.N. 2006a. Derivatization of starch granules as influenced by the presence of channels and reaction conditions. In: *Advances in Biopolymers, Molecules, Clusters, Networks and Interactions*, M.L. Fishman, P.X. Qi, and L. Wicker (Eds.). Washington, DC: American Chemical Society, pp. 165–184.

Han, X.-Z. and Hamaker, B.R. 2002. Association of starch granule proteins with starch ghosts and remnants revealed by confocal laser scanning microscopy. *Cereal Chem.* 79:892–896.

Han, X.-Z., Ao, Z., Janaswamy, S., Jane, J.-L., Chandrasekaran, R., and Hamaker, B.R. 2006b. Development of a low glycemic maize starch: Preparation and characterization. *Biomacromolecules.* 7:1162–1168.

Han, X.-Z., Benmoussa, M., Gray, J.A., BeMiller, J.N., and Hamaker, B.R. 2005. Detection of proteins in starch granule channels. *Cereal Chem.* 82:351–355.

Han, X.-Z., Campanella, O.H., Mix, N.C., and Hamaker, B.R. 2002. Consequence of starch damage on rheological properties of maize starch pastes. *Cereal Chem.* 79:897–901.

Hauber, R.J., BeMiller, J.N., and Fannon, J.E. 1992. Swelling and reactivity of maize starch granules. *Starch/Stärke.* 47:323–327.

Heinze, T., Haack, V., and Rensing, S. 2004. Starch derivatives of high degree of functionalization. 7. Preparation of cationic 2-hydroxypropyltrimethylammonium chloride starches. *Starch/Stärke.* 56:288–296.

Higley, J.S. 2005. Elucidation of starch granule surface composition and reactivity aided by 5-(4,6-dichlorotriazinyl)aminofluorescein (DTAF) derivatization and gel permeation chromatography coupled with fluorescence detection. M.S. Thesis, University of Idaho.

Hirsch, J.B. and Kokini, J.L. 2002. Understanding the mechanism of cross-linking agents (POCl$_3$, STMP, and EPI) through swelling behavior and pasting properties of cross-linked waxy maize starches. *Cereal Chem.* 79:102–107.

Hood, L.F. and Mercier, C. 1978. Molecular structure of unmodified and chemically modified manioc starches. *Carbohydrate Res.* 247:279–290.

Hoover, R. 2000. Acid-treated starches. *Food Rev. Int.* 16:369–392.

Hoover, R. and Manuel, H. 1996a. The effect of heat-moisture treatment on the structure and physicochemical properties of normal maize, waxy maize, dull waxy maize and amylomaize V starches. *J. Cereal Sci.* 23:153–162.

Hoover, R. and Manuel, H. 1996b. Effect of heat-moisture treatment on the structure and physicochemical properties of legume starches. *Food Res. Int.* 29:731–750.

Hoover, R. and Sosulski, F. 1986. Effect of cross-linking on functional properties of legume starches. *Starch/Stärke.* 38:149–155.

Hoover, R. and Vasanthan, T. 1994. The flow properties of native, heat-moisture treated, and annealed starches from wheat, oat, potato and lentil. *J. Food Biochem.* 18:67–82.

Hoover, R., Swamidas, G., and Vasanthan, T. 1993. Studies on the physicochemical properties of native, defatted, and heat-moisture treated pigeon pea (*Cajanus cajan L*) starch. *Carbohydrate Res.* 246:185–203.

Howling, D. 1980. The influence of the structure of starch on its rheological properties. *Food Chem.* 6:51–61.

Huang, Z.-Q., Lu, J.-P., Li, X.-H., and Tong, Z.-F. 2007. Effect of mechanical activation on physico-chemical properties and structure of cassava starch. *Carbohydrate Polym.* 68:128–135.

Huber, K.C. and BeMiller, J.N. 1997. Visualization of channels and cavities of corn and sorghum starch granules. *Cereal Chem.* 74:537–541.

Huber, K.C. and BeMiller, J.N. 2000. Channels of maize and sorghum starch granules. *Carbohydrate Polym.* 41:269–276.

Huber, K.C. and BeMiller, J.N. 2001. Location of sites of reaction within starch granules. *Cereal Chem.* 78:173–180.

Jackowski, R., Czuchajowska, Z., and Baik, B.-K. 2002. Granular cold water gelling starch prepared from chickpea starch using liquid ammonia and ethanol. *Cereal Chem.* 79:125–128.

Jacobasch, G., Dongowski, G., Schmiedl, D., and Mueller-Schmehl, K. 2006. Hydrothermal treatment of Novelose 330 results in high yield of resistant starch type 3 with beneficial prebiotic properties and decreased secondary bile acid formation in rats. *Brit. J. Nutr.* 95:1063–1074.

Jacobs, H. and Delcour, J.A. 1998. Hydrothermal modifications of granular starch, with retention of the granular structure: A review. *J. Agric. Food Chem.* 46:2895–2905.

Jacobs, H., Eerlingen, R.C., Clauwaert, W., and Delcour, J.A. 1995. Influence of annealing on the pasting properties of starches from varying botanical sources. *Cereal Chem.* 72:480–487.

Jacobs, H., Mischenko, N., Koch, M.H.J., Eerlingen, R.C., Delcour, J.A., and Reynaers, H. 1998. Evaluation of the impact of annealing on gelatinisation at intermediate water content of wheat and potato starches: A differential scanning calorimetry and small angle X-ray scattering study. *Carbohydrate Res.* 306:1–10.

Jane, J.-L. 1995. Starch properties, modifications, and applications. *J. Macromol. Sci. Pure Appl. Chem.* A32:751–757.

Jane, J.-L., Craig, S.A.S., Seib, P.A., and Hoseney, R.C. 1986. A granular cold-water starch gives a V-type X-ray diffraction pattern. *Carbohydrate Res.* 150:C5–C6.

Jane, J.-L., Xu, A., Radosavljevic, M., and Seib, P.A. 1992. Location of amylose in normal starch granules. I. Susceptibility of amylose and amylopectin to cross-linking reagents. *Cereal Chem.* 69:405–409.

Ji, Y., Ao, A., Han, J.-A., Jane, J.-L., and BeMiller, J.N. 2004. Waxy maize starch sub-populations with different gelatinization temperatures. *Carbohydrate Polym.* 57:177–190.

Jyothi, A.N., Moorthy, S.N., and Rajasekharan, K.N. 2006. Effect of cross-linking with epichlorohydrin on the properties of cassava (*Manihot esculenta* Crantz) starch. *Starch/Stärke.* 58:292–299.

Kalambur, S. and Rizvi. S.S.H. 2005. Biodegradable and functionally superior starch-polyester nanocomposites from reactive extrusion. *J. Appl. Polym. Sci.* 96:1072–1082.

Kalambur, S. and Rizvi, S.S.H. 2006. An overview of starch-based plastic blends from reactive extrusion. *J. Plast. Film Sheet.* 22:39–58.

Karam, L.B., Ferrero, C., Martino, M.N., Zaritsky, N.E., and Grossmann, M.V.E. 2006. Thermal, microstructural and textural characterisation of gelatinized corn, cassava, and yam starch blends. *Int. J. Food Sci. Technol.* 41:805–812.

Karam, L.B., Grossmann, M.V.E., Silva, R.S.S.F., Ferrero, C., and Zaritsky, N.E. 2005. Gel textural characteristics of corn, cassava, and yam starch blends: A surface response methodology approach. *Starch/Stärke.* 57:62–70.

Kasemsuwan, T. and Jane, J.-L. 1994. Location of amylose in normal starch granules. II. Locations of phosphodiester cross-linking revealed by phosphorus-31 nuclear magnetic resonance. *Cereal Chem.* 71:282–287.

Kaur, L., Singh, J., McCarthy, O.J., and Singh H. 2007. Physico-chemical, rheological and structural properties of fractionated potato starches. *J. Food Eng.* 82:383–394.

Kaur, L., Singh, J., and Singh, N. 2006. Effect of cross-linking on some properties of potato (*Solanum tuberosum* L.) starches. *J. Sci. Food Agric.* 86:1945–1954.

Kaur, L, Singh, N., and Singh, J. 2004. Factors influencing the properties of hydroxypropylated potato starches. *Carbohydrate Polym.* 55:211–223.

Kavitha, R. and BeMiller J.N. 1998. Characterization of hydroxypropylated potato starch. *Carbohydrate Polym.* 37:115–121.

Kawabata, A., Takase, N., Miyoshi, E., Sawayama, S., Kimura, T., and Kudo, K. 1994. Microscopic observation and x-ray diffractometry of heat/moisture-treated starch granules. *Starch/Starke.* 46:463–469.

Kerr, R.W. and Cleveland, F.C., Jr. 1962. Thickening agent. U.S. Patent 3,021,222; *Chem. Abstr.* 56:55759.

Kesselmans, R.P.W., Venema, B., and Hadderingh, E. 2001. Reactive processing by extrusion of high-amylopectin starch. Eur. Pat. Appl. 1,148,067; *Chem Abstr.* 135:319744.

Khalil, M.I. and Farag, S. 1994. Preparation and utilization of some ion-exchange starches. *Starch/Stärke.* 46:17–22.

Khalil, M.I. and Farag, S. 1998. Preparation of some cationic starches using the dry process. *Starch/Stärke.* 50:267–271.

Khalil, M.I., Farag, S., Aly, A.A., and Hebeish, A. 2002. Some studies on starch-urea-acid reaction mechanism. *Carbohydrate Polym.* 48:255–261.

Kim, H.-S. and Huber, K.C. 2008. Channels within soft wheat starch A- and B-type granules. *J. Cereal Sci.* 48:159–172.

King, J.M. and Tan, S.Y. 2005. Resistant starch with cooking properties similar to untreated starch. PCT Int. Patent Appl. WO 2005025327; *Chem. Abstr.* 142:315706.

Kiseleva, V.I., Genkina, N.K., Tester, R., Wasserman, L.A., Popov, A.A., and Yuryev, V.P. 2004. Annealing of normal, low and high amylose starches extracted from barley cultivars grown under different environmental conditions. *Carbohydrate Polym.* 56:157–168.

Kitahara, K., Hamasuna, K., Nozuma, K., Otani, M., Hamada, T., Shimada, T., Fujita, K., and Suganuma, T. 2007. Physicochemical properties of amylose-free and high-amylose starches from transgenic sweet potatoes modified by RNA interference. *Carbohydrate Polym.* 69:233–240.

Knutson, C.A. 1990. Annealing of maize starches at elevated temperatures. *Cereal Chem.* 67:376–384.

Koch, V.H., Bommer, H.D., and Koppers, J. 1982. Analytical investigations on phosphate cross-linked starches. *Starch/Stärke.* 24:16–21.

Komolprasert, V. 1990. Starch liquefaction by thermostable alpha-amylase during reactive extrusion. Ph.D. Dissertation, Michigan State University.

Kosan, B., Meister, F., Liebert, T., and Heinze, T. 2006. Hydrophobic modification of starch via grafting with an oxazoline-derivative. *Cellulose.* 13:105–113.

Kuhn, K. and Schlauch, S. 1994. Comparative study about commercially available starches for high shear and high temperature applications in food. *Starch/Stärke.* 46:208–218.

Kurakake, M., Noguchi, M., Fujioka, K., and Komaki, T. 1997. Effects on maize starch properties of heat-treatments with water-ethanol mixtures. *J. Cereal Sci.* 25:253–260.

Kweon, M.R., Bhirud, P.R., and Sosulski, F.W. 1996. An aqueous alcoholic-alkaline process for cationization of corn and pea starches. *Starch/Stärke.* 48:214–220.

Kweon, M.R., Sosulski, F.W., and Bhirud, P.R. 1997. Cationization of waxy and normal corn and barley starches by an aqueous alcohol process. *Starch/Stärke.* 49:59–66.

Kweon, M.R., Sosulski, F.W., and Han, H.S. 1997. Effect of aqueous ethanol cation-ization on functional properties of normal and waxy starches. *Starch/Stärke*. 49:202–207.

Lai, L.-S. and Liao, C.-L. 2002. Steady and dynamic shear rheological properties of starch and decolorized Hsian-tsao leaf gum composite systems. *Cereal Chem.* 79:58–63.

Lammers, G., Stamhuis, E.J., and Beenackers, A.A.C.M. 1993. Kinetics of the hydroxypropylation of potato starch in aqueous solution. *Ind. Eng. Chem. Res.* 32:835–842.

Larsson, A. 1998. Conformational changes of cationic potato amylopectin; influence by salt, influence by nanosized silica particles. *Colloids Surf. B Biointerfaces*. 12:23–34.

Larsson, I. and Eliasson, A.C. 1991. Annealing of starch at an intermediate water content. *Starch/Stärke*. 43:227–231.

Lawal, O.S. and Adebowale, K.O. 2005. An assessment of changes in thermal and physico-chemical parameters of jack bean (*Canavalia ensiformis*) starch fol-lowing hydrothermal modifications. *Eur. Food Res. Technol.* 221:631–638.

Leeman, A.M., Karlsson, M.E., Eliasson, A.C., and Björck, I.M.E. 2006. Resistant starch formation in temperature treated potato starches varying in amylose/amylopectin ratio. *Carbohydrate Polym.* 65:306–313.

Lehmann, U. and Robin, F. 2007. Slowly digestible starch: Its structure and health implications: A review. *Trends Food Sci. Technol.* 18:346–355.

Lelievre, J. 1985. Gelatinization of crosslinked potato starch. *Starch/Stärke*. 37:267–269.

Leong, Y.H., Karim, A.A., and Norziah, M.H. 2007. Effect of pullulanase debranch-ing of sago (*Metroxylon sagu*) starch at subgelatinization temperature on the yield of resistant starch. *Starch/Stärke*. 59:21–32.

Lewandowicz, G., Jankowski, T., and Fornal, J. 1997. Effect of microwave radiation on physico-chemical properties and structure of cereal starches. *Carbohydrate Polym.* 42:193–199.

Li, J.H., Vasanthan, T., Hoover, R., and Rossnagel, B.G. 2004. Starch from hull-less barley: V. In-vitro susceptibility of waxy, normal, and high-amylose starches towards hydrolysis by alpha-amylases and amyloglucosidase. *Food Chem.* 84:621–632.

Li, X., Chen, L., and Li, L. 2007. Characterization of acetylated resistant starch for controlled and targeted releasing carriers. *Abstr. Papers, Amer. Chem. Soc. 233rd Annual Meeting*, CARB–039.

Lii, C.Y., and Chen, L.H. 1980. The factors in the gel forming properties of Hsian-tsao (*Mesona procumbens* Hemsl). I. Extraction conditions and different starches. *Proc. Natl. Sci. Council ROC* 4:438–442.

Lim, H.S., BeMiller, J.N., and Lim, S.-T. 2003. Effect of dry heating with ionic gums at controlled pH on starch paste viscosity. *Cereal Chem.* 80:198–202.

Lim, J.U., Mun, S.H., Shin, M.S., and Song, J.Y. 2004. Increasing content of resistant starch by acid treatment. KR Patent 2004065072; *Chem. Abstr.* 145:187442.

Lim, S. and Seib, P.A. 1993. Location of phosphate esters in a wheat starch by ^{31}P-nuclear magnetic resonance microscopy. *Cereal Chem.* 70:145–152.

Lim, S.-T., Han, J.-A., Lim, H.S., and BeMiller, J.N. 2002. Modification of starch by dry heating with ionic gums at controlled pH on starch paste viscosity. *Cereal Chem.* 80:198–202.

Lim, S.-T., Han, J.-A., Lim, H.S., and BeMiller, J.N. 2003. Effect of dry heating with ionic gums. *Cereal Chem.* 79:601–606.

Liu, H.J., Corke, H., and Ramsden, L. 2000. The effect of autoclaving on the acetylation of ae, wx, and normal maize starches. *Starch/Stärke.* 52:353–360.

Liu, H.J., Ramsden, L., and Corke, H. 1999. Physical properties of cross-linked and acetylated normal and waxy rice starch. *Starch/Stärke.* 51:249–252.

Loisel, C., Maache-Rezzoug, Z., Esneault, C., and Doublier, J.L. 2006. Effect of hydrothermal treatment on the physical and rheological properties of maize starches. *J. Food Eng.* 73:45–54.

Lorenz, K. and Kulp, K. 1982. Cereal and root starch modification by heat-moisture treatment. *Starch/Stärke.* 34:50–54.

Lu, S., Chen, C.Y., and Lii, C.Y. 1996. Gel-chromatography fractionation and thermal characterization of rice starch affected by hydrothermal treatment. *Cereal Chem.* 73:5–11.

Luo, Z., He, X., Fu, X., Luo, F., and Gao, Q. 2006. Effect of microwave radiation on the physicochemical properties of normal maize, waxy maize and amylomaize V starches. *Starch/Stärke.* 58:468–474.

Mackey, L.N., James, M.D., Ensign, D.E., Gordon, G.C., Buchanan, L.L., Heinzmanm, S.W., Forshey, P.A., and Aydore, S. 2003. Non-thermoplastic starch fibers, starch composition, and process for making starch composition. U.S. Patent Appl. WO 2002-62392; *Chem. Abstr.* 139:181467.

Mali, S. and Grossmann, M.V.E. 2001. Preparation of acetylated distarch adipates by extrusion. *Lebensm. Wiss. Technol.* 34:384–389.

Maliger, R.B., McGlashan, S.A., Halley, P.J., and Matthew, L.G. 2006. Compatibilization of starch-polyester blends using reactive extrusion. *Polym. Eng. Sci.* 46:248–263.

Manelius, R., Buleon A., Nurmi, K., and Bertoft, E. 2000. The substitution pattern in cationised and oxidized potato starch granules. *Carbohydrate Res.* 329:621–633.

Manelius, R., Nurmi, K., and Bertoft, E. 2000. Enzymatic and acid hydrolysis of cationized waxy corn starch granules. *Cereal Chem.* 77:345–353.

Manelius, R., Nurmi, K., and Bertoft, E. 2005. Characterization of dextrins obtained by enzymatic treatment of cationic potato starch. *Starch/Stärke.* 57:291–300.

Manelius, R., Nurmi, K., Maaheimo, H., and Bertoft, E. 2002. Characterisation of fractions obtained by isoamylolysis and ion-exchange chromatography of cationic waxy maize starch. *Starch/Stärke.* 54:58–65.

Maningat, C.C., Bassi, S.D., Woo, K.S., Lasater, G.D., and Seib, P.A. 2008. Wheat starch: production, properties, modification, and uses. In *Starch: Chemistry and Technology,* 3rd ed., J.N. BeMiller and R.L. Whistler (Eds.). Orlando, FL: Elsevier, Chapter 10.

Martınez-Bustos, F., Lopez-Soto, M., San Martın-Martınez, E., Zazueta-Morales, J.J., and Velez-Medina, J.J. 2007. Effects of high energy milling on some functional properties of jicama starch (*Pachyrhizus erosus* L. Urban) and cassava starch (*Manihot esculenta Crantz*). *J. Food Eng.* 78:1212–1220.

Martinez-Bustos, F., Lopez-Soto, M., Zazueta-Morales, J.J., and Morales-Sanchez, E. 2005. Preparation and properties of pregelatinized cassava (*Manihot esculenta* Crantz) and jicama (*Pachyrhizus erosus*) starches using ohmic heating. *Agrociencia.* 39:275–283.

Maruta, I., Kurahashi, Y., Takano, R., Hayashi, K., Yoshino, O., Komaki, T., and Hara, S. 1994. Reduced-pressurized heat-moisture treatment: A new method for heat-moisture treatment of starch. *Starch/Stärke.* 46:177–181.

Meuser, F., Gimmler, N., and Oeding, J. 1990. System analysis of the derivatization of starch with a cooking extruder as reactor. *Starch/Stärke.* 42:330–336.

Miladinov. V.D., and Hanna, M.A. 2000. Starch esterification by reactive extrusion in food products. *Ind. Crop. Prod.* 11:51–57.

Miyoshi, E. 2002. Effects of heat-moisture treatments and lipids on gelatinization and retrogradation of maize and potato starches. *Cereal Chem.* 79:72–77.

Modig, G., Nilsson, P.O., and Wahlund, K.G. 2006. Influence of jet-cooking temperature and ionic strength on size and structure of cationic potato amylopectin starch as measured by asymmetrical flow field-flow fractionation multi-angle light scattering. *Starch/Stärke.* 58:55–65.

Mok, C. and Dick, J.W. 1991. Response of starch of different wheat classes to ball milling. *Cereal Chem.* 68:409–412.

Morrison, W.R. and Tester, R.F. 1994. Properties of damaged starch granules. IV. Composition of ball-milled wheat starches and fractions obtained on hydration. *J. Cereal Sci.* 20:69–77.

Morrison, W.R., Tester, R.F., and Gidley, M.J. 1994. Properties of damaged starch granules. II. Crystallinity, molecular order and gelatinization of ball-milled starches. *J. Cereal Sci.* 19:209–217.

Muhammad, K., Hussin, F., Man, Y.C., Ghazali, H.M., and Kennedy, J.F. 2000. Effect of pH on phosphorylation of sago starch. *Carbohydrate Polym.* 42:85–90.

Muhrbeck, P. and Svensson, E. 1996. Annealing properties of potato starches with different degrees of phosphorylation. *Carbohydrate Polym.* 31:263–261.

Muhrbeck, P. and Wischmann, B. 1998. Influence of phosphate esters on the annealing properties of starch. *Starch/Stärke.* 50:423–426.

Mussulman, W.C. and Wagoner, J.A. 1968. Electron microscopy of unmodified and acid-modified corn starches. *Cereal Chem.* 45:162–171.

Nabeshima, E.H. and Grossman, M.V.E. 2001. Functional properties of pregelatinized and crosslinked cassava starch obtained by extrusion with sodium trimetaphosphate. *Carbohydrate Polym.* 45:347–353.

Nakazawa, Y. and Wang, Y.-J. 2003. Acid hydrolysis of native and annealed starches and branch structure of their Naegeli dextrins. *Carbohydrate Res.* 338:2871–2882.

Narayan, R., Balakrishnan, S., and Shin, B.-Y. 2006. Biodegradable starch-vegetable oil graft copolymers and their biofiber composites, and a process for their manufacture. U.S. Patent Appl. 2005-124491; *Chem. Abstr.* 145:473365.

Narayan, R., Balakrishnan, S., Nabar, Y., Shin, B.-Y., Dubois, P., and Raquez, J.-M. 2006. Chemically-modified plasticized starch compositions by extrusion processing. U.S. Patent Appl. 2006/07945; *Chem. Abstr.* 144:490569.

Narkrugsa, W., Berghofer, E., and Camargo, L.C.A. 1992. Production of starch derivatives by extrusion cooking. *Starch/Stärke.* 44:81–90.

Nayouf, M., Loisel, C., and Doublier, J.L. 2003. Effect of thermomechanical treatment on the rheological properties of crosslinked waxy corn starch. *J. Food Eng.* 59:209–219.

Nehmer, W.L., Nobes, G.A.R., and Yackel, W.C. 2006. Production of enzyme-resistant starch by extrusion. U.S. Patent Appl. Publ. 2006272634; *Chem. Abstr.* 146:28982.

Nguyen, C.C., Martin, V.J., and Pauley, E.P. 1990. Starch graft polymer dispersions for paper coating. PCT Int. Patent Appl. WO 9009406; *Chem. Abstr.* 114:45303.

Novello-Cen, L. and Betancur-Ancona, D. 2005. Chemical and functional properties of *Phaseolus lunatis* and *Manihot esculenta* starch blends. *Starch/Stärke.* 57:431–441.

Obanni, M. and BeMiller, J.N. 1997. Properties of some starch blends. *Cereal Chem.* 74:431–436.

Onyango, C., Bley, T., Jacob, A., Henle, T., and Rohm, H. 2006. Influence of incubation temperature and time on resistant starch type III formation from autoclaved and acid-hydrolysed cassava starch. *Carbohydrate Polym.* 66:494–499.

Oosten, B.J. 1982. Tentative hypothesis to explain how electrolytes affect the gelatinization temperature of starches in water. *Starch/Stärke.* 34:233–239.

Ortega-Ojeda, F.E. and Eliasson, A.C. 2001. Gelatinisation and retrogradation behaviour of some starch mixtures. *Starch/Stärke.* 53:520–529.

Perera, C. and Hoover, H. 1998. The reactivity of porcine pancreatic alpha-amylase towards native, defatted and heat-moisture treated potato starches before and after hydroxypropylation. *Starch/Stärke.* 50:206–213.

Perera, C., Hoover R., and Martin, A.M. 1997. The effect of hydroxypropylation on the structure and physicochemical properties of native, defatted and heat-moisture treated potato starches. *Food Res. Int.* 30:235–247.

Pitchon, E., O'Rourke, J.D., and Joseph, T.H. 1981. Process and apparatus for cooking or gelatinizing materials. U.S. Patent 4,280,851; *Chem. Abstr.* 136:342846.

Preiss, J. 2008. Biochemistry and molecular biology of starch biosynthesis. In: *Starch: Chemistry and Technology,* 3rd ed., J.N. BeMiller and R.L. Whistler (Eds.). Orlando, FL: Elsevier. Chapter 4.

Pukkahuta, C., Shobsngob, S., and Varavinit, S. 2007. Effect of osmotic pressure on starch: New method of physical modification of starch. *Starch/Stärke.* 58:78–90.

Qu, D. and Wang, S.S. 2002. Modeling extrusion conversion of starch in a single-screw extruder. *J. Chinese Inst. Chem. Eng.* 33:33-51; *Chem. Abstr.* 136, 342846.

Radosta, S., Vorwerg, W., Ebert, A., Haji Begli, A., Grülc, D., and Wastyn, M. 2004. Properties of low-substituted cationic starch derivatives prepared by different derivatisation processes. *Starch/Stärke.* 56:277–287.

Rajagopalan, S. and Seib, P.A. 1992. Granular cold-water-soluble starches prepared at atmospheric-pressure. *J. Cereal Sci.* 16:13–28.

Rapaille, A. and Vanhemelrjck, J. 1997. Modified starches. In: *Thickening and Gelling Agents for Food.* London: Blackie Academic and Professional, pp. 199–229.

Richardson, S., Nilsson, G., Cohen, A., Momcilovic, D., Brinkmalm, G., and Gorton, L. 2003. Enzyme-aided investigation of the substituent distribution in cationic potato amylopectin starch. *Anal. Chem.* 75:6499–6508.

Rizvi, S.S.H. and Kalambur, S. 2005. Biodegradable starch-polyester blends produced by reactive extrusion. U.S. Patent Appl. 2004-567646; *Chem. Abstr.* 144:23535.

Röper, H. and Elvers, B. 2008. Starch. 3. Economic Aspects. In: *Ullman's Encyclopedia of Industrial Chemistry,* 7th ed., New York: John Wiley & Sons, pp. 21–22.

Rudnik, E. and Zukowska, E. 2004. Studies on preparation of starch succinate by reactive extrusion. *Polimery* (Warsaw) 49:132–134; *Chem. Abstr.* 141:397157.

Rutenberg, M.W. and Solarek, D. 1984. Starch derivatives: Production and uses. In: *Starch Chemistry and Technology*, R.L. Whistler, J.N. BeMiller, and E.F. Paschall (Eds.). San Diego: Academic Press, pp. 311–388.

Sajilata, M.G., Singhal, R.S., and Kulkarni, P.R. 2006. Resistant starch: A review. *Compr. Rev. Food Sci. Food Safety.* 5:1–17.

Sajjan, S.U. and Rao, M.R.R. 1987. Effect of hydrocolloids on the rheological properties of wheat starch. *Carbohydrate Polym.* 7:395–402.

Salomonsson, B.A., Fransson, G.M.B., and Theander, O. 1991. The cationic distribution in a cationised potato starch. *Starch/Stärke.* 43:81–82.

San Martin-Martinez, E., Aguilar-Mendez, M.A., Espinosa-Solares, T., Pless, R.C., and Quintana, Z.D. 2004. Starch phosphates produced by extrusion: Physical properties and influence on yogurt stability. *Starch/Stärke.* 56:199–207.

Sang, Y. and Seib, P.A. 2006. Resistant starches from amylose mutants of corn by simultaneous heat-moisture treatment and phosphorylation. *Carbohydrate Polym.* 63:167–175.

Sanguanpong, V., Chotineeranat, S., Piyachomkwan, K., Oates, C.G., Chinachoti, P., and Sriroth, K. 2003. Hydration and physicochemical properties of small-particle cassava starch. *J. Sci. Food Agric.* 83:123–132.

Sasaki, T., Yasui, T., and Matsuki, J. 2000. Influence of non-starch polysaccharides isolated from wheat flour on the gelatinization and gelation of wheat starches. *Food Hydrocoll.* 14:295–303.

Seguchi, M. 1984. Oil-binding capacity of heat-treated wheat starch. *Cereal Chem.* 61:241–244.

Seguchi, M. and Yamada, Y. 1988. Hydrophobic character of heat-treated wheat starch. *Cereal Chem.* 65:375–376.

Seidel, C., Kulicke, W.-M., Hess, C., Hartmann, B., Lechner, M.D., and Lazik, W. 2001. Influence of the cross-linking agent on the gel structure of starch derivatives. *Starch/Stärke.* 53:305–310.

Seker, M. and Hanna, M.A. 2005. Cross-linking starch at various moisture contents by phosphate substitution in an extruder. *Carbohydrate Polym.* 59:541–544.

Seker, M. and Hanna, M.A. 2006. Sodium hydroxide and trimetaphosphate levels affect properties of starch extrudates. *Ind. Crop. Prod.* 23:249–255.

Seker, M., Sadikoglu, H., and Hanna, M.A. 2004. Properties of cross-linked starch produced in a single screw extruder with and without a mixing element. *J. Food Proc. Eng.* 27:47–63.

Seker, M., Sadikoglu, H., Ozdemir, M., and Hanna, M.A. 2003. Cross-linking of starch with reactive extrusion and expansion of extrudates. *Int. J. Food Prop.* 6:473–480.

Seow, C.C. and Teo, C.H. 1993. Annealing of granular rice starches: Interpretation of the effect on phase transitions associated with gelatinizations. *Starch/Stärke.* 45:345–351.

Shey, J., Holtman, K.M., Wong, R.Y., Gregorski, K.S., Klamczynski, A.P., Orts, W.J., Glenn, G.M., and Imam, S.H. 2006. The azidation of starch. *Carbohydrate Polym.* 65:529–534.

Shi, X. and BeMiller, J.N, 2000. Effect of sulfate and citrate salts on derivatization of amylose and amylopectin during hydroxypropylation of corn starch. *Carbohydrate Polym.* 43:333–336.

Shi, X. and BeMiller, J.N. 2002. Effects of food gums on viscosities of starch suspensions during pasting. *Carbohydrate Polym.* 50:7–18.

Shi, Y.-C. and Trzasko, P.T. 1996. Process for producing amylase-resistant granular starch with high content of dietary fiber. Eur. Patent Appl. 747397; *Chem. Abstr.* 126:61804.

Shi, Y.-C., Cui, X., and Birkett, A.M. 2003a. Slowly digesting starch product. Eur. Patent Appl. 1362517; *Chem. Abstr.* 139:363947.

Shi, Y.-C., Cui, X., and Birkett, A.M. 2003b. Preparation of a slowly digestible starch product by enzymic debranching. Eur. Patent Appl. 1362919; *Chem. Abstr.* 139:363703.

Shim, J.Y. and Mulvaney, S.J. 1999. Effect of cooking temperature and stirring speed on rheological properties and microstructure of cornstarch and oat flour gels. *Cereal Foods World* 44:349–356.

Shin, M., Woo, K., and Seib, P.A. 2003. Hot-water solubilities and water sorptions of resistant starches at 25°C. *Cereal Chem.* 80:564–566.

Shin, S.I., Byun, J., Park, K.H., and Moon, T.W. 2004a. Effect of partial acid hydrolysis and heat-moisture treatment on formation of resistant tuber starch. *Cereal Chem.* 81:194–198.

Shin, S.I., Choi, H.J., Chung, K.M., Hamaker, B.R., Park, K.H., and Moon, T.W. 2004b. Slowly digestible starch from debranching waxy sorghum starch: Preparation and properties. *Cereal Chem.* 81:404–408.

Shin, S.I., Kim, H.J., Ha, H.J., Lee S.H., and Moon, T.W. 2005. Effect of hydrothermal treatment on formation and structural characteristics of slowly digestible non-pasted granular sweet potato starch. *Starch/Stärke.* 57:421–430.

Shinde, S.V., Nelson, J.E., and Huber, K.C. 2003. Soft wheat starch pasting behavior in relation to A- and B-type granule content and composition. *Cereal Chem.* 80:91–98.

Shiroza, T., Furihata, K., Endo, T., Seto, H., and Otake, N. 1982. The structures of diethylaminoethylated glucose and oligosaccharides derived from cationic starch. *Agric. Biol. Chem.* 46:1425–1427.

Siau, C.L., Karin, A.A., Norziah, M.H., and Wan Rosli, W.D. 2004. Effects of cationization on DSC thermal profiles, pasting and emulsifying properties of sago starch. *J. Sci. Food Agric.* 84:1722–1730.

Singh, J. and Singh, N. 2003. Studies on the morphological and rheological properties of granular cold-water soluble corn and potato starches. *Food Hydrocoll.* 17:63–72.

Singh, N. and Kaur, L. 2004. Morphological, thermal, rheological and retrogradation properties of potato starch fractions varying in granule size. *J. Sci. Food Agric.* 84:1241–1252.

Singh, S., Raina, C.S., Bawa, A.S., and Saxena, D.C. 2005. Effect of heat-moisture treatment and acid-modification on rheological, textural, and differential scanning calorimetry characteristics of sweet potato starch. *J. Food Sci.* 70:E373–E378.

Singh, V., Ali, S.Z., and Divakar, S. 1993. ^{13}C CP/MAS NMR spectroscopy of native and acid modified starches. *Starch/Stärke.* 45:59–62.

Solarek, D.B. 1986. Chemistry, properties and uses of modified starches. *Cereal Foods World* 31:597–597.

Stanley, K.D., Richmond, P.A., Yackel, W.C., Harris, D.W., Eilers, T.A., Marion, E.A., and Stanley, E.D. 2006. Enzyme-resistant starch and a method for its production. U.S. Patent Appl. Publ. 2006078667; *Chem. Abstr.* 144: 368799.

Stapley, J.A. and BeMiller, J.N. 2003. Hydroxypropylated starch: Granule subpopulation reactivity. *Cereal Chem.* 80:550–552.

Steeneken, P.A.M. 1984. Reactivity of amylose and amylopectin in potato starch. *Starch/Stärke.* 36:13–18.

Steeneken, P.A.M. and Smith, E. 1991. Topochemical effects of the methylation of starch. *Carbohydrate Res.* 209:239–249.

Steeneken, P.A.M. and Woortman, A.-J.J. 1994. Substitution patterns in methylated starch as studied by enzymatic degradation. *Carbohydrate Res.* 258:207–221.

Stevenson, D.G., Biswas, A., and Inglett, G.E. 2005. Thermal and pasting properties of microwaved corn starch. *Starch/Stärke.* 57:347–353.

Stute, R. 1992. Hydrothermal modification of starches: The difference between annealing and heat-moisture treatment. *Starch/Stärke.* 44:205–214.

Stute, R. and Kern, H. 1994. Starch mixtures as pudding starches. U.S. Patent 5,324,532.

Subramanian, K. and Hanna, M.A. 1996. Glycol glucosides process synthesis by reactive extrusion with a static mixer as postextruder reactor. *Cereal Chem.* 73:179–184.

Sun, T., Laerke, H.N., Jorgensen, H., and Knudsen, K.E.B. 2006. The effect of extrusion cooking of different starch sources on the in vitro and in vivo digestibility in growing pigs. *Anim. Feed Sci. Technol.* 131:66–86.

Takahash, R. and Ojima, T. 1969. Pregelatinization of wheat starch in a drum drier. *Starch/Stärke.* 21:318–321.

Takaya, T., Sano, C., and Nishinari, K. 2000. Thermal studies on the gelatinisation and retrogradation of heat-moisture treated starch. *Carbohydrate Polym.* 41:97–100.

Tamaki, S., Hisamatsu, M., Teranishi, K., Adachi, T., and Yamada, T. 1998. Structural change of maize starch granules by ball-mill treatment. *Starch/Stärke.* 50:342–348.

Tamaki, S., Hisamatsu, M., Teranishi, K., and Yamada, T. 1997. Structural change of potato starch granules by ball-mill treatment. *Starch/Stärke.* 49:431–438.

Tara, A., Berzin, F., Tighzert, L., and Vergnes, B. 2004. Preparation of cationic wheat starch by twin-screw reactive extrusion. *J. Appl. Polym. Sci.* 93:201–208.

Tester, R.F. and Debon, S.J.J. 2000. Annealing of starch: A review. *Int. J. Biol. Macromol.* 27:1–12.

Tester, R.F. and Morrison, W.R. 1994. Properties of damaged starch granules. V. Composition and swelling of fractions in wheat starch in water at various temperatures. *J. Cereal Sci.* 20:175–181.

Tester, R.F., Debon, S.J.J., and Karkalas, J. 1998. Annealing of wheat starch. *J. Cereal Sci.* 28:259–272.

Tester, R.F., Debon, S.J.J., and Sommerville, M.D. 2000. Annealing of maize starch. *Carbohydrate Polym.* 42:287–299.

Thompson, D.B. 2007. Resistant starch. In: *Functional Food Carbohydrates*, C.G. Biliaderis and M.S. Izydorczyk (Eds.). Boca Raton, FL: CRC Press, pp. 73–95.

Tomasik, P., Wang, Y.-J., and Jane, J.-L. 1995. Facile route to anionic starches. Succinylation, maleination and phthalation of corn starch on extrusion. *Starch/Stärke.* 47:96–99.

Tomka, I. 1998. Thermoplastic processable starch or starch derivative polymer mixtures for extruded or injection-molded articles. PCT Int. Patent Appl. WO 9806755; *Chem. Abstr.* 128:181802 .

Tran, T., Piyachomkwan, K., and Sriroth, K. 2007. Gelatinization and thermal properties of modified cassava starches. *Starch/Stärke.* 59:46–55.

Trimnell, D., Swanson, C.L., Shogren, R.L., and Fanta, G.F. 1993. Extrusion processing of granular starch-g-poly(methyl acrylate): Effect of extrusion conditions on morphology and properties. _J. Appl. Polym. Sci._ 48:1665–1675.

Tsai, M.-L., Li, C.-F., and Lii, C.-Y. 1997. Effects of granular structure on the pasting behaviors of starches. _Cereal Chem._ 74:750–757.

Turneport, L.J., Salomonsson, B.A., and Theander, O. 1990. Chemical characterization of bromine oxidized potato starch. _Starch/Stärke._ 42:413–417.

Tuting, W., Wegemann, K., and Mischnick, P. 2004. Enzymatic degradation and electrospray tandem mass spectrometry as tools for determining the structure of cationic starches prepared by wet and dry methods. _Carbohydrate Res._ 339:637–648.

Vaca-Garcia, C., Borredon, M.E., and Gaset, A. 2000. Method for making a cellulose or starch fatty ester by esterification or transesterification. France Patent Appl. 99-2220; _Chem. Abstr._ 133:194826.

Van Beynum, G.M.A. and Joels, J.A. 1985. _Starch Conversion Technology._ New York: Marcel Dekker.

van der Burgt, Y.E.M., Bergsma J., Bleeker I.P., Mijland, P.J.H.C., Kamerling, J.P., and Vliegenthart, J.F.G. 2000. Structural studies on methylated starch granules. _Starch/Stärke._ 52:40–43.

van der Burgt, Y.E.M., Bergsma J., Bleeker I.P., Mijland, P.J.H.C., van der Kerk-van Hoof, A., Kamerling, J.P., and Vliegenthart, J.F.G. 1998. Distribution of methyl substituents over branched and linear regions in methylated starches. _Carbohydrate Res._ 312:201–208.

Van Hung, P. and Morita, N. 2005a. Effects of granule sizes on physicochemical properties of cross-linked and acetylated wheat starches. _Starch/Stärke._ 57:413–420.

Van Hung, P. and Morita, N. 2005b. Physicochemical properties of hydroxypropylated and cross-linked starches from A-type and B-type wheat starch granules. _Carbohydrate Polym._ 59:239–246.

Van Hung, P., Maeda T., and Morita N. 2007. Study on physicochemical characteristics of waxy and high-amylose wheat starches in comparison with normal wheat starch. _Starch/Stärke._ 59:125–131.

van Warners, A., Stamhuis, E.J., and Beenakers, A.A.C.M. 1994. Kinetics of the hydroxyethylation of starch in alkaline salt-containing aqueous slurries. _Ind. Eng. Chem. Res._ 33:981–992.

Vasanthan, T., Sosulski, F.W., and Hoover, R. 1995. The reactivity of native and autoclaved starches from different origins towards acetylation and cationization. _Starch/Stärke._ 47:135–143.

Vergnes, B. and Berzin, F. 2004. Modelling of flow and chemistry in twin screw extruders. _Plast. Rubber Compos._ 33:409–415.

Vermeire, A., Kiekens, F., Corveleyn, S., and Remon, J.P. 1999. Evaluation of the emulsifying properties of some cationic starches. _Drug Dev. Ind. Pharm._ 25:367–371.

Vermeylen, R., Goderis, B., and Delcour, J.A. 2006. An X-ray study of hydrothermally treated potato starch. _Carbohydrate Polym._ 64:364–375.

Vihervaara, T., Bruun, H.H., Backman, R., and Paakkanen, M. 1990. The effect of different methods of cationisation on the starch granule and its gelatinisation product. _Starch/Stärke._ 42:64–68.

Villwock, V.K. and BeMiller, J.N. 2005. Effects of salts on the reaction of normal corn starch with propylene oxide. _Starch/Stärke._ 57:281–290.

Waduge, R.N., Hoover, R., Vasanthan, T., Gao, J., and Li, J. 2006. Effect of annealing on the structure and physicochemical properties of barley starches of varying amylose content. *Food Res. Int.* 39:59–77.

Waliszewski, K.N., Aparicio, M.A., Bello, L.A., and Monroy, J.A. 2003. Changes of banana starch by chemical and physical modification. *Carbohydrate Polym.* 52:237–242.

Wang, J., Jin, Z., and Yuan, X. 2006. Preparation of resistant starch from starch-guar gum extrudates and their properties. *Food Chem.* 101:20–25.

Wang, L., Shogren, R.L., and Willett, J.L. 1997a. Preparation of starch succinates by reactive extrusion. *Starch/Stärke.* 49:116–120.

Wang, L., Shogren, R.L., and Willett, J.L. 1997b. Preparation of starch succinates by reactive extrusion. *Polym. Mater. Sci. Eng.* 76:416–417.

Wang, W.J., Powell, A.D., and Oaks, C.G. 1997. Effect of annealing on the hydrolysis of sago starch granules. *Carbohydrate Polym.* 33:195–202.

Wang, Y.-J. and Wang, L. 2000. Effects of modification sequence on structures and properties of hydroxypropylated and crosslinked waxy maize starch. *Starch/ Stärke.* 52:406–412.

Wanous, M.P. 2004. Texturizing and stabilizing, by gum. *Prepared Foods.* 173:108–118.

Wasserman, L.A., Signorelli, M., Schiraldi, A., Yuryev, V., Boggini, G., Bertini, S., and Fessas, D. 2007. Preparation of wheat resistant starch: Treatment of gels and DSC characterization. *J. Therm. Anal. Calorim.* 87:153–157.

Watson, S.A. and Eckhoff, S.R. 2008. Corn and sorghum starches: production. In: *Starch: Chemistry and Technology*, 3rd ed., J.N. BeMiller and R.L. Whistler (Eds.). Orlando, FL: Elsevier. Chapter 9.

Whistler, R.L. 1984. History and future expectation of starch use. In: *Starch Chemistry and Technology*, R.L. Whistler, J.N. BeMiller, and E.F. Paschall (Eds.). San Diego: Academic Press, pp. 1–9.

Whistler, R.L. and Spencer, W.W. 1960. Distribution of substituents in corn starch granules with low degrees of substitution. *Arch. Biochem. Biophys.* 87:137–139.

Whistler, R.L., Madson, M.A., Zhao, J., and Daniel, J.R. 1998. Surface derivatization of corn starch granules. *Cereal Chem.* 75:72–74.

White, P.J. and Tziotis, A. 2004. New corn starches. In: *Starch in Food*, A.C. Eliasson (Ed.). Cambridge, UK: Woodhead, pp. 295–320.

Wilke, O. and Mischnick, P. 1997. Determination of the substitution pattern of cationic starch esters. *Starch/Stärke.* 49:453–458.

Willett, J.L. 2008. Starch in polymer compositions. In: *Starch: Chemistry and Technology*, 3rd ed., J.N. BeMiller and R.L. Whistler (Eds.). Orlando, FL: Elsevier, Chapter 19.

Willett, J.L. and Finkenstadt, V.L. 2003. Preparation of starch-graft-polyacrylamide copolymers by reactive extrusion. *Polym. Eng. Sci.* 43:1666–1674.

Willett, J.L. and Finkenstadt, V.L. 2004. Reactive extrusion of starch-polyacrylamide graft copolymers. *Ann. Tech. Conf., Soc. Plastics Eng. 62nd*, Vol. 1, pp. 336–339.

Willett, J.L. and Finkenstadt, V.L. 2006a. Initiator effects in reactive extrusion of starch-polyacrylamide graft copolymers. *J. Appl. Polym. Sci.* 99:52–58.

Willett, J.L. and Finkenstadt, V.L. 2006b. Reactive extrusion of starch-polyacrylamide graft copolymers using various starches. *J. Polym. Environ.* 14:125–129.

Wing, R.E. and Willett, J.L. 1997. Water soluble oxidized starches by peroxide reactive extrusion. *Ind. Crop. Prod.* 7:45–52.

Wongsagonsup, R., Shobsngob, S., Oonkhanond, B., and Varavinit, S. 2005. Zeta potential and pasting properties of phosphorylated or crosslinked rice starches. *Starch/Stärke.* 57:32–37.

Woo, K.S. and Seib, P.A. 1997. Crosslinking of wheat starch and hydroxypropylated wheat starch in alkaline slurry with sodium trimetaphosphate. *Carbohydrate Polym.* 33:263–271.

Woo, K.S. and Seib, P.A. 2002. Cross-linked resistant starch: Preparation and properties. *Cereal Chem.* 79:819–825.

Woo, K.S., Bassi, S., and Maningat, C.C. 2005. Oxidized reversibly swelling granular starch with capability of rapid hydration and resistance of digestion. U.S. Patent Appl. Publ. 2005256306; *Chem. Abstr.* 143:479637.

Wu, Y. and Seib, P.A. 1990. Acetylated and hydroxyprpoylated distarch phosphates from waxy barley: paste properties and freeze-thaw stability. *Cereal Chem.* 67:202–208.

Wurzburg, O.B., Ed. 1986a. *Modified Starches: Properties and Uses.* Boca Raton, FL: CRC Press.

Wurzburg, O.B. 1986b. Introduction. In: *Modified Starches: Properties and Uses*, O.B. Wurzburg (Ed.). Boca Raton, FL: CRC Press, pp. 3–16.

Wurzburg, O.B. 1986c. Converted starches. In: *Modified Starches: Properties and Uses*, O.B. Wurzburg (Ed.). Boca Raton, FL: CRC Press, pp. 17–40.

Wurzburg, O.B. and Szymanski, C.D. 1970. Modified starches in the food industry. *J. Agric. Food Chem.* 18:997–1001.

Xie, F., Yu, L., Liu, H., and Chen, L. 2006. Starch modification using reactive extrusion. *Starch/Stärke* 58:131–139.

Xie, S.X., Liu, Q,. and Cui, S.W. 2005. Starch modification and applications. In: *Food Carbohydrates: Chemistry, Physical Properties, and Applications*, S.W. Cui (Ed.). Boca Raton, FL: Taylor & Francis Group, pp. 357–418.

Yamamori, M., Kato, M., Yui, M., and Kawasaki, M. 2006. Resistant starch and starch pasting properties of a starch synthase IIa-deficient wheat with apparent high amylose. *Aust. J. Agric. Res.* 57:531–535.

Yang, C.Z., Shu, X.L., Zhang, L.L., Wang, W.Y., Zhao, H.J., Ma, C.X., and Wu, D.X. 2006. Starch properties of mutant rice high in resistant starch. *J. Agric. Food Chem.* 54:523–528.

Yao, Y., Zhang, J., and Ding, X. 2003. Retrogradation of starch mixtures containing rice starch. *J. Food Sci.* 68:260–265.

Yeh, A. and Yeh, S. 1993. Some characteristics of hydroxypropylated and cross-linked rice starch. *Cereal Chem.* 70:596–601.

Yeh, A. and Yeh, S. 1996. Kinetics of phase transition of native, cross-linked, and hydroxypropylated rice starches. *Starch/Stärke.* 48:17–21.

Yeo, L.L., Woo, K.S., and Seib, P.A. 2002. Cross-linked resistant starch: Preparation and properties. *Abstr. Papers, Amer. Chem. Soc. 223rd Annual Meeting*, CARB-047.

Yoneya, T., Ishibashi, K., Hironaka, K., and Yamamoto, K. 2003. Influence of cross-linked potato starch treated with $POCl_3$ on DSC, rheological properties and granule size. *Carbohydrate Polym.* 53:447–457.

Yoo, Y.D., Kim, Y.W., and Cho, W.Y. 1994. Biodegradable polyethylene compositions chemically bonded with starch and a simplified process for their manufacture. UK Patent Appl. 2,272,700; *Chem. Abstr.* 122:82927.

Yook, C., Pek, U.-H., and Park, K.-H. 1993. Gelatinization and retrogradation characteristics of hydroxypropylated and cross-linked rices. *J. Food Sci.* 58:405–407.

Yook, C., Sosulski, F., and Bhirud, P.R. 1994. Effects of cationization on functional properties of pea and corn starches. *Starch/Stärke.* 46:393–399.

Yoon, K.J., Carr, M.E., and Bagley, E.B. 1992. Reactive extrusion vs. batch preparation of starch-g-polyacrylonitrile. *J. Appl. Polym. Sci.* 45:1093–1100.

Yoshimura, M., Takaya, T., and Nishinari, K. 1996. Effects of konjac-glucomannan on the gelatinization and retrogradation of corn starch as determined by rheology and differential scanning calorimetry. *J. Agric. Food Chem.* 44:2970–2976.

Zhang, G., Venkatachalam, M., and Hamaker, B.R. 2006. Structural basis for the slow digesting property of native cereal starches. *Biomacromolecules.* 7:3259–3266.

Zhao, K., Zhang, S., Fang, G., and Yang, C. 2005. Preparation techniques for resistant starch from mung bean. *Shipin Gongye Keji* 26:121–123; *Chem. Abstr.* 144:330135.

Zheng, G.H., Han, H.L., and Bhatty, R.S. 1999. Functional properties of cross-linked and hydroxypropylated waxy hull-less barley starches. *Cereal Chem.* 76:182–188.

Zhu, Q. and Bertoft, E. 1997. Enzymic analysis of the structure of oxidized potato starches. *Int. J. Biol. Macromol.* 21:131–135.

Zhu, Q., Sjoholm, R., Nurmi, K., and Bertoft, E. 1998. Structural characterization of oxidized potato starch. *Carbohydrate Res.* 309:213–218.

Ziegler, G.R., Thompson, D.B., and Casasnovas, J. 1993. Dynamic measurement of starch granule swelling during gelatinization. *Cereal Chem.* 70:247–251.

chapter nine

Starch-based nanocomposites

Eliton S. Medeiros
EMBRAPA Instrumentação Agropecuária

Alain Dufresne
Ecole Internationale du Papier, de la Communication Imprimée des Biomatériaux
Institut Polytechnique de Grenoble

William J. Orts
Bioproduct Chemistry and Engineering Unit, U.S. Department of Agriculture

Contents

9.1 Introduction

In recent years there has been a growing effort to develop new biodegradable materials from environmentally friendly and renewable resources whose feasibility in suiting their properties to a particular application can result in easily tailored composite materials. The utilization of natural polymers such as starch, lignin, cellulose, and proteins in the plastics industry is considered a viable approach to reduce surplus agricultural products and to develop biodegradable materials. Natural polymers derived from renewable resources have these environmental advantages when compared with petroleum-derived ones. Starch is one of the most studied and promising raw materials for the production of biodegradable materials because it is a widely abundant, relatively low-cost, and renewable natural polysaccharide obtained from a great variety of crops (Roper and Koch, 1990; Scott and Gilead, 1995; Chandra and Rustgi, 1998).

Composites have been used since ancient times. Archaeological evidence of the earliest civilizations reveals mud and straw "composites," bricks that are applied not unlike our modern composites today (Durant, 1963; Scherman, 2001). Natural composite materials range from wood to blood vessels (Shalaby and Latour, 1997), all of which show the advantage of reinforced structures. Modern nanocomposites advanced rapidly, in part when researchers from the automobile industry in the early 1990s created composites using nanostructured clay reinforcements (i.e., clay–polymer nanocomposites) that provided significant improvements in dimensional stability, stiffness, and heat distortion temperature relative to pure polymer. Nanocomposites utilize low levels of fillers, thereby minimizing the change in rheological properties during processing compared to traditional composites (Orts et al., 2005). Studies of characterization techniques and synthesis of new nanomaterials have opened up many research avenues, leading to a vast number of polymer matrices and fillers that can be used to produce nanocomposites.

One noteworthy example was introduced in the mid-1990s by Favier, Chanzy, and Cavaille (1995), who reported organic nanocomposites based on polymers reinforced with cellulose nanofibrils. In this research, crystalline cellulose needles (at 3–6% cellulose loadings) were dispersed in a copolymer acrylate latex film, increasing the dynamic storage modulus more than threefold. Other researchers (Favier, Chanzy, and Cavaille, 1995; Greene and Imam, 1998) have continued work on natural cellulose nanoscale microfibrils highlighting advantages of these natural composites. Natural fibers such as wood are essentially cellulose-reinforced fibrils bound together by lignin and hemicellulose matrix, whereas bones, the basic structures of all vertebrates, are collagen fibrils embedded in an inorganic apatite matrix (Chawla, 1998; Callister, 2006).

There are three significant distinctions between clay-reinforced and cellulose nanofibril-reinforced nanocomposites. First, clays are often flat platelets, whereas typical cellulosic nanofibrils are long crystalline "needles" usually ranging in size from 10–20 nm in width, with an average ratio of 20–100. Cellulose surfaces provide the potential for significant surface modification using well-established carbohydrate chemistry, some of which makes them ideally suited to interact with starch, a similar carbohydrate-based polymer. Sources of cellulose microfibrils, including wood, straw, bagasse, bacteria, and sea organisms, are widely diverse, providing a wide range of potential nanoparticle properties. The most-studied matrix for these cellulose-based nanocrystal composites has been starch, in part because starch is processed under aqueous conditions.

This chapter discusses these nanocomposites, the starch matrix and reinforcements—organic and inorganic materials—that have at least one of their dimensions in the range of 1–100 nm, the generally accepted definition of nanomaterials (Fishbine, 2002).

9.2 Nanocomposites

Nanotechnology principles and characterization techniques of materials have been revisited in search of new properties and applications. Nanocomposites not only exhibit enhancement in mechanical properties (tensile strength, elastic modulus, etc.) and physical properties (better barrier properties, reduced flammability, etc.), but they also show improved optical transparency when compared to traditional composites (Fischer, 2003). A low amount of added filler (less than 5% w/w) overcomes rheological implications, limiting the uses of many traditional composites (Sanadi et al., 1995; Bledzki and Gassan, 1999; Sanadi et al., 2001; Medeiros et al., 2005).

The nature and effectiveness of interactions at the interface phases are main variables acting on mechanical and physical properties of nanocomposites. Both interfacial region and particle dispersion play an important role in nanocomposite properties, mainly when fillers and reinforcements have a large contact surface area. Table 9.1 compares surface areas for several micro- and nanomaterials. A strong interface assures that any applied load can be easily transferred from the matrix to reinforcements, avoiding premature failure. A poor interface is also a drawback in situations other than external mechanical loading; for example, because of differential thermal expansions of reinforcement and matrix, premature failure can occur at a weak interface when the composite is subjected to thermal stress (Chawla, 1998; Callister, 2006).

Table 9.1 Surface Areas of Main
Materials Used as Reinforcements and
Fillers in Composites and
Nanocomposites

Fibers/fillers	Surface area (m^2/g)
Natural fibers	0.5
Glass fibers	1
Paper fibers	4
Cellulose nanofibrils	250
Graphite	25–300
Fumed silica	100–400
Exfoliated clays	500
Carbon nanotubes	300–1000

Sources: Dresselhaus, Dresselhaus and Avouris, 2001; Balaposzhinimaev, 2003; Bismarck, Mishra, and Lampke, 2005; Yang, Kim, and Lee, 2005; Simonsen, 2008.

9.2.1 Starch as matrix in nanocomposites

Several thermoplastic and elastomeric polymers have been used as matrices for nanocomposites (Giannellis, 1996; LeBaron, Wang, and Pinnavaia, 1999; Ajayan, Schadler, and Braun 2003; Gao, 2004). As in conventional composites, the matrix role is to support and protect the filler materials, transmitting and distributing the applied load between reinforcing agents (Giannellis, 1996; Chawla, 1998; LeBaron, Wang, and Pinnavaia, 1999; Callister, 2006). Because the range of particle sizes is on the nanoscale, the large surface area of the filler material makes the particle–matrix interfacial region even more important. Matrices support the filler surface by chemical reactions or adsorption, providing the strength of interfacial adhesion. In certain cases, the interface may be composed of an additional constituent such as bonding agents or interlayers between the two components of the composite.

The choice of a matrix depends on several factors such as application, compatibility of the components, processing, and costs. However, with the recent advances in development of biodegradable polymers, the search for biodegradable matrices such as starch, poly(lactic acid), poly(ε-caprolatones), polyhydroxybutyrate, poly(butylene succinate), and polyhydroxyalkanoates has received attention (Shalaby and Latour, 1997; Avérous, Fringant, and Moroa, 2001; Fischer, 2003; Ray and Okamoto, 2003; Okamoto, 2005). For many years starch has been considered a natural polymer with high potential for applications in biodegradable plastics because of its renewability, biodegradability, low cost, availability, and mechanical properties (Avérous, Fringant, and Moroa, 2001; Corradini et al., 2006). Composites based on the opposite polysaccharide nanocrystals are often obtained by a casting/evaporation method. Increasing the affinity between the polysaccharide filler and the host matrix results in less favorable mechanical performance. This behavior is ascribed to the uniqueness of the reinforcing phenomenon of polysaccharide nanocrystals resulting from the formation of a percolating network due to hydrogen bonding forces (Dufresne, 2007). The microstructure of the matrix and the resulting competition between matrix–filler and filler–filler interactions also affect the mechanical behavior of the polysaccharide nanocrystal-reinforced nanocomposites. However, the major drawbacks associated with the use of starch-based materials in packaging applications are due to their hydrophilic nature, poor processability and brittleness, and postprocessing crystallization, resulting in poor mechanical properties (Whistler and Paschall, 1965, 1967; Aichholzer and Fritz, 1998; Rhim and Ng, 2007).

Significant research on starch as a matrix for nanocomposites has been done, with the matrix mechanical properties changing dramatically according to the starch botanical origin, plasticizer addition, and filler choice. Strong interactions between cellulose crystallites from cottonseed lintners and glycerol plasticized starch matrix play a key role in reinforcing properties (Lu, Weng, and Cao, 2005). In plasticized starch-based nanocomposites reinforced with cellulose whiskers (Anglès and Dufresne, 2001), loss of mechanical properties was observed with adding reinforcement, an effect similar to that observed for surface chemically modified chitin whiskers reinforced natural rubber (Gopalan Nair et al., 2003).

A transcrystallization resulting in a broader interphase and in a broadening of the main relaxation process of the matrix on plasticized starch reinforced with cellulose whiskers (Anglès and Dufresne, 2001) suggests that the strong loss of performance was due to the importance of the accumulation of the main plasticizer (glycerol) in the interfacial zone resulting in restricting filler–filler interactions (Anglès and Dufresne, 2000).

9.2.2   Nanocomposite classification

There are several methods to classify composites and nanocomposites, which are based on the nature and origin of the matrix, shape and size of the reinforcements, and uses of the obtained material (Chawla, 1998; Callister, 2006). According to filler particle shape, nanocomposites can be classified into particulate, elongated particle, and layered nanocomposites.

Particulate nanocomposites generally use isodimensional particles as reinforcement. The reinforcing effect is moderate and the general purpose of using these reinforcements is to enhance composite resistance to flammability or to decrease permeability or costs.

Elongated particle nanocomposites are composed of reinforcement particles, such as cellulose nanofibrils and carbon nanotubes. These nanocomposites present better mechanical properties due to the particles' higher aspect ratio.

Layered particle-reinforced polymer nanocomposites, also known as layered polymer nanocomposites (LPN), are classified into three subcategories according to how the particles are dispersed in the matrix. *Intercalated nanocomposites* are formed by intercalated polymer chains on sheets of layered nanoparticles, *exfoliated nanocomposites* are obtained by separation of individual layers, and *flocculated* or *phase-separated nanocomposites* present no separation between the layers due to particle–particle interactions. This last class of composites is often called microcomposites because the individual laminae do not separate, thus acting as microparticles dispersed in the polymeric matrix. Their mechanical and physical properties are poorer than those of exfoliated and intercalated nanocomposites (Ajayan, Schadler, and Braun, 2003; Ray and Okamoto, 2003; Orts et al., 2005). Figure 9.1 shows the structure of layered nanocomposites.

9.2.3   Fillers

Fillers are generally defined as organic or inorganic particulate material (spheres, plates, flakes, sheets, fibers, fibrils, and whiskers) used in composites to reduce costs. They are generally the least expensive of the major ingredients. In addition, they often modify mechanical or physical properties, such as processability, flammability, conductivity, shrinkage, weight, and visual appearance. Depending on the application, fillers are given specific terms, such as smoke suppressors, nucleating agents, and UV stabilizers among others (Shalaby and Latour, 1997; Chawla, 1998; Medeiros et al., 2001, 2002; Ray and Okamoto, 2003). Nanocomposite fillers can provide multiple advantages even if added in small amounts.

Fillers can also be classified into three major categories, according to their particle size and shape. *Isodimensional* or *zero-dimensional particles* have the same size in all directions (examples include spherical silica, nanoclusters, metallic nanoparticles, carbon black, and fullerenes) and their aspect ratio is usually close to unity (L/D ~1). The second category of particles, also known as *cellulose nanofibrils, microcrystalline cellulose, cellulose nanocrystals,* or *nanowhiskers,* consists of elongated particles or fibrils with diameter between 1 and 100 and length of several hundreds of nanometers. The third category is comprised of particles shaped as a lamina or sheet, with width and thickness ranging from a few angstroms to several hundreds of nanometers and length of thousands of nanometers, called layered particles. These particles are composed of stacks of laminae; usually found in nature, they can also be produced synthetically (Grim, 1962, 1988; Ajayan, Schadler, and Braun, 2003).

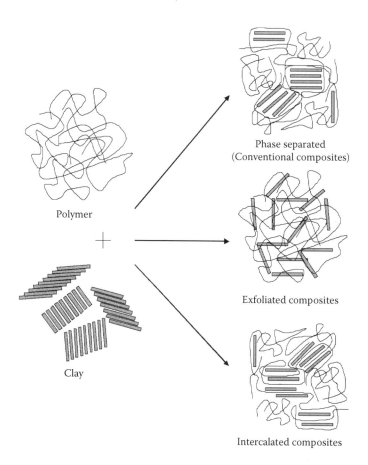

Figure 9.1 Different types of nanocomposites from polymers and clays. Conventional composites are also known as microcomposites.

Due to their high aspect ratios which confer a higher enhancement of the mechanical properties of the nanocomposites, nanomaterials of the last two categories are the most popular. The most common fillers used to obtain nanocomposites are carbon nanotubes, carbon black, fumed silica and clays (inorganic fillers), and cellulose-based fibrils (organic fillers). Carbon nanotubes are molecular-scale tubes formed by rolled-up graphene sheets with outstanding mechanical and electrical properties, with the stiffest and strongest fibers (Balaposzhinimaev, 2003; Popov, 2004).

9.2.3.1 Clays

Clays are composed of colloidal fragments of primary silicates, also known as clay minerals. The structure of a clay mineral is made up of tetrahedral and octahedral sheets of small cations, such as aluminum or

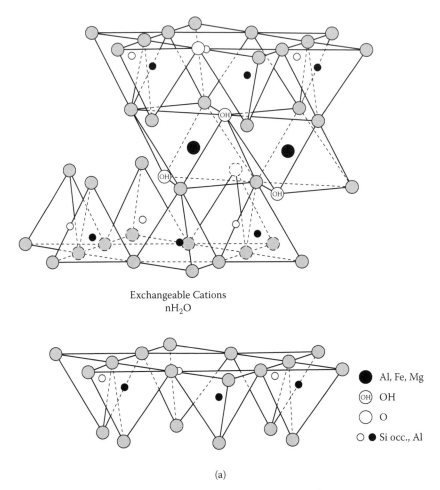

Exchangeable Cations
nH$_2$O

● Al, Fe, Mg
(OH) OH
○ O
○ ● Si occ., Al

(a)

Figure 9.2 Structures of some of the clay minerals used as fillers in nanocomposites: (a) montmorillonite, (b) muscovite, (c) kaolinite. Note the layered structures composed of tetrahedrons and octahedrons with atoms positioned at both interstitial and corner positions (Grim, 1962).

magnesium coordinated by oxygen atoms. Clays are classified based on the way that tetrahedral (n) and octahedral (m) sheets are packaged into layers as n:m clays. For example, 1:1 clays are formed by one tetrahedral and one octahedral group in each layer. Figure 9.2 shows the tetrahedral–octahedral structures of some of the clay minerals used in nanocomposites. One important feature of clays is that the space between layers contains hydrated cations such as Na$^+$ or K$^+$ that can undergo exchange reaction with organic as well as inorganic cations (Grim, 1962, 1988). Addition of clay (as sodium montmorillonite) to a matrix of plasticized starch and glycerol results in improved properties when compared with the starch

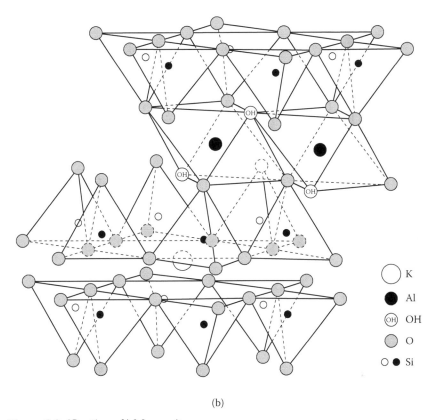

(b)

Figure 9.2 (Continued.) Muscovite.

matrix (Chen and Evans, 2005; Park et al., 2003; Qiao, Jiang, and Sun, 2005; Chiou et al., 2007; Magalhães and Andrade, 2009).

Cation exchange reactions in clays are very important to impart functionality and compatibility with polymers as shown schematically in Figure 9.3 (Fischer, 2003; Ray and Okamoto, 2003).

9.2.3.2 Cellulose

Cellulose is the most abundant organic polymer in nature with an estimated annual production of 1.5×10^{12} tons (Crawford, 1981). Chemically, cellulose is a linear polymer made up of several hundred to over 10,000 repeating units of β(1–4) linked D-glucose (Figure 9.4) (Klemm et al., 2005). Physically, cellulose is found in the form of microfibrils constituted of amorphous and crystalline domains in combination with other substances such as lignin, hemicelluloses, and proteins making up the basic structural unit of plant cell walls as shown in Figure 9.5. Cotton, wood, leaves, and stalks are examples of raw materials from plants that can be

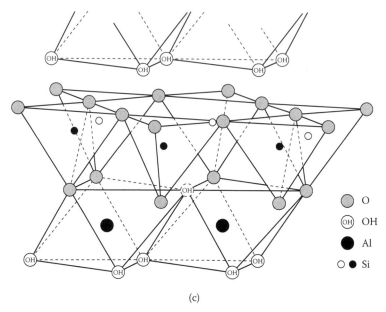

(c)

Figure 9.2 (Continued.) Kaolinite.

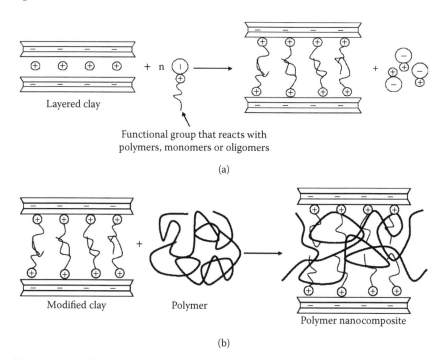

Figure 9.3 (a) Cation exchange reaction to produce organically modified clays; (b) compounding with a polymer matrix to make nanocomposites.

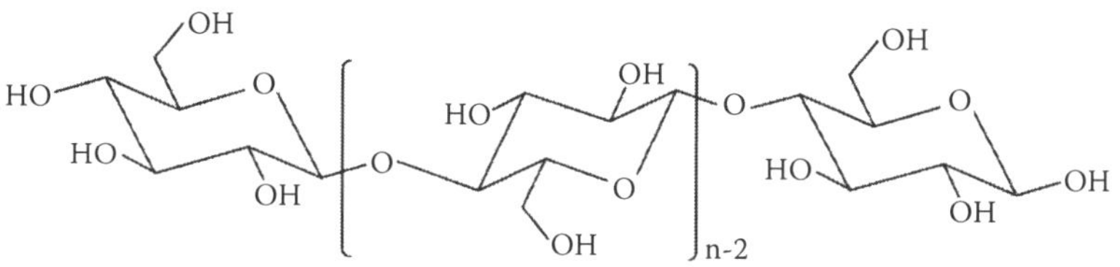

Figure 9.4 Chemical structure of cellulose.

sources of cellulose, but it can also be produced by bacteria, algae, and fungi (Klemm et al., 2005).

Crystalline cellulose can be isolated by treatment of cotton, sisal, and wood with strong acids such as hydrochloric and sulfuric to remove the amorphous parts, yielding fibrils with diameters in the range of 5 to 20 nm and aspect ratio of about 1 to 100 times (Figure 9.6) (Klemm et al., 2005).

As a result of the acid treatment used to obtain cellulose, nanofibrils often have electronegative surface charges on their surface. Particle–particle or particle–solvent–particle interactions form three-dimensional networks giving rise to a rheological behavior reminiscent of clay nanocomposite behavior (Crawford, 1981; Lima and Borsali, 2002; Klemm et al., 2005). However, acid hydrolysis also decreases thermal stability of nanofibrils, which determines the range of starch processing temperatures (Tadmor and Gogos, 1979; Morton-Jones, 1989; Glasser et al., 1999). The thermal stability can be recovered to a certain extent by modifying the extraction process, such as by the use of low acid concentrations, low acid-to-cellulose ratio, and short reaction times (Roman and Winter, 2004), or post-extraction processes such as restabilization by partly neutralizing the sulfuric acid groups with strong bases such as sodium hydroxide (Wang, Ding, and Cheng, 2007).

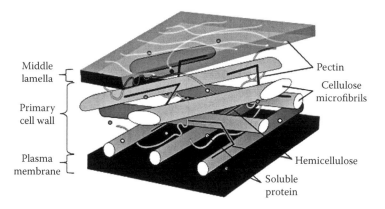

Figure 9.5 Physical structure of cellulose showing the location and arrangement of cellulose microfibrils in plant cell walls (Wikipedia, 2008).

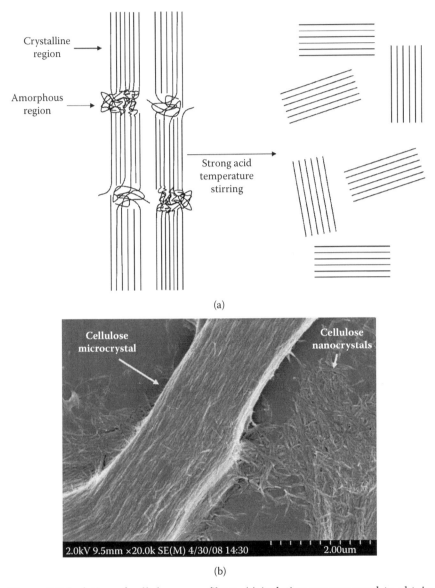

(a)

(b)

Figure 9.6 Isolation of cellulose nanofibers: (a) isolation process used to obtain cellulose microfibrils; (b) disintegration of microfibrils by acid attack forming nanofibrils (Rosa et al., 2008).

Although cellulose nanofibrils have been studied and used industrially since the 1960s (Battista, 1975), recently these nanostructures have gained more scientific attention due to the application of nanomaterials from renewable resources. These nanofibrils possess several advantages in engineering material, such as low cost and density, renewability, bio-degradability, nontoxicity, formation of stable aqueous suspension, and remarkable mechanical properties—they are capable of improving the mechanical performance of polymers at very low fiber concentrations (Orts et al., 2005; Marcovich et al., 2006). Consequently, crystalline cellulose has a modulus rivaling steel, therefore representing an appropriate material for nanocomposites. With the above-mentioned benefits, cellulose-based nanocomposites are organic alternatives to clay-based particulates, with the added advantage of utilizing well-understood cellulose chemistry. Moreover, due to cellulose polyfunctionality, chemical reactions can be carried out on the surfaces of the nanofibrils, enhancing their interaction with a vast number of polymeric matrices (Klemm et al., 2005; Orts et al., 2007).

9.2.4   Preparation, processing, and characterization of nanocomposites

9.2.4.1   Preparation

One of the most important steps in the preparation of polymer nanocomposites is the mixing process. In order for the properties of a compatible filler to be fully exploited, it has to be well distributed and well dispersed, as shown in Figure 9.7. This is important in nanocomposites because filler content is low and because high surface area nanoparticles have a natural tendency to agglomerate rather than disperse in the matrix (Tadmor and Gogos, 1979; Morton-Jones, 1989; Fischer, 2003; Ray and Okamoto, 2003,).

Nanocomposites are basically prepared by solution and melt dispersion (Ray and Okamoto, 2003), sol–gel synthesis (Ajayan, Schadler, and Braun, 2003), in situ polymerization, and self-assembly (Decher and Schlenoff, 2002; Medeiros, Paterno, and Mattoso, 2006).

Solution dispersion is a process whereby the filler is dispersed in an organic solvent, followed by swelling. Meanwhile, the polymer is dissolved in the same solvent for further mixing. Both the mixing intensity and duration are critical to ensure a good distribution and dispersion of filler particles. The major limitations of these processes are that the polymer needs to be soluble in the same solvent used to disperse and swell the filler, and the solvent removal is an additional step that limits large-scale industrial processing.

Melt dispersion is probably the most practical process of nanocomposite preparation, inasmuch as traditional methods of composite mixing such as twin- and single-screw extruders, rheometers, and mixers can be used, providing a good mixing efficiency (Tadmor and Gogos, 1979; Morton-Jones, 1989). Melt dispersion consists of mixing the filler with a polymer melt to assure a good dispersion by high shear and temperature.

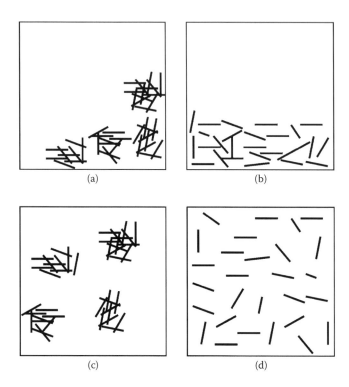

Figure 9.7 Degrees of dispersion and distribution of fibrillar reinforcements in a polymeric matrix: (a) poorly distributed and poorly dispersed; (b) poorly distributed and well dispersed; (c) poorly dispersed and well distributed; and (d) well dispersed and well distributed.

Shear prevents the filler particles from aggregating and higher temperatures promote polymer chain diffusion between separated particles. The thermodynamical driving forces change according to the polarity of the fillers, chemical compatibility with the polymeric matrix, processing temperature, and polymer molecular weight (Giannellis, 1996; LeBaron, Wang, and Pinnavaia, 1999).

In situ polymerization comprises filler dispersion in a liquid monomer or a monomer dissolved in a suitable solvent, allowing monomers to diffuse between the layers. Upon further addition of initiator or exposure of appropriate source of light or heat, the polymerization takes place *in situ* producing the nanocomposite.

9.2.4.2　Processing

Although they can be carried out separately, mixing and processing generally take place at the same time to reduce costs and time. The quality of final products depends strongly on this step, because it can impart defects, inefficient dispersion, and undesired orientation of the reinforcement, resulting in premature failure and other environmental stress cracking processes (Tadmor and Gogos, 1979; Morton-Jones, 1989; Chawla, 1998; Shackelford, 1999; Callister, 2006).

The nature of the matrix (thermoplastic, thermoset, or elastomeric), the nature of the fillers, and their thermal resistance should be the main factors considered when choosing a process. The most common methods are extrusion, injection molding, casting, and compression molding; reactive extrusion is a potential option (Tadmor and Gogos, 1979; Fischer, 2003; Ray and Okamoto, 2003). More information on polymer processing operations can be found in the specialized literature (Tadmor and Gogos, 1979; Morton-Jones, 1989; Rosato, 1997; Begishev and Malkin, 1999; Brydson, 1999; Crawford, 1999).

9.2.4.3 Characterization

Actually, there are several techniques of nanocomposite characterization (Brundle, Evans, and Wilson, 1992). The choice of the most appropriate technique to characterize nanocomposites depends strongly on the nature of the materials and their applications.

Particle size, distribution and dispersion, degree of interaction, and the filler–matrix interface are some parameters evaluated by direct or indirect techniques. Direct techniques include microscopic observations (scanning electron microscopy, transmission electron microscopy, scanning probe microscopy, and x-ray diffraction). Indirect techniques provide data about nanocomposite characterization, correlating with mechanical, physical, rheological, thermal, and thermomechanical properties; they include rheometric and calorimetric analysis.

Wide angle x-ray diffraction (WAXD) is used most broadly in nanocomposite characterization. This technique, which is based on changes in the Bragg peaks, provides details about the crystalline orientation and interplanar spaces. WAXD is especially useful for polymer-layered nanocomposites to evaluate the exfoliation degree on clays and the interplanar expansion after chemical surface treatments. WAXD has been used in cellulose nanocomposites to study changes in crystalline degree and structure of cellulose nanofibrils and polymeric matrices after fiber extraction and composite formation or processing (Edgar and Gray, 2003; Nishino, Matsuda, and Hirao 2004; Huang et al., 2005). Figure 9.8 illustrates application of WAXD on samples of starch/montmorillonite composites from 1.01 to 1.25 and 2.08 nm.

Transmission electron microscopy (TEM), field emission scanning electron microscopy (FESEM), and scanning electron microscopy (SEM) are the most commonly used methods. Details about microscopy in characterization of starch and starch-based materials are provided in Chapter 2. Figure 9.9 compares ranges of sizes of structures that can be analyzed

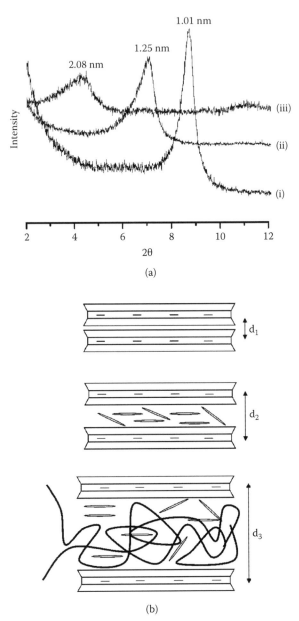

(a)

(b)

Figure 9.8 (a) WAXD patterns of (i) untreated montmorillonite (MMT), (ii) activated montmorillonite (EMMT), and (iii) thermoplastic starch/activated montmorillonite nanocomposites (TPS/EMMT) (Huang et al., 2005); and (b) changes in the interplanar spacing of montmorillonite (d_1) with activation (d_2), and intercalation (d_3).

by microscopic techniques. Transmission electron microscopy is by far the most used imaging technique in nanocomposite characterization. It reveals spatial distribution of the various phases and defects through direct visualization of ultra-thin microtomed layers. In polymer clay nanocomposites, TEM allows one to observe the degree of dispersion, particle size distribution of cellulose nanofibrils (Figure 9.10) (Rodriguez, Thielemans, and Dufresne, 2006), and clay exfoliation, intercalation, or flocculation (Figure 9.11). However, special care must be taken during sample preparation, because a sample must be thin enough to be transparent to the electron beam (Ray and Okamoto, 2003).

Scanning electron microscopy is particularly useful for studying the fracture surfaces of composites, identifying the various modes of failure, the interfacial region, and therefore identifying filler–matrix interaction. Figure 9.12 shows SEM micrographs of the structures of the nanocomposites of thermoplastic starch (TPS) and cellulose nanofibers. It can be observed that the nanofibers (white spots) are well dispersed and covered by the matrix with no fiber pull-out or debonding, confirming the improved adhesion between the nanofibers and the polymeric matrix (Chawla, 1998; Medeiros et al., 2005). Scanning probe microscopy (SPM) is also used on surface characterizations when roughness modification

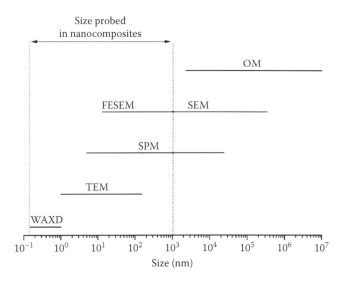

Figure 9.9 Comparison of the size ranges of structures that can be analyzed by TEM, FESEM, SEM, and OM. Field emission scanning electron microscopy (FESEM) is an improved SEM in terms of spatial resolution and minimized sample charging and damage.

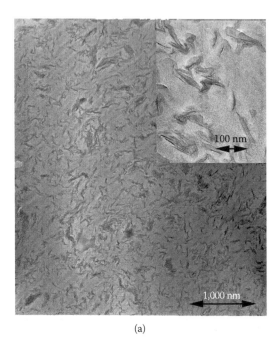

(a)

Figure 9.10 TEM micrographs showing (a) particles of montmorillonite dispersed in a starch–poly(vinyl alcohol) matrix at low and high magnification (upper right corner); and (b) cellulose nanofibrils obtained from sisal and a histogram for the distribution of diameters (upper right corner) (Rodriguez, Thielemans, and Dufresne, 2006).

by extraction or surface treatments and nanoindentation tests of the nanofibrils are required (Edgar and Gray, 2003), being particularly useful to investigate self-assembled nanocomposite structures (Decher and Schlenoff, 2002). Conversely, optical microscopy (OM) has little or no practical use for nanocomposites because the dimensions of the reinforcing agents are on the nanoscale, which lies below the resolution limit of optical microscopes.

Measurement of mechanical, physical, dynamic–mechanical, and thermal properties is often used to evaluate the impact of filler on nanocomposites (Orts et al., 2005). For example, additions of stiff cellulose nanofibrils, carbon nanotubes, and clays affect rheological, mechanical (tensile, flexural, and impact), and dynamic mechanical properties of nanocomposites, which can be measured by dynamic–mechanical thermal analysis. Inorganic fillers impart high thermal stability; thus measurements of thermal stability (thermal mechanical analysis, TMA, and thermogravimetry, TGA), heat distortion temperature (HDT), and flame retardancy and smoke suppression (flammability tests) are all useful in nanocomposite

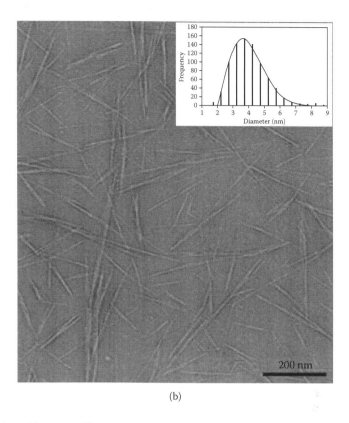

(b)

Figure 9.10 (Continued.)

characterization (Alexandre and Dubois, 2000; Bower, 2002; Ray and Okamoto, 2003; Jansson and Järnström, 2005; Sperling, 2006).

Thermal analysis by differential scanning calorimetry (DSC) has been applied to nancomposites to follow matrix properties after filler addition, allowing us to understand polymer chain mobility and, hence, the nucleating process (Bower, 2002; Sperling, 2006). Knowledge of crystallization kinetics of crystalline matrices is fundamental in setting up processing parameters such as molding temperature and time. Thermal characterization is an important tool because retrogradation and crystallization of thermoplastic starch during aging lead to undesired changes in thermomechanical properties, the main drawback for application of starch-based plastics. Increase of the degree of crystallinity of the thermoplastic starch matrix has been reported (Angellier et al., 2006b), implying that crystallization of the matrix occurs at the interface between the starch nanocrystals filler and the matrix based on to the similar chemical structure of both components.

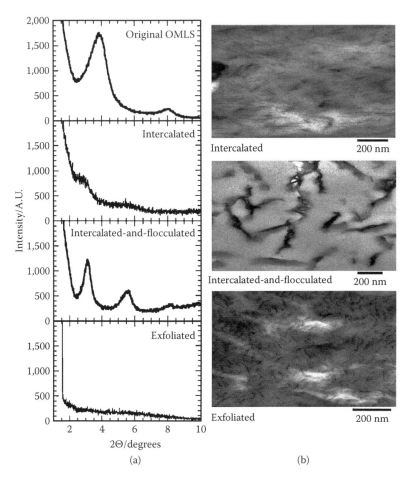

Figure 9.11 Comparison of WAXD patterns and TEM images for organically modified layered silicates (OMLSs) and nanocomposites displaying three different morphologies (exfoliated, flocculated, and intercalated) (Ray and Okamoto, 2003).

Fourier transform infrared spectroscopy (FTIR) and nuclear magnetic resonance (NMR) analyses are used to detect functional groups to study the structures of nanocomposites, mainly to evaluate cross-linking, compatibilization, and surface treatment of the fillers. For example, NMR had been used to study nanocomposites based on cross-linked starch and cellulose nanofibrils (Orts et al., 2007). Details about applying NMR to characterize starch and starch gels are discussed in Chapter 4.

Dynamic mechanical thermal analysis (DMTA) measures the response of a material to oscillatory deformation as a function of temperature and

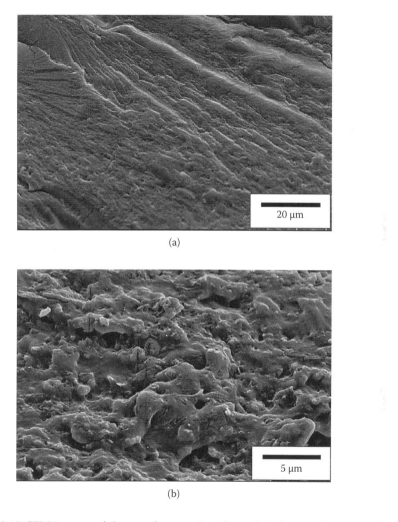

(a)

(b)

Figure 9.12 SEM images of the cryofractured surface of (a) thermoplastic starch and (b) cellulose nanofibers/starch nanocomposite (Alemdar and Sain, 2008).

decomposes this response into two components: elastic, measured by the storage modulus (E'), and viscous, measured by the loss modulus (E"), providing information on changes in molecular mobility with the addition of rigid fillers. It is a very suitable technique to determine mechanical properties of nanocomposites and the influence of filler addition. Moreover, because in many cases there is a need to add a third component to improve mechanical properties, glass transition measurements can also be used to evaluate the degree of compatibilization and interactions of multiple components (Bower, 2002; Sperling, 2006).

Gas permeability and changes in barrier properties are other important properties of nanocomposites. The migration of gases through packaging reduces the shelf life of food and beverages and durability of many other industrialized products (Alexandre and Dubois, 2000; Sanadi et al., 2001; Ray and Okamoto, 2003; Jansson and Järnström, 2005). Some nanocomposites have shown barrier properties because nanofillers, such as exfoliated clays, act as barriers that prevent gas diffusion. Gases need to follow tortuous paths around the nanoparticles, thus making the diffusion path longer as implied in Figure 9.13. Measurement of barrier properties can give useful insights on the nature of particle dispersion in nanocomposites.

There is no simple predictable trend in increasing or decreasing a specific nanocomposite property with the addition of nanostructured reinforcements. Several combinations must be tried to offset competing tendencies, depending on interactions among the components, filler size, surface properties, shape and aspect ratio, and nature of the polymeric matrix. Table 9.2 summarizes some of the trends in mechanical, physical, and rheological properties of the polymer nanocomposite as a function of the interaction among the components and the crystalline or amorphous nature of the matrix (Jordan et al., 2005).

9.3   Starch-based nanocomposites

As previously mentioned, starch is one of the most studied natural polymers with many potential applications in biodegradable packaging. However, due to major drawbacks associated with the use of starch-based materials, such as hydrophilic nature, poor processability, brittleness, and postprocessing crystallization, the number of commercially available starch-based products is still limited.

Several attempts to improve starch-based materials by traditional composite formulation resulted in materials with moderately improved properties, but limited processability due to high viscosity. Recently, however, starch-based nanocomposites with the addition of low quantities of fillers have given rise to large-scale improvements in physical, mechanical, and barrier properties (Ruiz-Hitzky, Darder, and Aranda, 2005; Rhim and Ng, 2007; Yang, Wang, and Wang, 2007). This section discuss some examples of starch nanocomposites reinforced with inorganic clays or carbon nanotubes and organic (cellulosic) fillers.

9.3.1   Starch nanocomposites with inorganic reinforcements

Inorganic fillers when added to the starch matrix generally lead to composites with better overall properties. Filler surface properties can be modified by several techniques such as organophilization, silanization,

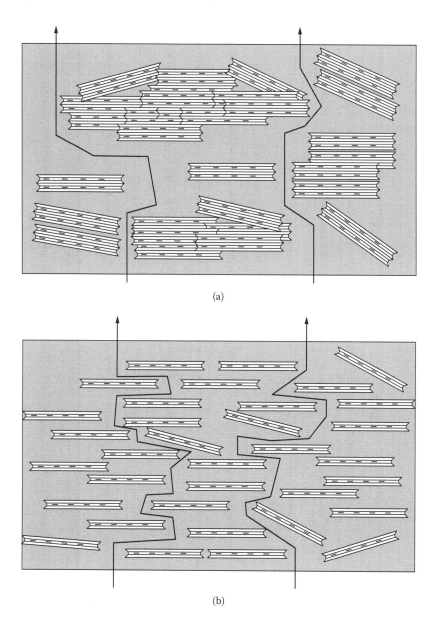

Figure 9.13 Possible diffusion paths of gas molecules in (a) a traditional composite and (b) exfoliated polymer clay nanocomposite. Note the increase in path length that ultimately results in decreased permeability.

Table 9.2 Polymer Nanocomposite Trends Based on Nature of Matrix and Its
Interaction with Nanostructured Fillers

| Property | Polymer matrix | | Interaction |
	Crystalline	Amorphous	
Elastic modulus	Increases with volume fraction of filler Increases or no change with decrease in size	Increases with volume fraction of filler Increases with decrease in size	Good
	Increases with volume fraction of filler Increases with decrease in particle size Greater increase than for good interaction	Increases with volume fraction of filler Increases with decrease in particle size	Poor
Yield stress/strain	Increases with volume fraction of filler Increases with decrease in particle size Decreases with addition of particles	Not applicable	Good
	Decreases with addition of particles	Decreases with addition of particles	Poor
Ultimate stress/strain	Increases with decrease in particle size No unified result for change in volume fraction	Nano > micro after 20 wt.%.	Good
	Lower than pure for small volume fraction	Decreases with addition of particles	Poor
Density/volume	Increases volume with decrease in size	Increases volume with decrease in size	Good
	N/A	N/A	Poor
Strain to failure	Decreases with addition of particles	Decreases with addition of particles Increases with decrease in size	Good
	Decreases with addition of particles	Increases with addition of particles	Poor
Glass transition temperature	Decreases with addition of particles	Increases with addition of particles	Good

Table 9.2 Polymer Nanocomposite Trends Based on Nature of Matrix and Its Interaction with Nanostructured Fillers (Continued)

| Property | Polymer matrix | | |
	Crystalline	Amorphous	Interaction
	Decreases with addition of particles	Level until 0.5%, drops off level from 1–10 wt.	Poor
Crystallinity	No major effect	Not applicable	Good
	No major effect	Not applicable	Poor
	Increases with volume fraction of filler		Good
	Increases with decrease in particle size		
Viscoelasticity	N/A	Increases with addition of particles; drop at 1 wt.% flowed by increase in this property	Poor

N/A = Not applicable.
Source: Jordan et al., 2005.

alkylation, and surface etching that improve interface properties, processability, and particle dispersion, thereby producing nanocomposites with better properties.

Cao et al. (2007) studied the utilization of multiwalled carbon nanotubes (MWCNT) as filler reinforcements to improve the performance of thermoplastic starch. Details about TPS structure and applications are discussed in Chapter 6. Physical properties of TPS/MWCNT nanocomposites improved with an increase in MWCNT content. For 0 to 3.0 wt.%, the composites showed an increase in tensile strength by 66% (from 2.85 to 4.73 MPa), Young's modulus by 89% (from 20.74 to 39.18 MPa), and T_g by 53% (from 16.5 to 25.3°C). The incorporation of MWCNT into the TPS matrix also led to a decrease in water uptake up to 14%. This enhancement was attributed to a hydrogen bond between carbon nanotubes and TPS molecules. In a similar study, Ma, Yu, and Wang (2008) found only a modest increase in Young's modulus and tensile strength and a slight decrease in elongation at break as MWCNT content increased, with interfacial adhesion between the components of the nanocomposite.

Nanocomposites of glycerol-plasticized corn starch and hydrated kaolin (10, 20, 30, 40, 50, and 60 w/w parts of kaolin/100 parts of TPS) showed an increase in tensile properties as a consequence of strong bonding between kaolin and the matrix. For instance, nanocomposites filled

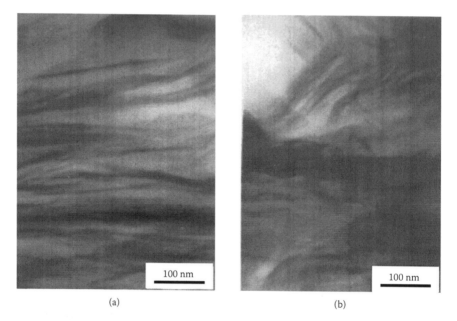

(a)   (b)

Figure 9.14 TEM micrographs of nanocomposites of montmorillonite and thermoplastic starch (MMT/TPS) with (a) 5% untreated MMT, (b) 10% untreated MMT, (c) 5% organophilized MMT, and (d) 10% organophilized MMT (Park et al., 2003). Note the differences between dispersions of natural (hydrophilic) and organophilized (hydrophobic) clays.

with 50 wt.% of kaolin showed an increase in the tensile strength and Young's modulus, respectively, from 5 to 7.5 MPa and 120 to 290 MPa, whereas tensile strain at break decreased from 30 to 14 wt.%. Moreover, kaolin led to a reduction in water uptake, although thermal resistance—measured by the onset of the thermal decomposition curves obtained by thermogravimetry—did not suffer any changes, as would be expected by the addition of much more thermally stable mineral fillers (Carvalho, Curvelo, and Agnelli, 2001).

In a similar study carried out by Chen and Evans (2005), nanocomposites of potato starch reinforced with four different types of clays (natural sodium montmorillonite, natural hectorite, a hectorite modified with 2-methyl, 2-hydrogenated tallow quaternary ammonium chloride, and kaolinite) obtained by melt processing and clay content ranging from 0 to 12 wt.%, showed that treated hectorite and kaolinite formed conventional composites, with little intercalation whereas natural smectic clays, montmorillonite, and hectorite with starch readily formed intercalated nanocomposites. In all cases, clay increased the elastic modulus of TPS. For nanocomposites, montmorillonite generally provided a slight improvement in the modulus over untreated hectorite.

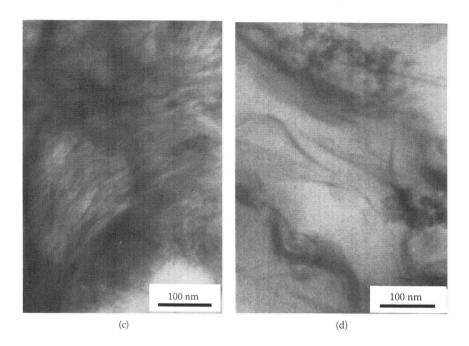

(c)   (d)

Figure 9.14 (Continued.)

Characterization of nanocomposites obtained by injection molding of thermoplastic potato starch, natural montmorillonite modified with quaternary ammonium salt (Na⁺MMT), and organically modified (MMT) with methyl bis-2-hydroxyethyl ammonium cations located in the silicate gallery and glycerol as plasticizer showed that composites using unmodified clays had higher tensile strength, thermal stability, and better barrier properties to water vapor than the those containing organophilized MMT. It is likely that the hydrophilic TPS had better affinity to nonorganophilized clays which helped form intercalated nanostructures with the untreated clay (Figure 9.14). This result implied that higher interaction led to better dispersion of the filler in the matrix, resulting in better mechanical, thermal, and barrier properties (Park et al., 2003). Similar results were reported by Chiou et al. (2005), by incorporating montmorillonite clays (both hydrophilic and hydrophobic clays) into wheat, potato, corn, and waxy corn starches. Composites reinforced with hydrophilic MMT had improved mechanical properties due to greater intercalation in the gelatinized starch matrices. Composites from wheat and corn starches showed similar elastic modulus values during gelatinization, whereas both potato and waxy corn starch samples had elastic modulus values that decreased rapidly at higher temperatures. These results could be explained by the high swelling of potato starch, with "softer" granules leading to lower elastic modulus. On the other hand, results from waxy corn could be explained by the low amylose content of this particular starch. The lack of amylose would imply fewer physical cross-links of leached amylose, starch granule, and clay, leading to a lower elastic modulus.

Effects of plasticizers on nanocomposites have been widely studied (Pandey and Singh 2005; Qiao, Jiang, and Sun, 2005; Chiou et al., 2007). Nanocomposites with organically modified montmorillonite (OMMT) and acetylated starch (TPAS) plasticized with glycerol showed a better reinforcement than composites with montmorillonite (MMT), resulting in a decrease in the hydrophilic nature of acetylated starch. This improved the dispersion of OMMT in TPAS when compared with unmodified MMT (Qiao, Jiang, and Sun, 2005). These results corroborated those obtained by Park et al. (2003), where better interaction was observed in unmodified MMT with unmodified starch because of their similar hydrophilic nature.

Studies of nanocomposites based on native corn starch (75 wt.%) with sodium montmorillonite with different addition sequences of glycerol (20 wt.%) as plasticizer and clay (5 wt.%) showed that the sequence of component addition had a significant effect on the nature of composites formed and, accordingly, the final properties. In these experiments, solution dispersion was obtained by (i) gelatinization of starch followed by plasticization and addition of clay dispersion and boiling for 30 min; (ii) mixture of clay dispersion with starch in water and boiling for 30 min with constant vigorous stirring, followed by the addition of plasticizer; (iii) starch, clay dispersion, and glycerol mixed together and boiled for 30 min; and (iv) glycerol mixed with clay dispersion and stirred for 5 h at room temperature followed by addition of starch and heating for 30 min (Pandey and Singh, 2005). The best mechanical properties can be obtained if plasticizer is added after mixing clay in the starch matrix.

Nanocomposite foams obtained by melt processing of TPS with urea as plasticizer and ammonium-treated montmorillonite (NH_4 MMT) showed enhanced dispersion due to the NH_4 MMT in the TPS, producing exfoliated TPS–clay nanocomposites (Chen, Chen, and Evans, 2005). Synergistic effects of the urea as plasticizer with ammonium enhanced clay dispersion and foaming. Nanocomposites of glycerol-plasticized starch reinforced with montmorillonite exhibited a significant enhancement in mechanical properties, showing significant improvement in Young's modulus with the addition of 5 wt.% of clay (Cyras et al., 2008).

9.3.2   *Nanocomposites based on starch and organic reinforcements*

Cellulose nanofibrils have been the most studied organic reinforcements in starch-based nanocomposites because of their remarkable mechanical properties. The affinity between starch and cellulose due to their structural similarity can be exploited, not only to enhance the mechanical properties of composites, but also to produce biodegradable materials (Avérous, 2004; Zhao, Torley, and Halley, 2008).

There are three significant differences between clay-reinforced and cellulose nanofibril-reinforced nanocomposites: (i) typical cellulosic nanofibrils are long crystalline "needles" ranging in size from 10–20 nm in width and an average aspect ratio of 20–100, providing contrast to the lamellar structures of most clays; (ii) cellulose surfaces provide a greater potential for surface modification using well-established carbohydrate chemistry; and (iii) sources of cellulose microfibrils, including wood, straw, bagasse, bacteria, and sea animals (tunicates), are widely diverse, providing a wide range of potential nanoparticle properties (Orts et al., 2005). Moreover, cellulose nanofibrils are known to align in the magnetic and electrical fields (Sugiyama, Chanzy, and Maret, 1992; Yoshiharu et al., 1997), opening up the possibility to control the degree of orientation during processing. One example is the extrusion blow molding of packaging films and electric or magnetic devices with orientation in film processing.

In plasticized materials, the mechanical properties are strongly related to moisture content and humidity conditions, and the addition of nanofibrils on nanocomposites can reduce water uptake. The distribution of compounds in the matrices also changes the mechanical properties. Anglès and Dufresne (2000, 2001), observing nanocomposites from glycerol-plasticized waxy maize starch and cellulose whiskers extracted from tunicates by casting and evaporation under vacuum, concluded that glycerol and water redistributed themselves within the matrix, diffusing toward the cellulose–amylopectin interface, and changing properties in a short time period. Mechanical properties of these nanocomposites are more dependent on plasticizer and moisture content than on the addition of whiskers. In this example, accumulation of plasticizer at the interface increased the ability of amylopectin chains to crystallize, leading to the formation of a transcrystalline zone around the whiskers. Such crystalline zones accounted for the lower water uptake of the nanocomposites with increasing filler content. A very low reinforcing effect was observed upon the addition of tunicin whiskers as a consequence of this plasticizer accumulation at the interfacial zone.

Nanocomposites from cellulose microfibrils gelatinized with potato starch (Dufresne and Vignon, 1998) reduce water sensitivity, increasing thermomechanical stability. Mechanical properties of nanocomposites based on glycerol-plasticized waxy maize starch and cellulose whiskers (Anglès and Dufresne, 2000, 2001; Dufresne, Dupeyre, and Vignon, 2000) show a relationship between plasticizer content and relative humidity conditions during storage. The reinforcing effect is more significant in plasticized starch due to the decrease in glass transition temperature of the matrix below room temperature.

Consequently, the reinforcing effect of cellulose nanofibrils in plasticized starch matrices plays a fundamental role in moisture, the amount and type of plasticizer, and the specific reinforcement (Mathew and Dufresne, 2002). Glycerol-plasticized nanocomposites are hampered by accumulation of glycerol on the cellulose whisker surface, giving rise to antiplasticization effects (Anglès and Dufresne, 2000; Dufresne, Dupeyre, and Vignon, 2000), and thus poor mechanical properties.

Nanocomposites from waxy maize starch plasticized with sorbitol and tunicin whiskers exhibited a single glass transition without any evidence of preferential migration of plasticizers toward the cellulose or transcrystallization of amylopectin on the cellulose surface. Fractured surfaces of the composites showed uniform distribution of whiskers in the matrices for all the compositions (Figure 9.15). Glass transition temperature of the plasticized amylopectin matrix increased up to a whisker content of ~10–15 wt.%, and a significant increase in crystallinity was also observed in the composites by increasing either moisture content or whiskers content. Water uptake of the composites remained roughly constant upon whisker addition (Mathew and Dufresne, 2002).

Addition of cellulose microfibrils extracted from cotton, softwood, or bacterial cellulose at low concentrations to wheat or potato starch blended with pectin has a significant effect on their mechanical properties (Orts et al., 2005). Young's modulus of wheat starch nanocomposites reinforced with cotton nanofibrils increased fivefold with the addition of only 2.1 wt.% of nanofibrils (Table 9.3). In this example, the source of cellulose nanofibrils influences the mechanical properties of the composites; composites reinforced with cotton and wood-derived microfibrils are indistinguishable, whereas composites reinforced with bacterial cellulose exhibit Young's modulus and elongation at maximum load significantly lower. On the other hand, the addition of 5 wt.% of cellulose nanofibrils to a 50–50 wt.% starch–pectin blend results in a decrease in mechanical properties. These results corroborate previous works on starch-based nanocomposites (Anglès and Dufresne, 2000, 2001; Dufresne, Dupeyre, and Vignon, 2000), where the addition of a third component can give rise to complex interactions between the components, often resulting in poorer mechanical properties.

In studies with glycerol-plasticized starch nanocomposites reinforced with cellulose nanofibrils from wheat straw, Alemdar and Sain (2008) found that the addition of 10 wt.% of cellulose nanofibrils improved the tensile strength and Young's modulus of the composites. Nanocomposites based on wheat starch plasticized with glycerol and reinforced with cellulose nanofibrils extracted from ramie fibers by acid hydrolysis (Lu, Weng, and Cao, 2006) showed improvement in water resistance, good dispersion,

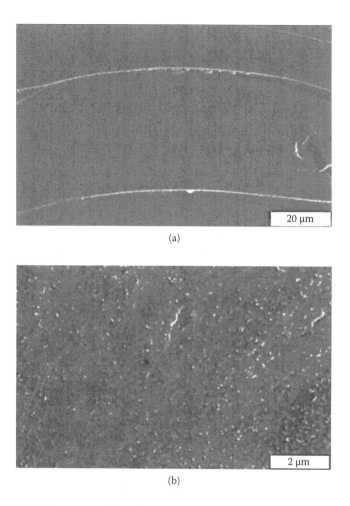

(a)

(b)

Figure 9.15 SEM micrographs of the fractured surfaces of (a) unfilled sorbitol plasticized starch matrix and related composites filled with (b) 5, (c) 15, and (d) 25 wt.% tunicin whiskers (white dots) (Mathew and Dufresne, 2002).

good adhesion of components, an increase in Young's modulus (from 56 to 480 MPa), and improvement in tensile strength (from 2.8 to 6.9 MPa) with increasing filler content from 0 to 40 wt.% (Figure 9.16).

Mechanical, dynamic mechanical, and thermal properties of nano-composites of potato starch reinforced with either cellulose nanofibrils (CNF) or layered silicate (LS) plasticized with water and sorbitol with 5 wt.% of either CNF or synthetic hectorite, showed well-distributed reinforcements in the starch matrix (Kvien et al., 2007) with a significant improvement in tensile properties compared to the pure matrix (Table 9.4). Significant improvement of the storage modulus at elevated

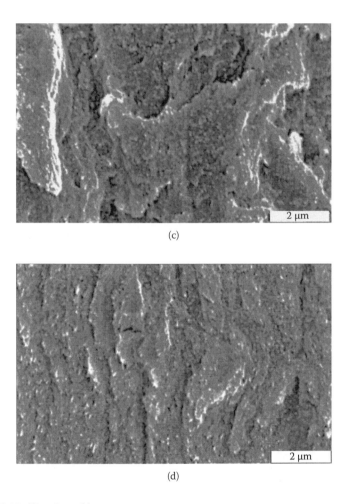

(c)

(d)

Figure 9.15 (Continued.)

temperatures on layered silicate nanocomposite was observed, which could be expected taking into account the high thermal resistance and surface area of layered silicates.

Although these results provide interesting insights into composites reinforced with organic (CNF) and inorganic (LS) reinforcements, more work is required to understand the nature of starch-based nanocomposites, especially the fundamental surface interactions between matrix and filler. This is especially true given the differences among the various types of clays and nanofibers, because there is a wide range of clay materials and cellulose nanofibers. Moreover, nanocomposite properties are strongly dependent on the interfacial region adhesion which also depends on particle size, superficial energy, surface modification, and so on.

Table 9.3 Young's Modulus (E) and Elongation at
Maximum Load (ε_m) of Composites of Wheat
Starch and Cellulose Nanofibrils Extracted from
Cotton as a Function of Fiber Content

Nanofibrils (%)	E (Gpa, %)	ε_m (%)
0	1.39 (0)*	2.7 (0)
2.1	5.09 (266)	3.9 (44)
5	9.34 (572)	8.4 (210)
10.3	12.45 (796)	8.8 (226)

Source: Orts et al., 2005.

Considering that there is no commercial process for large-scale isolation of cellulose nanofibrils, the applicability of highly loaded cellulose-based nanocomposites is presently limited. The production of plasticized-starch-based nanocomposites with improved properties can be feasible, implying that large-scale commercial isolation may be worth considering. Moreover, the average deformation of plasticized starch nanocomposites is higher than for nonplasticized ones, thereby opening up the possibility of developing less brittle starch-based composites with potential applications in packaging materials.

9.3.3 Starch-based nanocomposite blends: Nanocomposites of starch in combination with other polymers

Blends of starch with natural and synthetic polymers have been extensively studied in order to overcome aging effects (i.e., recrystallization after processing), to impart better mechanical properties, and to reduce the hydrophilicity of starch (Chandra and Rustgi, 1998, Ray and Okamoto, 2003; Okamoto, 2005). The addition of natural biodegradable polymers, nevertheless, has gained more attention with the intent of producing fully biodegradable blends (Corradini et al., 2006).

Although this chapter is concerned mainly with starch nanocomposites as matrix components where the addition of organic or inorganic nanoparticles improves mechanical or physical properties, it must be considered that starch blends with other polymers and nanostructure reinforcements (polymer–nanocomposite blends) can provide another viable approach toward improving even more starch-based nanocomposites.

Examples of nanocomposite blends with enhanced properties are increasingly cited in the literature. For example, improved blends were obtained by polyethylene–octene elastomer, starch and montmorillonite (Liao and Wu, 2005); commercial starch–PCL blend and montmorillonite (Pérez et al., 2007); starch, poly(vinyl alcohol), and montmorillonite (Dean

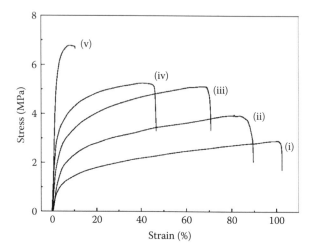

Figure 9.16 Stress versus strain curves for nanocomposites of starch reinforced with ramie nanofibrils: (i) glycerol-plasticized starch, and composites with (ii) 5, (iii) 10, (iv) 25, and (v) 40 wt.% of ramie nanofibrils (Lu, Weng, and Cao, 2006).

et al., 2008); and starch, poly(lactic acid), and montmorillonite (Lee, Chen, and Hanna, 2008).

Small amounts of MMT are also used to obtain biodegradable blends of starch and polyester with significantly improved tensile properties. Nanocomposite blends, with different starch–polyester ratios and organophilized with montmorillonite at 1.5 and 5 wt.%, prepared by twin screw extrusion showed that the addition of MMT improved both the processing and tensile properties when compared to original starch blends. The type of nanocomposite produced (intercalated or exfoliated) depended on the amount of clay added and the ratio of starch to polyester, which also influenced mechanical and thermal properties (McGlashan and Halley, 2003).

Table 9.4 Tensile and Dynamic Mechanical Properties of Starch and Starch Nanocomposites Reinforced with Cellulose Nanofibrils (S-CNF) and Layered Silicate (S-LS)

Materials	E(MPa)[a]	σ_y(MPa)	ε_{max} (%)	E' at 25°C (MPa)	E' at 60°C (MPa)
Starch	370 ± 35	11.3 ± 1.0	25 ± 11	370 ± 97	69 ± 7
S-CNF	460 ± 10	13.7 ± 1.3	32 ± 10	502 ± 10	182 ± 20
S-LS	460 ± 45	12.5 ±1.3	31 ± 12	1,020 ± 20	286 ± 40

[a] E = Young's modulus; σ_y = yield strength; ε_{max} = elongation at break; E' = storage modulus.
Source: Kvien et al., 2007.

Nanocomposite blends are easier to process than the base blends using a film-blowing tower (McGlashan and Halley, 2003; Kalambur and Rizvi, 2005, 2006). Reactive extrusion to develop biodegradable starch–polycaprolactone (PCL) nanocomposite blends (using Fenton's reagent to oxidize starch and initiate cross-linking between oxidized starch and PCL) showed enhanced properties and improved interfacial adhesion between starch and PCL (Scheme 9.1). These biodegradable nanocomposite blends contained up to 40% starch, resulting in tough materials with elongational properties comparable to those of 100% polyester (Kalambur and Rizvi, 2005, 2006).

Less known is the ability to use starch nanocrystals as reinforcement in polymer matrices. Starch nanocrystals with dimensions of a few nanometers are formed from acid hydrolysis of starch granules. These starch crystals are mainly formed of crystalline amylose, because acid hydrolysis removes the amorphous domains comprised mostly of amylopectin (Dufresne, Cavaillé, and Helbert, 1996; Dufresne and Cavaillé, 1998).

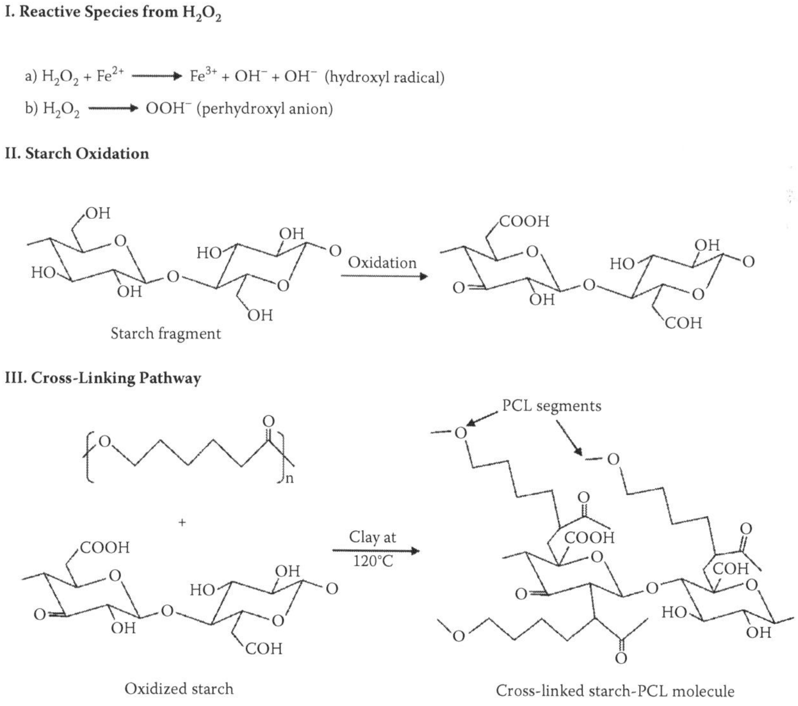

I. Reactive Species from H_2O_2

a) $H_2O_2 + Fe^{2+} \longrightarrow Fe^{3+} + OH^- + OH^-$ (hydroxyl radical)

b) $H_2O_2 \longrightarrow OOH^-$ (perhydroxyl anion)

II. Starch Oxidation

Starch fragment   Oxidation

III. Cross-Linking Pathway

PCL segments

Oxidized starch   Clay at 120°C   Cross-linked starch-PCL molecule

Scheme 9.1 Oxidation and cross-linking steps used during reactive extrusion to produce starch–polyester–montmorillonite nanocomposite blends with improved properties (Kalambur and Rizvi, 2005).

Such starch nanocrystal-reinforced nanocomposites have been prepared with poly(β-hydroxyoctanoate) (Dubief, Cavaillé, and Helbert, 1999), natural rubber (Angellier et al., 2005; Angellier, Molina-Boisseau, and Dufresne, 2005, 2006), poly(styrene-co-butyl acrylate) (Angellier et al., 2005c), waxy maize starch (Angellier et al., 2006b), sorbitol-plasticized pullulan (Kristo and Biliaderis, 2007), and poly(vinyl alcohol) (Chen et al., 2008). Moreover, the possibility of surface modification by grafting can lead to reinforcements compatible with hydrophilic and hydrophobic matrices (Thielemans, Belgacem, and Dufresne, 2006; Labet, Thielemans, and Dufresne, 2007). Considering a glycerol plasticized starch matrix, the increase of relaxation temperature can be associated with the glass–rubber transition of amylopectin-rich domains as increasing the starch nanocrystal content (Angellier et al., 2006b). The reduction in the molecular mobility of matrix amylopectin chains for filled materials is explained by the establishment of hydrogen bonding of both components. This increase of T_g led to a considerable slowing down of the retrogradation of the matrix, perhaps corresponding to a reduction in chain mobility (Chapter 5).

Oxygen and water permeability of natural rubber reinforced with waxy maize starch nanocrystals decreases continuously upon addition of starch nanoparticles (Angellier et al., 2005). It shows that the hydrophilic nature of starch nanocrystals does not increase the permeability of natural rubber to water vapor. Starch nanocrystals exhibit reduced oxygen diffusion when adding the filler and accounting for oxygen permeability. This phenomenon could be mainly due to the structural modification of the film and not to the decrease of the solubility of oxygen. Both observations were ascribed to the nanoscale platelet-like morphology of starch nanocrystals, which increases difficulty of the diffusion path in matrices. Conversely, no significant differences in water vapor permeability were observed in samples with sorbitol and pullulan plasticized starch for reinforced unfilled films compared to those containing up to 20 wt.% of nanocrystals (Kristo and Biliaderis, 2007). Moreover, the possibility of surface modification by grafting can lead to reinforcements compatible with hydrophilic and hydrophobic matrices (Angellier et al., 2005d). Alkenyl succinic anhydride (ASA), used for acylating the surfaces of starch nanocrystals, has been widely employed as a sizing agent in papermaking on pulp fibers in aqueous systems. The surface chemical modification of waxy maize starch nanocrystals with ASA is done by a toluene–(dimethylammo)pyridine system. Isocyanates, such as phenyl isocyanate (PI), can also be used to chemically modify the surface of starch nanocrystals (Angellier et al., 2005d). The lower polarity of the modified nanocrystals is demonstrated by their migration to the methylene chloride phase from the aqueous phase as can be seen in Figure 9.17.

Figure 9.17 Wettability tests: (a) a drop of an aqueous suspension of waxy maize starch nanocrystals in dichloromethane, (b) migration of unmodified starch nanocrystals in distilled water, and migration of (c) alkenyl succinic anhydride(ASA)-modified and (d) phenyl isocyanate(PI)-modified starch nanocrystals modified with (c) ASA and (d) PI in dichloromethane (Angellier et al., 2005d).

A promising way of processing new nanocomposites consists of transforming starch nanocrystals into a co-continuous material through long-chain surface chemical modification. Nanoparticles are surface modified by grafting agents bearing a reactive endgroup and a long hydrophobic tail. The general objective of this surface chemical modification is to increase the apolar character of the nanoparticle. From this procedure, biocomposite materials based on fully renewable resources can then be processed by classical methods such as hot-pressing, extrusion, injection molding, or thermoforming. Very few reports in the literature have applied this very promising methodology.

Surface modifications on nanocrystals from chemically modified starch with poly(ethylene glycol) methyl ether (PEGME) and stearic acid chloride obtained by grafting methods provide unique properties to the nanocrystals, especially their ability to avoid aggregation due to reduced hydrogen bonding and polar interactions among individual particles (Thielemans, Belgacem, and Dufresne, 2006).

Starch nanoparticles grafted with polytetrahydrofuran, polycaprolactone, and poly(ethylene glycol) monobutyl ether chains using toluene 2,4-diisocyanate as linking agent resulted in individualization of nanoparticles or the formation of a film and, as expected, the grafting efficiency decreased with the lengths of the polymeric chains (Labet, Thielemans, and Dufresne, 2007).

Long-chain surface modifications can yield some extraordinary possibilities. The surface modifications can act as binding sites for active agents in drug delivery systems or for leaching out toxins in purifying and treatment systems. These surface modifications may also be able to interdiffuse, upon heating, to form the polymer matrix phase. However, the covalent linkage between reinforcement and matrix will result in near-perfect stress transfer at the interface with exceptional mechanical properties of the composite as a result. The large surface area, inherent in nanoparticles, guarantees a large surface activity and a high grafting per unit mass of particles. Their small size also reduces the required chain length for interdiffusion to assure sufficient matrix phase cohesion.

9.3.4 *Methods of improving filler and matrix interaction in starch-based nanocomposites*

Carbohydrate chemistry is a well-established field with proven techniques to modify starch and cellulose functionality. Nanocomposites of formamide–ethanolamine plasticized thermoplastic starch (FETPS) as the matrix and ethanolamine activated montmorillonite (EMMT) as the reinforcing phase can be obtained in two steps: (i) mixture of starch with formamide and ethanolamine followed by extrusion in a single screw extruder and pelletization; and (ii) mixture of formamide FETPS with 0–10 wt.% of added MMT previously activated with ethanolamine followed by extrusion. Intercalated FETPS and EMMT layers can improve interaction through modification, resulting in suitable mechanical properties (Figure 9.18), thermal stability, and water resistance (Huang et al., 2005).

Improvements in mechanical properties of biodegradable nanocomposites from thermoplastic corn starch (TPCS) plasticized with urea and

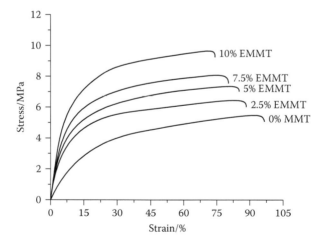

Figure 9.18 Effect of the addition of ethanolamine-activated montmorillonite (EMMT) on stress–strain curves of nanocomposites with thermoplastic starch plasticized with formamide and ethanolamine (FETPS) (Huang et al., 2005).

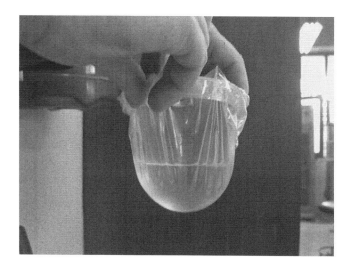

Figure 9.19 Starch/MMT/chitosan nanocomposite film with improved hydrophobic nature (Kampeerapappun et al., 2007).

formamide and citric acid-activated montmorillonite can also be obtained by melt intercalation (Huang, Yu, and Ma, 2006). Addition of chitosan nanoparticles into cassava starch–montmorillonite nanocomposites can increase surface hydrophobicity of films (Kampeerapappun et al., 2007). For example, at 5 wt.% montmorillonite, the contact angles of composite films increased from 62° to 72, 75, 78, and 83° with increases in chitosan content of 0, 5, 10, 15, and 20 wt.%, respectively. Decrease in hydrophilicity of the nanocomposite films displayed a lower transmission rate of water vapor as can be seen in Figure 9.19.

Nanocomposites of starch and cellulose nanofibrils modifed by azide cross-linking (Orts et al. 2007) improved the compatibility between both the polymer matrix and the cellulose nanofibrils. Azide derivatives of different starches were synthesized as depicted in Scheme 9.2. ^{13}C NMR and FT-IR spectra confirmed the azide derivative can be selectively cross-linked with heat or application of UV light. Nanofibrils derived from sulfuric acid hydrolysis of cotton cellulose added to extruded wheat starch gels at concentrations of 2–10 wt.% had a significant effect on mechanical properties of the nanocomposites, mainly on Young's modulus and elongation at maximum load increased (Figure 9.20).

9.4   Final remarks

Efforts to develop new materials from environmentally friendly, biodegradable, and renewable resources have grown in recent years. Among these new materials, starch is one of the most studied and promising

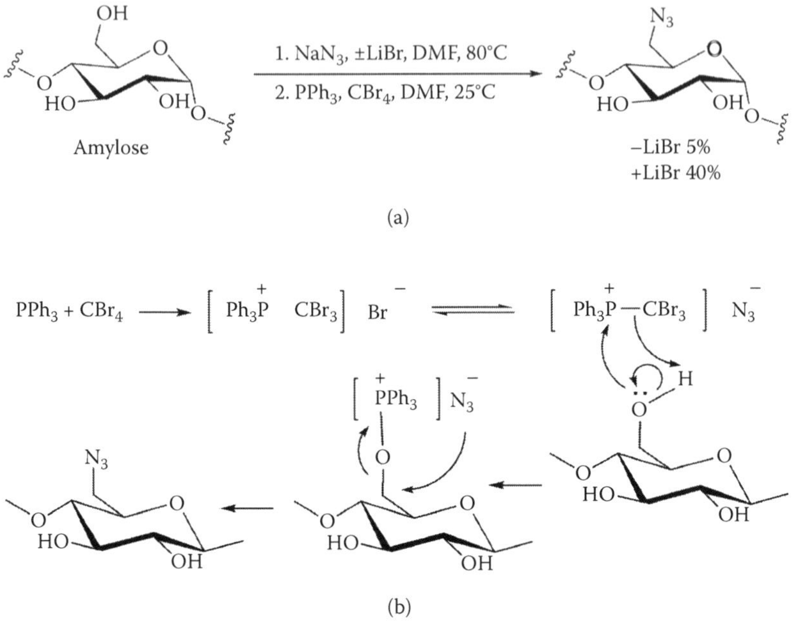

(a)

(b)

Scheme 9.2 Proposed azidation reactions for (a) starch: synthesis of 6-azido-6-deoxyamylose; and (b) cellulose: synthesis of 6-azido-6-deoxycellulose (Orts et al., 2007).

raw materials for the production of biodegradable plastics because it is a widely abundant and relatively low-cost natural polysaccharide obtained from a great variety of crops. Moreover, its use in the plastics industry has been considered a viable approach to reduce surplus agricultural products and develop biodegradable materials with a more favorable carbon footprint than petroleum-derived feedstocks.

The advent of new techniques for fabrication and characterization of materials on the nanoscale has made possible the production of starch-based nanocomposites with enhanced mechanical and physical properties, and greatly improved their optical transparency when compared to traditional composites. These properties can be easily tailored to a particular application by addition of nanostructured fillers or reinforcements such as clays, cellulose nanofibrils, and carbon nanotubes.

Much has been done to improve composite properties by examining the nature of individual components, their mechanisms of interaction, reinforcement dispersibility, processability, hydrophilicity, costs, and other economical and environmental aspects. However, despite all efforts toward better starch-based nanocomposites, effectiveness of interactions at the interfacial region between reinforcements and starch matrix

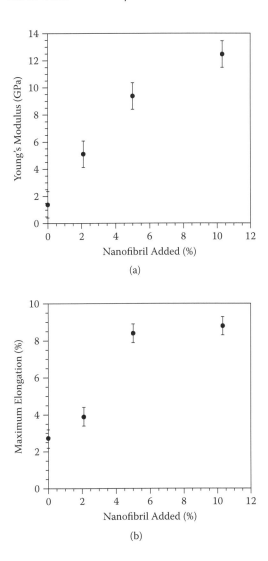

Figure 9.20 Effect of nanofibril concentration on (a) Young's modulus (GPa) and (b) maximum elongation to break (%) of wheat starch thermoplastic strings reinforced with cotton-derived nanofibrils (Orts et al., 2007).

durability in outdoor applications and moisture sensitivity continue to limit the large-scale production, use, and commercialization of biodegradable nanocomposites based on starch.

In summary, the technology reviewed here shows a growing effort to produce starch-based nanocomposites with improved properties and to scale up nanocomposite production and applications in many sectors of the plastics industry. This review shows how biodegradable materials can be exploited to produce composites with tailored properties by the correct choice of components, their relative proportions, and methods of modification.

References

Aichholzer W. and Fritz, H.-G. 1998. Rheological characterization of thermoplastic starch materials. *Starch/Stärke.* 50:77–83.

Ajayan, P.M., Schadler, L.S., and Braun, P.V. (Eds.) 2003. *Nanocomposite Science and Technology.* Weinheim: Wiley-VCH.

Alemdar, A. and Sain, M. 2008. Biocomposites from wheat straw nanofibers: Morphology, thermal and mechanical properties. *Comp. Sci. Tech.* 68:557–565.

Alexandre, M. and Dubois, P. 2000. Polymer-layered silicate nanocomposites: Preparation, properties and uses of a new class of materials. *Mat. Sci. Eng.* 28:1–63.

Angellier, H., Molina-Boisseau, S., and Dufresne, A. 2005. Mechanical properties of waxy maize starch nanocrystal reinforced natural rubber. *Macromolecules.* 38:9161–9170.

Angellier, H., Molina-Boisseau, S., and Dufresne, A. 2006. Waxy maize starch nanocrystals as filler in natural rubber. *Macromol. Symp.* 233:132–136.

Angellier, H., Molina-Boisseau, S., Belgacem, M.N., and Dufresne, A. 2005d. Surface chemical modification of waxy maize starch nanocrystals. *Langmuir.* 21:2425–2433.

Angellier, H., Molina-Boisseau, S., Dole, P., and Dufresne, A. 2006b Thermoplastic starch-waxy maize starch nanocrystals nanocomposites. *Biomacromolecules.* 7:531–539.

Angellier, H., Molina-Boisseau, S., Lebrun, L., and Dufresne, A. 2005. Processing and structural properties of waxy maize starch nanocrystals reinforced natural rubber. *Macromolecules.* 38:3783–3792.

Angellier, H., Putaux, J. L., Molina-Boisseau, S., Dupeyre, D., and Dufresne, A. 2005c . Starch nanocrystal fillers in an acrylic polymer matrix. *Macromol. Symp.* 221:95–104.

Anglès, M.N. and Dufresne, A. 2000. Plasticized starch/tunicin whiskers nanocomposites. 1. Structural analysis. *Macromolecules.* 33:8344–8353.

Anglès M.N. and Dufresne, A. 2001. Plasticized starch/tunicin whiskers nanocomposite materials. 2. Mechanical behavior. *Macromolecules.* 34:2921–2931.

Avérous, L. 2004. Biodegradable multiphase systems based on plasticized starch: A review. *J. Macromol. Sci. Polym. Rev.* C44: 231–274.

Avérous, L., Fringant, C., and Moroa, L. 2001. Starch-based biodegradable materials suitable for thermoforming packaging. *Starch/Stärke.* 53:368–371.

Balaposzhinimaev, B.S. 2003. Glass fiber based catalysts for environment protection. *Science and Technology. Proceedings KORUS 2003*, 4:115–122.

Battista, O.A. 1975. *Microcrystalline Polymer Science.* New York: McGraw-Hill.

Begishev, V.P. and Malkin, A.Y. 1999. *Reactive Processing of Polymers.* Toronto: ChernTec.

Bismarck, A., Mishra, S., and Lampke, T. 2005. Plant fibers as reinforcements for green composites. In: *Natural Fibers, Biopolymers, and Their Biocomposites*, A. K. Mohanty, M. Misra, and L.T. Drzal (Eds.). Boca Raton, FL: CRC Press.

Bledzki, A.K. and Gassan J. 1999. Composites reinforced with cellulose based fibers. *Prog. Polym. Sci.* 24:221–274.

Bower, D.I. 2002. *An Introduction to Polymer Physics.* Cambridge, MA: Cambridge University Press.

Brundle, C.R., Evans, Jr., C.A., and Wilson, S. (Eds.) 1992. *Encyclopedia of Materials Characterization, Surfaces, Interfaces, Thin Films.* Boston: Butterworth-Heinemann.

Brydson J.A. 1999. *Plastics Materials,* 7th ed. Oxford: Butterworth-Heinemann.

Callister, Jr., W.D. 2006. *Materials Science and Engineering: An Introduction.* New York: Wiley Interscience.

Cao, X., Chen, Y. Chang, P.R., and Huneault, M.A. 2007. Preparation and properties of plasticized starch/multiwalled carbon nanotubes composites. *J. Appl. Polym. Sci.* 106:1431–1437.

Carvalho, A.J.F., Curvelo, A.A.S., and Agnelli, J.A.M. 2001. A first insight on composites of thermoplastic starch and kaolin. *Carbohydrate Polym.* 45:189–194.

Chandra, R. and Rustgi, R. 1998. Biodegradable polymers. *Prog. Polym. Sci.* 23:1273–1335.

Chawla, K.K. 1998. *Composite Materials, Science and Engineering.* New York: Springer.

Chen, B. and Evans, J.R.G. 2005. Thermoplastic starch–clay nanocomposites and their characteristics. *Carbohydrate Polym.* 61:455–463.

Chen, M., Chen, B., and Evans, J.R.G. 2005. Novel thermoplastic starch–clay nanocomposite foams. *Nanotechnology.* 16:2334–2337.

Chen, Y., Cao, X., Chang, P.R., and Huneault, M.A. 2008. Comparative study on the films of polyvinyl alcohol/pea starch nanocrystals and polyvinyl alcohol/native pea starch. *Carbohydrate Polym.* 73:8-17.

Chiou, B.-S., Wood, D., Yee, E., Imam, S.H., Glenn, G.M., and Orts, W.J. 2007. *Polym. Eng. Sci.,* 47:1898–1904.

Chiou, B.-S., Yee, E., Glenn, G.M., and Orts, W.J. 2005. Rheology of starch-clay nanocomposites. *Carbohydrate Polym.* 59:467–475.

Corradini, E., Medeiros, E.S., Carvalho, A.J.F., Curvelo, A.A.S., and Mattoso, L.H.C. 2006. Mechanical and morphological characterization of starch/zein blends plasticized with glycerol. *J. Appl. Polym. Sci.* 101:4133–4139.

Crawford, R.J. 1999. *Plastics Engineering.* 3rd ed. Oxford: Butterworth-Heinemann.

Crawford, R.L. 1981. *Lignin Biodegradation and Transformation.* New York: John Wiley & Sons.

Cyras, V.P., Manfredi, L.B., Ton-That, M.T., and Vázquez, A. 2008. Physical and mechanical properties of thermoplastic starch/montmorillonite nanocomposite films. *Carbohydrate Polym.* 73:55–63.

Dean, K.M., Do, M.D., Petinakis E., and Yu, L. 2008. Key interactions in biodegradable thermoplastic starch/polyvinyl alcohol/montmorillonite micro- and nanocomposites. *Comp. Sci. Tech.* 68:1453–1462.

Decher, G. and Schlenoff, J.B. (Eds.) 2002. *Multilayer Thin Films.* Berlin: Wiley-VCH.

Dresselhaus, M.S., Dresselhaus G., and Avouris, P. (Eds.) 2001. *Carbon Nanotubes: Synthesis, Structure, Properties, and Applications.* Munich: Springer.

Dubief, D., Samain, E., and Dufresne, A. 1999. Polysaccharide microcrystals reinforced amorphous poly-β-hydroxyoctanoate nanocomposite materials. *Macromolecules.* 32:5765–5771.

Dufresne, A. 2007. Biopolymers on nanocomposites. In: *Biopolymers Technology,* A.C. Bertolini (Ed.). São Paulo: Cultura Academica.

Dufresne, A. and Cavaillé, J.-Y. 1998. Clustering and percolation effects in micro-crystalline starch-reinforced thermoplastic. *J. Polym. Sci. B: Polym. Phys.* 36:2211–2224.

Dufresne, A. and Vignon, M. R. 1998. Improvement of starch film performances using cellulose microfibrils. *Macromolecules.* 31:2693–2696.

Dufresne, A., Cavaillé, J.-Y. and Helbert, W. 1996. New nanocomposite mate-rials: Microcrystalline starch reinforced thermoplastic. *Macromolecules.* 29:7624–7626.

Dufresne, A., Dupeyre, D., and Vignon, M.R. 2000. Cellulose microfibrils from potato tuber cells: Processing and characterization of starch-cellulose micro-fibril composites. *J. Appl. Polym. Sci.* 76:2080–2092.

Durant, W. 1963. *The Story of Civilization: Our Oriental Heritage*, Vol. 1. New York: Simon & Schuster.

Edgar, C.D. and Gray, D.G. 2003. Smooth model cellulose I surfaces from nano-crystal suspensions. *Cellulose* 10:299–306.

Favier, V., Chanzy, H., and Cavaille, J.Y. 1995. Polymer nanocomposites reinforced by cellulose whiskers. *Macromolecules.* 28:6365–6367.

Fischer, H. 2003. Polymer nanocomposites: From fundamental research to specific applications. *Mat. Sci. Eng. C* 23:763–772.

Fishbine, G. 2002. *The Investor's Guide to Nanotechnology and Micromachines.* New York: John Wiley & Sons.

Gao, F. 2004. Clay/polymer composites: The story. *Mat. Today.* Nov:50–55.

Giannellis, E.P. 1996. Polymer layered silicate nanocomposites. *Adv. Mater.* 8:29–35.

Glasser, W.G., Taib, R., Jain, R.K., and Kander, R. 1999. Fiber-reinforced cellulosic thermoplastic composites. *J. Appl. Polym. Sci.* 73:1329–1340.

Gopalan Nair, K., Dufresne, A., Gandini, A., and Belgacem, M.N. 2003. Crab shells chitin whiskers reinforced natural rubber nanocomposites. 3. Effect of chemical modification of chitin whiskers. *Biomacromolecules.* 46:1835–1842.

Greene, R.V. and Imam, S.H. (Eds.) 1998. *Biopolymers: Utilizing Natures Advanced Materials.* New York: ACS.

Grim, R.E. 1962. Clay mineralogy. *Science.* 135:890–898.

Grim, R.E. 1988. The history of the development of clay mineralogy. *Clays Clay Min.* 36:97–101.

Huang, M., Yu, J., and Ma, X. 2006. High mechanical performance MMT-urea and formamide-plasticized thermoplastic cornstarch biodegradable nanocom-posites. *Carbohydrate Polym.* 63:393–399.

Huang, M., Yu, J., Ma, X., and Jin, P. 2005. High performance biodegradable ther-moplastic starch—EMMT nanoplastics. *Polymer.* 46:3157–3162.

Jansson, A. and Järnström, L. 2005. Barrier and mechanical properties of modified starches. *Cellulose.* 12:423–433.

Jordan, J., Jacob, K.I., Tannenbaum, R., Sharaf, M.A., and Jasiuk, I. 2005. Experimental trends in polymer nanocomposites—A review. *Mat. Sci. Eng. A.* 393:1–11.

Kalambur, S. and Rizvi, S.S.H. 2005. Biodegradable and functionally superior starch–polyester nanocomposites from reactive extrusion. *J. Appl. Polym. Sci.* 96:1072–1082.

Kalambur, S. and Rizvi, S.S.H. 2006. Rheological behavior of starch–polycaprolac-tone nanocomposite melts synthesized by reactive extrusion, *Polym. Eng. Sci.* 46:650–658.

Kampeerapappun, P., Aht-ong, D., Pentrakoon, D., and Srikulkit, K. 2007. Preparation of cassava starch/montmorillonite composite film. *Carbohydrate Polym.* 67:155–163.

Klemm, D., Heublein, B., Fink, H.-P., and Boh, A. 2005. Cellulose: Fascinating biopolymer and sustainable raw material. *Angew. Chem. Int. Ed.* 44:3358–3393.

Kristo, E. and Biliaderis, C.G. 2007. Physical properties of starch nanocrystal-reinforced pullulan films. *Carbohydrate Polym.* 68:146–158.

Kvien, I., Sugiyama, J., Votrubec, M., and Oksman, K. 2007. Characterization of starch based nanocomposites, *J. Mater. Sci.* 42:8163–8171.

Labet, M., Thielemans, W., and Dufresne, A. 2007. Polymer grafting onto starch nanocrystals, *Biomacromolecules.* 8:2916–2927.

LeBaron, P.C., Wang, Z., and Pinnavaia, T.J. 1999. Polymer-layered silicate nanocomposites: an overview. *Appl. Clay Sci* .15:11–29.

Lee, S.Y., Chen, H., and Hanna, M.A. 2008. Preparation and characterization of tapioca starch–polylactic acid nanocomposite foams by melt intercalation based on clay type. *Ind. Crop. Prod.* 110:2337–2344.

Liao, H.T. and Wu, C.S. 2005. Synthesis and characterization of polyethylene-octene elastomer/clay/biodegradable starch nanocomposites. *J. Appl. Polym. Sci.* 97:397–404.

Lima, M.M.S. and Borsali, R. 2002. Static and dynamic light scattering from polyelectrolyte microcrystal cellulose. *Langmuir.* 18:992–996.

Lu, Y., Weng, L., and Cao, X. 2005. Biocomposites of plasticized starch reinforced with cellulose crystallites from cottonseed linter. *Macromol. Biosci.* 5:1101–1107.

Lu, Y., Weng, L., and Cao, X. 2006. Morphological, thermal and mechanical properties of ramie crystallites—reinforced plasticized starch biocomposites. *Carbohydrate Polym.* 63:198–204.

Ma, X., Yu, J., and Wang, N. 2008. Glycerol plasticized-starch/multiwall carbon nanotube composites for electroactive polymers. *Comp. Sci. Tech.* 68:268–273.

Magalhães, N.F. and Andrade, C.T. 2009. Thermoplastic corn starch/clay hybrids: Effect of clay type and content on physical properties. *Carbohydrate Polym.* 75:712–718. .

Marcovich, N.E., Auad, M.L., Bellesi, N.E., Nutt, S.R., and Aranguren, M.I. 2006. Cellulose micro/nanocrystals reinforced polyurethane. *J. Mater. Res.* 21:870–881.

Mathew, A.P. and Dufresne, A. 2002. Morphological investigation of nanocomposites from sorbitol plasticized starch and tunicin whiskers. *Biomacromolecules.* 3:609–617.

McGlashan, S.A. and Halley, P.J. 2003. Preparation and characterization of biodegradable starch-based nanocomposite materials. *Polym. Int.* 52:1767–1773.

Medeiros, E.S., Agnelli, J.A.M., Joseph, K., Carvalho, L.H., and Mattoso, L.H.C. 2005. Mechanical properties of phenolic composites reinforced with jute/cotton hybrid fabrics. *Polym. Comp.* 6:1–11.

Medeiros, E.S., Paterno, L.G., and Mattoso, L.H.C. 2006. In: *Encyclopedia of Sensors,* C.A. Grimes, E.C. Dickey, and M.V. Pishko (Eds.). Valencia: American Scientific.

Medeiros, E.S., Tocchetto, R.S., Carvalho, L.H., Conceição, M.M., and Souza, A.G. 2002. Nucleating effect and dynamic crystallization of a polypropylene/attapulgite. *J. Therm. Anal. Calorim.* 67:279–285.

Medeiros, E.S., Tocchetto, R.S., Carvalho, L.H., Santos, I.M.G., and Souza, A.G. 2001. Nucleating effect and dynamic crystallization of a polypropylene/talc system. *J. Therm. Anal. Calorim.* 66:523–531.

Morton-Jones, D.H. 1989. *Polymer Processing*, London: Chapman & Hall.

Nishino, T., Matsuda, I., and Hirao, K. 2004. All-cellulose composite. *Macromolecules.* 37:7683–7687.

Okamoto, M. 2005. Biodegradable polymer/layered silicate nanocomposites: A review. In: *Handbook of Biodegradable Polymeric Materials and Their Applications.* Vol. 1, S. Mallapragada and B. Narasimhan (Eds.). Valencia: American Scientific, Chapter 8.

Orts, W.J., Baker, D.A., Shey, J., Chiou, B.-S., Imam, S.H., Glenn, G.M., and Mattoso, L.H.C. 2007. Starch-fiber nanocomposites: Reinforcement through improved processing and azide cross-linking. In: *Biopolymers and Technology*, A.C. Bertolini (Ed.). São Paulo: Cultura Academica Editora.

Orts, W.J., Shey J., Imam, S.H., Glenn, G.M., Guttman, M.E., and Revol, J.-F. 2005. Application of cellulose microfibrils in polymer nanocomposites. *J. Polym. Environ.* 13:301–306.

Pandey, J.K. and Singh, R.P. 2005. Green nanocomposites from renewable resources: Effect of plasticizer on the structure and material properties of clay-filled starch. *Starch/Stärke.* 57:8–15.

Park, H.M., Lee, W.K., Park, C.Y., Cho, W.J., and Ha, C.S. 2003. Environmentally friendly polymer hybrids Part I. Mechanical, thermal, and barrier properties of thermoplastic starch/clay nanocomposites, *J. Mat. Sci.* 38:909–915.

Pérez, C.J., Alvarez, V.A., Mondragón, I., and Vázquez, A. 2007. Mechanical properties of layered silicate/starch polycaprolactone blend nanocomposites. *Polym. Int.* 56:686–693.

Popov, V.N. 2004. Carbon nanotubes: Properties and application. *Mat. Sci. Eng. R.* 43:61–102.

Qiao, X., Jiang, W., and Sun, K. 2005. Reinforced thermoplastic acetylated starch with layered silicates. *Starch/Stärke.* 57:581–586.

Ray, S.S. and Okamoto, M. 2003. Polymer/layered silicate nanocomposites: A review from preparation to processing. *Prog. Polym. Sci.* 28:1539–1641.

Rhim J.-W. and Ng, P.K.W. 2007. Natural biopolymer-based nanocomposite films for packaging applications. *Crit. Rev. Food Sci. Nutr.* 47:411–433.

Rodriguez, N.L.G., Thielemans, W., and Dufresne, A. 2006. Sisal cellulose whiskers reinforced polyvinyl acetate nanocomposites. *Cellulose.* 13:261–270.

Roman, M. and Winter, W.T. 2004. Effect of sulfate groups from sulfuric acid hydrolysis on the thermal degradation behavior of bacterial cellulose. *Biomacromolecules.* 5:1671–1677.

Roper, H. and Koch, H. 1990. The role of starch in biodegradable thermoplastic materials. *Starch/Stärke.* 42:123–130.

Rosa, M.F., Chiou, B.-S., Medeiros, E.S., Wood, D.F., Williams, T.G., Mattoso, L.H.C., Orts, W.J., and Imam, S.H. 2008. Effect of fiber treatments on tensile and thermal properties of starch/ethylene vinyl alcohol copolymers/coir biocomposites. *Bioresource Technol.* 100: 5196–5202.

Rosato, D.V. 1997. *Plastics Processing Data Handbook*, 2nd ed. London: Chapman & Hall.

Ruiz-Hitzky, E., Darder, M., and Aranda, P. 2005. Functional biopolymer nanocomposites based on layered solids. *J. Mater. Chem.* 15:3650–3662.

Sanadi A.R., Caulfield, D.F., Jacobson R.E., and Rowell, R.M. 1995. Renewable agricultural fibers as reinforcing fillers in plastics: Mechanical properties of kenaf fiber–polypropylene composites. *Ind. Eng. Chem. Res.* 34:1889–1896.

Sanadi, A.R., Hunt, J.F., Caulfield, D.F., Kovacsvolgyi, G., and Destree, B. 2001. High fiber-low matrix composites: Kenaf fiber/polypropylene. *The 6th International Conference on Woodfiber-Plastic Composites*, pp. 121–124.

Scherman, N. 2001. *Tanach: The Stone Edition.* New York: Mesorah.

Scott, G. and Gilead, D. 1995. *Degradable Polymers: Principles & Applications.* London: Chapman & Hall.

Shackelford, J. F. 1999. *Introduction to Materials Science for Engineers*, 5th ed. Upper Saddle River, NJ: Prentice Hall.

Shalaby, W.S. and Latour, R.A. 1997. *Handbook of Composites.* Berlin: Springer.

Simonsen, J. 2008. Bio-bio-based nanocomposites: Challenges and opportunities. http://www.swst.org/meetings/AM05/simonsen.pdf/ accessed October 23.

Sperling, L.H. 2006. *Introduction to Physical Polymer Science*, 4th ed. Hoboken, NJ: John Wiley & Sons.

Sugiyama, J., Chanzy, J.H., and Maret, G. 1992. Orientation of cellulose microcrystals by strong magnetic fields. *Macromolecules.* 25:4232–4234.

Tadmor, Z. and Gogos, C.G. 1979. *Principles of Polymer Processing.* New York: John Wiley & Sons.

Thielemans, W., Belgacem, M.N., and Dufresne, A. 2006. Starch nanocrystals with large chain surface modifications. *Langmuir.* 22:4804–4810.

Wang, N., Ding, E., and Cheng, R. 2007. Thermal degradation behaviors of spherical cellulose nanocrystals with sulfate groups. *Polymer.* 48:3486-3493.

Whistler, R.L. and Paschall, E.F. 1965. *Starch: Chemistry and Technology*, Vol. 1. New York: Academic Press.

Whistler, R.L. and Paschall, E.F. 1967. *Starch: Chemistry and Technology*, Vol. 2. New York: Academic Press,

Yang, C.M., Kim, D.Y., and Lee, Y.H. 2005. Single-Walled Carbon Nanotube Network with Bimodal Pore Structures of Uniform Microporosity and Mesoporosity. *J. Nanosci. Nanotechnol.* 5:970–974.

Yang, K.K., Wang, X.L., and Wang, Y.Z. 2007. Progress in nanocomposite of biodegradable polymer. *J. Ind. Eng. Chem.* 13:485–500.

Yoshiharu, N., Shigenori, K., Masahisa, W., and Takeshi, O. 1997. Cellulose microcrystal film of high uniaxial orientation. *Macromolecules.* 30:6395–6397.

Zhao, R., Torley, P., and Halley, P.J. 2008. Emerging biodegradable materials: Starch- and protein-based bio-nanocomposites, *J. Mat. Sci.* 43:3058–3071.

chapter ten

Biodegradation of starch blends

Derval dos Santos Rosa
Universidade Federal do ABC

Cristina das Graças Fassina Guedes
Universidade São Francisco

Contents

10.1 Introduction

Synthetic polymers have been common applications on products that are used daily as substitutes for expensive materials such as steel, aluminum, paper, and glass. Some of these polymers, such as polystyrene (PS), polypropylene (PP), and polyethylene (PE) are widely used for packaging products as well as in the biomedical field and agriculture (Kiatkamjornwong, Thakeow, and Sonsuk, 2001; El-Rehim et al., 2004). The main advantages of these materials are their convenience, low cost, light weight, durability, and ease of production (Kiatkamjornwong, Thakeow, and Sonsuk, 2001; Mehrabzadeh and Farahmand, 2001; El-Rehim, 2004; Ramis et al., 2004; Bonelli et al., 2005). However, these synthetic polymers generate industrial and post-consumer wastes (Dintcheva et al., 2001). To reduce the amount of plastic waste accumulated in landfills, it has been suggested to collect, separate, and recycle some of these materials (Ramis et al., 2004). However, the difficulties associated with the separation of plastic waste have generally made this process impractical for unidentifiable products (Bikiaris et al., 1997; Carvalho and Rosa, 2005).

Interest in reuse of these polymers has led to the development of various techniques of recycling and applications (Bonelli et al., 2005). During separation of the polymers for recycling, it is possible to obtain two fractions: a light fraction floating on water and a heavy fraction sinking in water. The light fraction is essentially made of low-density polyethylene (LDPE), high-density polyethylene (HDPE), and polypropylene (PP). These materials are used in the production of disposable products.

An extensively investigated alternative is to replace synthetic polymers with biodegradable ones, which have a shorter half-life in the environment under appropriate conditions of moisture, temperature, and oxygen availability (Rosa, Guedes, and Pedroso, 2004a; Rosa et al., 2004c; Rosa, Volponi, and Guedes, 2006; Rosa, Guedes, and Carvalho, 2007a; Chiellini et al., 2006; Calil et al., 2006; Krzan et al., 2006; Lei et al., 2007; Umare, Chandure, and Pandey, 2007). Half-lives of these biopolymers depend on their structural characteristics, molecular weight, and crystallinity.

The use of products derived from renewable resources reduces greenhouse gas emissions, energy consumption, and the exploitation of nonrenewable resources. This cycle of use is completed when the raw materials of agricultural origin return to the earth through biodegradation and composting, without releasing pollutants (Rosa, Volponi, and Guedes, 2006). The American Society for Testing and Materials (ASTM D6400, 1998) defines a biodegradable material as one that can be completely assimilated by microorganisms that transform biomaterials into elements that are not harmful to nature. A wide range of organic species can be used in this process, but the most important organisms are bacteria and fungi. However, the extent of biodegradation is highly influenced by the composition of the material, including its carbon content and the presence of additives, plasticizers, and compatibilizers (Meier et al., 2004).

In recent years, many attempts have focused on blending plastic materials with cheap and natural polymers to create new materials with specific properties (Koenig and Huang, 1995; Wu, 2003; Rosa and Pantano, 2003; Rosa et al., 2004b). There has been a growing interest in developing starch-based products because starch is abundant, low cost, and totally biodegradable (Westling et al., 1998; Preechawong, 2004; Chen et al., 2005; Rosa, Volponi, and Guedes, 2006).

Thermoformed containers and transparent films are widely used in the packaging industry. Biodegradable bags are used for waste collection and for disposable materials, bio-fillers, and inorganic replacers (Alonso et al., 1999; Karim, Norziah, and Seow, 2000; Rosa, Volponi, and Guedes, 2006).

There is a wide range of starch-based formulations produced by plastic processors, including biodegradable products in different environments and starch-based biopolymers, which are 100% biodegradable in water and carbon only by microorganism action. They are commercialized for use in bottle blowing, cast film, injection molding, and thermoforming, as short-life consumer plastics and food packaging, as well as for products in landscape gardening and in agriculture (BIOP, 2008).

The addition of starch in mixtures with polyolefin such as polyethylene or polypropylene increases the biodegradation rate; consumption of starch by microorganisms creates porosity and also increases the polymer's superficial area and the oxygen reactions (Abd et al., 2004). It has been used in an extensive number of mixtures with a variety of polymers, in order to accelerate the degradable rate of these polymers, such as polycaprolactone and LDPE. Often, to be used in blends, starch is converted into an essentially amorphous, homogeneous material by thermal processing (Myllarinen et al., 2002). The main limitations of starch are its low melting temperature (T_m 65°C), poor mechanical properties, and low stability caused by water absorption (Chen et al., 2005). Some of these problems can be overcome by physical or chemical modifications, including blending with other polymers (Singh et al., 2003; Chen et al., 2005; Pandey et al., 2005).

10.2   Factors influencing polymer biodegradation

It has been found that the resistance of conventional polymers to microorganisms is primarily due to two factors: (1) the low surface area and relative impermeability of plastic films and molded objects, and (2) the very high molecular weight of the plastic material. Microorganisms tend to attack the ends of large carbon-chain molecules and the number of ends is inversely proportional to the molecular weight. In order to make polymer biodegradable, it is necessary to break down the surface area and to reduce the molecular weight.

The polymer biodegradability mechanisms are generally associated with certain factors such as the action of microorganisms; polymer process aging; impact of external factors such as temperature (thermal degradation), light (photodegradation), morphology, species of fungi acting on polymer biodegradation, oxygen, pressure, ozone, and water. These factors have significant effects on the decomposition of biodegradable polymers, inducing the break of macromolecular chains, variations in the composition of separate units, and degradation of the surface layer of the polymer.

All these processes play an important role in the growth of biodegradable microorganisms (Semenov, Gumargalieva, and Zaikov, 2003). In the next section, the main factors involved in polymer biodegradation are discussed.

10.2.1   Thermal degradation

Temperature is a critical factor in polymer degradation. The increase of temperature often accelerates polymer biodegradation, but this effect is reversed at temperatures above 65°C (Chandra and Rustgi, 1998), which eliminates the microorganisms responsible for stabilizing the transformation of waste material in humus (Reed and Gilding, 1981).

High temperatures increase the rate of degradation by favoring the nonenzymatic hydrolysis of ester bonds (Reed and Gilding, 1981), and the presence of thermophilic microorganisms. Biodegradability tests for natural (starch, cellulose, PHB/HV) and synthetic (PCL, SG, PLA) polymers conducted at different temperatures (35 and 55°C) in cured compost indicated that cellulose and starch biodegradations were higher at 35°C, whereas on other polymers degradations were higher at 55°C. In the biodegradation test at 55°C, compost harvested right after the thermophilic degradation stage showed higher biodegradation activity than the cured compost for both the synthetic aliphatic polyester and cellulose (Joo, Shin, and Bae, 1999). However, evaluation of biodegradation on buried PCL, PHB, and PHB-V in simulated soil by mass retention at 46 and 24°C, presented faster degradation taking place at 46°C (Lotto et al., 2004).

Consequently, the greatest quantity of microbiological damage in materials is observed under tropical climate conditions. Research carried out at river and marine fleet facilities in various climatic zones indicates that the main foci of the mold fungi are rubber, fiber, plastics, and varnished fabric (Semenov, Gumargalieva, and Zaikov, 2003).

10.2.2   Photo-oxidation

Ultraviolet radiation (UV) is one of the most effective factors accelerating the degradation rate of organic materials (Osawa, 1992; Sanches, Ferreira, and Felisberti, 1999). Free-radical initiators of polymerization produce chemically incorporated fragments that are relatively stable under irradiation.

Materials synthesized by using transition metal catalysts almost always contain significant amounts of transition metal ions that may initiate photo-oxidation. Sunlight striking the surface of polymers is absorbed, oxidizing the materials and thus beginning their degradation (Osawa, 1992). UV radiation causes irreversible chemical modifications that affect mechanical properties of organic materials. This degradation can make the polymers brittle, with loss of resistance and changes of color (Berre and Lala, 1989). Isolated double-bond units absorb energy radiation λ>300 nm, and peroxides and hydroperoxides absorb energy at 300–360 nm. UV spectra obtained from starches show the presence of photo-sensitive compounds excited at 360 and 290 nm. The presence of these chromophores promotes UV and sunlight absorption, leading to starch depolymerization. Because chromophores are minor compounds present in starches, the botanical origin of the starch influences the extent of UV depolymerization. For example, cassava starch appears to be more sen-sitive to photodegradation than corn starch (Bertolini et al., 2001a). The photodepolymerization by UV irradiation results in a strong impact on the rheological properties of starches, changing the viscosity and macro-molecular structure (Bertolini, Mestres, and Colonna, 2000; Bertolini et al., 2001a). As does UV irradiation (Bertolini et al., 2001a), gamma irradiation also depolymerizes starches, inducing modifications due, at least in part, to a mechanism involving free radicals (Bertolini et al., 2001b).

Some strategies to reduce the molar mass of the chain of PP and PE to be consumed by the microorganisms are the incorporation of carbonyl groups by the addition of pro-oxidant (Bikiaris et al., 1997; Ramis et al., 2004) and by UV (Bikiaris et al., 1997). The UV radiation or the action of solar light also reduces the size of polymeric chains of the PE and forms oxidant groups such as carbonyls, carboxyls, and hydroxyls. The cetone groups of PE are degraded by UV, however, the hydroperoxide groups are degraded both by UV and by temperature (Abd et al., 2004). The effect of UV radiation on the biodegradation of PP–starch blends (Morancho et al., 2006) reduces the polymers' thermal stability, contributing to blend degradation in the soil. Polyhydroxybutyrate (PHB)–low-density poly-ethylene (LDPE) blends modified by the addition of oxidized polyethyl-ene wax (OPW) and manganese stearate (MnS) are more susceptible to photo-oxidation by sunlight, indicating the degradation of PHB followed by hydrolysis over the known bioplastic (Rosa et al., 2007b).

10.2.3 Polymer structure

It is well known that the biodegradation rate depends strongly on the polymer structure. The linkages of the polymer backbone, endgroups, and their chemical activity are important factors affecting biodegradation (Zee, 2005). Linkages involving hetero atoms, such as ester and amide, are considered susceptible to enzymatic degradation.

The in vivo degradation rate of a series of L lactic acid (L-LA)–glycolic acid (GA) homo- and co-polymers suggests that degradation rates change according to the proportion of LA to GA (Cutright et al., 1994).

In poly(β-caprolactone) (PCL)–starch blends the addition of starch reduces the tensile strength at break, the elongation at break, and Young's modulus. Blends containing starches with high amylose content present lower values of tensile strength and higher values of elongation at break, suggesting that the composition of these blends favors biodegradation (Rosa, Lopes, and Calil, 2007c).

10.2.4   Influence of morphology

The crystallinity of polymers influences their rate of hydrolysis, mainly due to water access on crystalline regions. Hydrolysis is therefore initially restricted to the amorphous phase and to the fringes of the crystallites. Consequently, the highly crystalline polymers are relatively resistant to oxidation (Scott and Gilead, 1995). For example, oxidation of polyolefins occurs almost exclusively in the amorphous region of the polymer because the crystallites are impermeable to oxygen. A semicrystalline polymer, such as granular starch, tends to limit the accessibility of the hydrophilic nature, confining the degradation to the amorphous regions of the polymer (Zee, 2005). Degradation of semicrystalline polymers occurs in two stages: the first one consists of water diffusion into the amorphous regions with random hydrolytic scission of ester bonds and the second one starts when most of the amorphous regions are degraded. Hydrolytic attack then progresses within the crystalline domains. For example, amorphous regions of PCL films are degraded first. Also bulk degradation caused by phosphate-buffered solution with enzymes from *Cryptococcus* and *Fusarium* in 40% methylamine (Jarrett et al., 1985) and thermal conditions (Pitt et al., 1981) are predominant on amorphous regions.

In polymer blends, interactions with other polymers also affect the biodegradation properties. An additional material may act as a barrier to prevent migration of microorganisms, enzymes, moisture, or oxygen into the polymer domains of interest. Conversely, blends of nonbiodegradable and biodegradable polymers, or grafting of biodegradable polymer onto a nonbiodegradable backbone polymer may result in a biodegradable system (Zee, 2005).

Starch is a hydrophilic polymer and the presence of this material in blends induces water absorption, promoting blend hydrolysis and biodegradation by microorganisms. In starch–PCL blends, cleavage of starch leads to the fragmentation of PCL and provides a large surface area for the enzyme to act on, thereby accelerating the degradation of the blends (Rosa, Volponi, and Guedes, 2006). In PCL–starch blends the action of α-amylase on starch results in fragmentation of PCL accelerating the blend's biodegradation (Araújo, Cunha, and Mota, 2004). These changes can accelerate the biodegradation promoted by microorganisms, leading to an increase of porosity, void formation, and the loss of integrity of the plastic matrix (Bastioli, 2005).

10.2.5 Water absorption

Absorption of water is defined as the amount of water absorbed by a material when immersed in or exposed to water for a stipulated period of time. All organic polymeric materials can absorb moisture to some extent, resulting in swelling, dissolving, leaching, plasticizing, or hydrolyzing, events which can result in discoloration, embrittlement, loss of mechanical and electrical properties, lower resistance to heat and weathering, and stress cracking.

The presence of water is necessary to promote the hydrolysis of the polymer, which takes part in the biodegradation mechanism. Therefore, hydrophilic polymers associated with amorphous regions present the advantage of water uptake and consequently make the biodegradation process easier.

The products derived from starch present higher permeability to humidity and they degrade quickly for many types of applications. The water acts as a disruptive agent to the starch, breaking the hydrogenated linkages between the chains.

From the molecular viewpoint, ester hydrolysis is a well-known reaction in organic chemistry:

$$RCOOR' + H_2O \rightleftarrows RCOOH + R'OH$$

This reaction can be catalyzed by either acids or bases. Also, the reaction product, RCOOH, is able to accelerate ester hydrolysis according to a phenomenon called autocatalysis. In the case of aliphatic polyesters, chain cleavage at the ester bond level is autocatalyzed by carboxyl endgroups initially present or generated by the degradation reaction:

$$\sim COO \sim + H_2O \rightarrow \sim \xrightarrow{\text{COOH}} + OH \sim$$

Hydrolytic degradation between PHB and a series of HB/V co-polymers shows an increase in surface energy in the presence of hydroxyl and carboxyl groups and consequently a faster "surface erosion process." However, surface erosion rapidly appeared in competition with a "bulk erosion process" which resulted from the diffusion of products of chain scission from the matrix (Holland et al., 1987).

10.2.6 pH condition

High alkalinity speeds up the degradation rate, indicating ester hydrolysis as the degradation mechanism (Holland et al.,1987). The biodegradability of poly-β-hydroxybutyrate (PHB) and polycaprolactone (PCL) in soil at pH

7.0, 9.0, and 11.0 is directly proportional to the soil alkalinity. This reaction produces RCOOH groups, which are also able to accelerate ester hydrolysis through autocatalysis. In the case of aliphatic polyesters, the cleavage of ester bonds is autocatalyzed by carboxyl groups present at the start or generated after the first stage of degradation (Rosa et al., 2002).

Hydrolytic degradation of a HB–HV co-polymer in buffered physiological saline at various pH shows that neutral and acidic solutions produce a diffuse surface degradation, whereas the alkaline solution seems to be more aggressive, with site-specific attacks causing deep points of "surface erosion" (Knowles and Hastings, 1991). It suggests that the hydrolysis mechanism occurs by different pathways, depending on the pH of the surrounding medium. In acidic and neutral solutions, hydrolysis proceeds by a protonation process, whereas in alkaline media, hydroxyl ions are attached to the carbonyl carbons.

10.2.7   Action of microorganisms

Biodegradation of polymers by microorganisms takes place in the soil layer from 5 to 15 cm deep. Several species of fungi are typically present during biodegradation of polymers, acting as microorganism degraders, and the most typical ones are: *Aspergillus awamori, Aspergillus niger, Aspergillus oryzae, Trichoderma* sp., *Aspergillus amstelodami, Aspergillus flavus, Shactomim globosum, Trichoderma lignorum, Cephalosporum aeremonium, Penicillium* sp., *Rhizopus nigricans,* and *Fusarium roseum* (Semenov, Gumargalieva, and Zaikov, 2003).

Degradation of poly(ε-caprolactone)–cellulose acetate blends buried in simulated soil show the presence of *Aspergillus fumigatus, Aspergillus niger, Aspergillus versicolor,* other *Aspergillus* spp., *Penicillium simplicissimum,* other *Penicillium* spp., and *Cladosporium cladosporioides* fungi (Rosa et al., 2008). These microorganisms have also been identified in previous studies of polymer degradation (Ishii et al., 2007), with ~20 fungal species, including several *Aspergillus* representatives, in degraded samples of poly(ethyl succinate) (PESu).

Blends composted by poly(ethylene terephtalate) (PET) under action of *Aspergillus niger* and nitric acid show an increase in surface roughness, swelling of the material, and decrease in molecular weight; these changes are considered fundamental to promoting polymer degradation (Marqués-Calvo et al., 2006). Surface erosion of PCL during biodegradation in compost, in the presence of plant residues under anaerobic conditions and in the presence of the fungus *Aspergillus fumigatus* at 23°C and 50°C, promotes longitudinal cuts or organized cracks on the polymer surface, whereas samples incubated under anaerobic conditions show no changes in surface morphology (Eldsäter et al., 2000), resulting in polymer surface erosion in PCL–CA blends (Rosa et al., 2008). Co-polymeric starch grafted with methyl vinyl ketone or methacrolein (MVK) or phenyl vinyl ketone (PVK) in the presence of *Aspergillus niger* is more susceptible to photo-degradation, inducing cuts and cracks on the surface and, consequently, favoring the action of other microorganisms (Choi et al., 1998).

Other than the species of microorganisms, the interactions between materials and microorganisms have a strong impact on polymer degradation (Semenov, Gumargalieva, and Zaikov, 2003). Adhesion of microorganisms is commonly characterized by the quantity of cells fixed at the specific surface and the force necessary for detaching them, which can be stimulated by molecular, chemical, capillary, and electric forces. The possibility and intensity of microbiological development are determined by substrate properties, nutritive medium, and microorganism genetic features. The biodegradation of polymers starts with the lag phase, which is characterized by a period of adaptation and selection of the degrading microorganisms. During this period, a complex of enzymes necessary to degrade the substrate is synthesized, accelerating the growth of the biomass and biochemical metabolic mechanisms in cells. The second phase of biodegradation occurs from the end of the lag phase until about 90% of the maximum level of biodegradation; this phase is characterized by an increase in the biomass. Finally, exhaustion of nutritive components occurs, limiting the development of the microorganism and (or) up to the formation of metabolites inhibiting its growth (Semenov, Gumargalieva, and Zaikov, 2003). The last phase extends from the end of the biodegradation phase until the end of the test (Rosa, Volponi, and Guedes, 2006).

10.3 Tests of polymer biodegradability

Reliable methods are essential to evaluate the polymer biodegradation rate in soils. The first biodegradation tests were adapted from industrial tests used in evaluation of leather and rubber biodegradation. Because they have similarities to disposal conditions, tests by burying in soil have been widely conducted in order to evaluate polymer biodegradation. They were based on procedures such as soil burial, zone of inhibition, and qualitative surface examination. However, certain disadvantages appear: the long time necessary to perform the tests and problems in their reproducibility due to the difficulty of controlling environmental conditions and microbial populations.

In 1977 the Organisation for Economic Cooperation and Development (OECD) launched the Chemical Testing Programme and Especial Programme on the Control of Chemicals, reporting that chemicals entered in the environment may undergo various changes by physical, chemical, and biological reaction in each compartment of the environment (Kitano and Yakabe, 1993). Table 10.1 summarizes the main factors responsi-

Table 10.1 Factors Responsible for Polymer Degradation

Environment	Degradation		Accumulation	
	Biotic	Abiotic	Biotic	Abiotic
Atmosphere		Photodegradation		
Water	Biodegradation	Photodegradation		
		Hydrolysis		
Soil/sediment	Biodegradation			Geoaccumulation
	Soil degradation			

ble for degradation and accumulation under different environmental conditions.

The strategy for biodegradability testing is based on three levels of testing: (1) ready biodegradability or screening, (2) inherent biodegradability, and (3) emulation of environmental compartments. There is also a requirement to examine the possible presence of stable intermediates at the second level if only partial mineralization occurs. The conditions provided by the tests at the first two levels are progressively less stringent (Kitano and Yakabe, 1993).

The American Standard Testing Methods (ASTM D6400, 1998) Committee D20.96 is developing standards for the terminology and tests of biodegradable polymers and environmentally degradable plastic materials. When degradable plastic materials are exposed to a given environment, several outcomes are possible, depending on the plastic material and its mode of degradation. They may remain unchanged in the environment and be termed recalcitrant, they may be degraded completely and removed from the environment through biodegradation processes, or they may be partially degraded and remain in the environment for an indefinite period of time. The importance of biodegradation in the environment and its role relative to other environmental degradation pathways of photodegradation, oxidative degradation, and hydrolytic degradation,

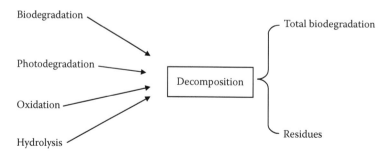

Figure 10.1 Interrelationships of environmentally degradable plastic materials.

is represented in Figure 10.1. It is noteworthy that biodegradation is the only pathway for the complete elimination of plastics and fragments in the environment.

Biodegradation of polymers depends on the soil conditions, species, and the number of microorganisms responsible for the degradation and their own optimal growth conditions in the soil. Considering burying attempts to measure polymer biodegradation, two different experimental soil conditions are often used:

1. Compost inocula. The compost inoculum is composed of three-month-old, well-aerated compost derived from the organic fraction of solid waste and sieved (screen mesh, 10 mm) to remove large inert material.
2. Simulated soil. The simulated soil is composed of 23% loamy silt, 23% organic matter, 23% sand, and 31% distilled water (w/w). Polymer samples are weighed and buried at room temperature (24°C) and biodegradation is monitored every 15 days for 180 days by measuring the mass retention. The buried samples are recovered, washed with distilled water, and dried at room temperature until they show no weight variation.

10.4   Biodegradability evaluation parameters

Most parameters used to measure polymer biodegradation are based on compounds resulting from degradation or physical–chemical changes in samples. Biodegradation of polymers is often measured by:

1. The carbon dioxide production or oxygen consumption
2. Weight loss of the sample
3. Changes in polymer properties

10.4.1   Carbon dioxide production

Biodegradation is a natural process initiated by the action of microorganisms on the polymer surface, where polymers are converted into water and carbon dioxide (CO_2) (Stevens, 1999; Yam et al., 2000). One of the most common biodegradation tests evaluates the carbon dioxide rate released by polymer samples in the presence of inoculated biodegrading microorganisms. In this test, analytical grade cellulose and LDPE are used as positive and negative controls, respectively. To assess the CO_2 production, four identical systems are placed in a reactor (Figure 10.2) and immersed in organic composted soil. The reactor (receptacle B) is maintained at 58°C and the system is monitored every 24 h for 46 days (ASTM D5338-98, 1998). The CO_2 produced by polymer degradation is then recovered in receptacle

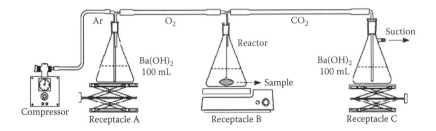

Figure 10.2 Scheme for monitoring CO_2 production.

C (Figure 10.2) and the extent of aerobic biodegradation during the test is measured by CO_2 titration.

Carbon dioxide production in the reactor (Figure 10.2) is accomplished by a short induction time, showing an increase in the carbon dioxide rates during induction, as can be seen in Figure 10.3, which shows the production of carbon dioxide from starch in simulated soil.

Evaluation of biodegradation of poly(ε-caprolactone) (PCL), cellulose acetate (CA) films, and their blends (PCL–CA, 60/40 and 40/60 wt.%) using an aerobic biodegradation resulted in higher CO_2 production and faster degradation for samples composed of 40PCL/60CA (Calil et al., 2006). In a Sturm test of PCL, CA, and their blends, the 40PCL/60CA blend showed higher CO_2 production in an aerobic medium and, consequently, faster biodegradation than the other blends.

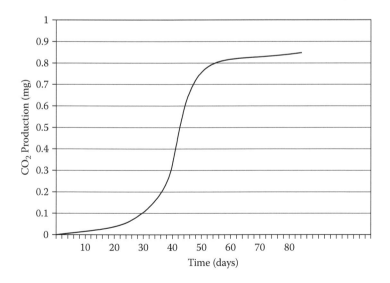

Figure 10.3 Biodegradation of a sample based on CO_2 production.

10.4.2   *Biodegradability based on mass retention of sample*

Some biodegradation measurements are based on weight loss of buried samples, in simulated soil at room temperature (24°C); biodegradation is monitored every 15 days for 180 days by measuring the mass retention. Biodegradation in simulated soil has been used to investigate the effect of corn starch degradation of blends with PCL, showing a decrease of mass retention samples often inversely proportional to starch content addition in the blends (Rosa, Volponi, and Guedes, 2006).

10.4.3   *Changes in polymer properties*

In degradation of polymers, changes in their physical and chemical properties can be observed. Table 10.2 shows evidence of possible degradation of LDPE and blends, measured by physical changes. For all LDPE–starch formulations, the tensile strength decreased with increasing starch content, indicating that corn starch behaved as a nonreinforcing filler. For blends containing LDPE, the tensile strength decreased by 57, 60, and 74% for the blends containing 30, 40, and 50 wt.% starch, respectively, in relation to pure LDPE. For blends containing reprocessed LDPE, tensile strength decreased by 50, 59, and 76%, for the blends containing 30, 40, and 50 wt.%, respectively, compared to pure reprocessed LDPE. A possible explanation for the decrease in tensile strength after the addition of starch could be the heterogeneous distribution of starch in LDPE and the low interfacial interaction between components of the blend, which resulted in mechanical rupture at the blend interface (Pedroso and Rosa, 2005).

The biodegradation rate of cellophane films was 41.2% in four months during winter whereas it degraded 76.5% in two months during summer, showing the influence of soil and environmental conditions on film biodegradation (Jeong, Park, and Shin, 1998).

Table 10.2 Tensile Strength for Virgin and Reprocessed LDPE and
LDPE–Starch Blends

Formulation LDPE/starch (wt.%)	LDPE	
	Virgin	Reprocessed
100/0	16 ± 1	12.1 ± 0.6
70/30	7.0 ± 0.2	6.1 ± 0.8
60/40	6.3 ± 0.6	5.0 ± 0.1
50/50	4.1 ± 0.2	3.0 ± 0.1

Source: Pedroso and Rosa, 2005.

After aging in soil composting medium at 46°C for 86 days, poly-β-(hydroxybutyrate) (PHB) and poly-β-(hydroxybutyrate-co-β-valerate) (PHB-V) polymers showed a decrease in tensile strength at break (76% for PHB and 74% for PHB-V) and their roughness increased faster in soil composting medium at 46°C (Rosa et al., 2004c). Biodegradation in blends composed of polyethylene graft glycidyl methacrylate (PE-g-GMA) and poly(ε-caprolactone) (PCL) and cellulose acetate (CA) (80/20, 60/40, 40/60, and 20/80 PCL/CA, w/w) showed the PE-g-GMA presented higher tensile strength retention during three months of aging in simulated soil. Samples of pure PCL were more porous, whereas PE-g-GMA decreased the mass loss of the polymers, possibly by enhancing the interaction between PCL and CA, with the formation of hydrogen bonds between the carbonyl groups of PCL and the hydroxyl groups of CA. The addition of PE-g-GMA decreased the rate of biodegradation of the polymers, increasing the tensile strength retention during the three months submitted in simulated soil, possibly by increasing the interaction between the constituent polymers in PCL/CA blends. The greater interaction may have resulted from the formation of hydrogen bonds between the carbonyl groups of PCL (carbonyl ester) and the hydroxyl groups in CA, with fewer free spaces among the polymer chains (Rosa et al., 2008).

10.5   Final remarks

Biodegradable polymers are materials of scientific and technological interest and their use in several applications has been proposed to solve environmental problems. Starch is a renewable source that has been identified as a raw material to be used in blends requiring biodegradable polymers. Starches are viable for use in development of new biodegradable polymers and applications in products with low lifetime due to their tendency to biodegrade in favorable environments. The development of methods to measure the biodegradability of polymers is essential to support research on new biodegradable materials. Studies aiming to improve the reproducibility of biodegradability tests simulating environmental conditions are crucial to evaluate potential natural polymers, such as starches, as replacements for synthetic materials.

References

Abd El-Rehim, H.A., Hegazy, E.-S.A., Ali, A.M., and Rabie, A.M. 2004. Synergistic effect of combining UV-sunlight-soil burial treatment on the biodegradation rate of LDPE/starch blends. *J. Photochem. Photobiol. A Chem.* 163:547–556.

Alonso, A.G., Escrig, A.J., Carrón, N.M., Bravo, L., and Calixto, F.S. 1999. Assessment of some parameters involved in the gelatinization and retrogradation of starch. *Food Chem.* 66:181–187.

Araújo, M.A., Cunha, A.M., and Mota, M. 2004. Enzymatic degradation of starch-based thermoplastic compounds used in protheses: Identification of the degradation products in solution. *Biomaterials.* 25:2687–2693.

ASTM Designation D5338–98. 1998. Standard test method for determining aerobic biodegradation of plastic materials under controlled composting conditions, 08.03:472.

ASTM Standard D6400 Standard Specification for Compostable Plastics, ASTM International, West Conshohocken, PA, www.astm.org

Bastioli, C. 2005. Starch-based technology. In: *Handbook of Biodegradable Polymers*, C. Bastiolli (Ed.). UK: Rapra Techology Limited.

Berre, B. and Lala, D. 1989. Investigation on photochemical dosimeters for ultraviolet radiation. *Solar Energy.* 42:405–416.

Bertolini, A.C., Mestres, C., and Colonna, P. 2000. Rheological properties of UV-irradiated starches. *Starch.* 52:340–344.

Bertolini, A.C., Mestres, C., Buléon, A., Lerner, D., Raffi, J., and Colonna, P. 2001a. Photodegradation of cassava and corn starches. *J. Agricult. Food Chem.* 49:675–682.

Bertolini, A.C., Mestres, C., Colonna, P., and Raffi, J. 2001b. Free radicals formation in the UV and gamma irradiated cassava starch. *Carbohydrate Polym.* 44:269–271.

Bikiaris, D., Prinos, J., Perrier, C., and Panayiotou, C. 1997. Thermoanalytical study of the effect of EAA and starch on the thermo-oxidative degradation of LDPE. *Polym. Degradation Stability* 57:313–324.

BIOP. 2008. Introducing Biop Biopolymer Technologies AG "Friendly by Nature" http://www.biopag.de/index.php (accessed December 2005).

Bonelli, C.M.C., Elzubair, A., Suarez, J.C.M., and Mano, E.B. 2005. Comportamento térmico, mecânico e morfológico de compósitos de polietileno de alta densidade reciclado com fibra de piaçava. *Polímeros: Ciência e Tecnologia.* 15:256–260.

Calil, M.R., Gaboardi, F., Guedes, C.G.F., and Rosa, D.S. 2006. Comparison of the biodegradation of poly(ε-caprolactone), cellulose acetate and their blends by the Sturm test and selected cultured fungi. *Polym. Test.* 25:597–604.

Carvalho, C.L. and Rosa, D.S. 2005. Gestão e Caracterização dos resíduos plásticos domésticos recicláveis oriundos de posto de entrega voluntária. *Revista Brasileira de Aplicações de Vácuo.* 24:43–48.

Chandra, R. and Rustgi, R. 1998. Biodegradable polymers. *Prog. Polym. Sci.* 23:1273–335.

Chen, L., Ni, Y., Bian, X., Qiu, X., Zhuang, X., Chen, X., and Jing, X. 2005. A novel approach to grafting polymerization of ε-caprolactone onto starch granules. *Carbohydrate Polym.* 60:103–109.

Chiellini, E., Corti, A., Sarto, G.D., and D'Antone, S. 2006. Effect of hydrolysis degree on biodegradation behaviour of poly(vinyl alcohol). *Polym. Degradation Stability.* 91:3397–3406.

Choi, W.M., Jung, I.D., Kwon, S.K., Ha, C.S., and Cho W.J. 1998. Syntheses and photobiodegradable properties of graft copolymers of vinyl ketones and starch. *Polym. Degradation Stability.* 61:15–20.

Cutright, D.E., Perez, B., Beasley, J.D., Larson, W.J., and Posey, W.R. 1974. Degradation rates of polymers and copolymers of polylactic and polyglycolic acids. *Oral Surg. Oral Med. Oral Pathol.* 37:142–152.

Dintcheva, N.T., La Mantia, F.P., Trotta, F., Luda, M.P., Camino, G., Paci, M., Di Maio, L., and Acierno, D. 2001. Effects of filler type and processing apparatus on the properties of the recycled "light fraction" from municipal postconsumer plastics. *Polym. Adv. Technol.* 12:552–560.

Eldsäter, C., Erlandsson, B., Renstad, R., Albertsson, A.-C., and Karlsson, S. 2000. The biodegradation of amorphous and crystalline regions in film-blown poly(ε-caprolactone). *Polymer.* 41:1297–1304.

El-Rehim, H.A.A., Hegazy, E.-S.A., Ali, A.M., and Rabie, A.M. 2004. Synergistic effect of combining UV-sunlight–soil burial treatment on the biodegradation rate of LDPE/starch blends. *J. Photochem. Photobiol. A Chem.* 163:547–556.

Holland, S.J., Jolly, A.M., Yasin, M., and Tighe, B.J. 1987. Polymer for biodegradable medical devices. II. Hydroxybutyrate-hydroxyvalerate copolymer: Hydrolytic degradation studies. *Biomaterials.* 8:289–295.

Ishii, N., Inoue, Y., Shimada, K., Tezuka, Y., Mitomo, H., and Kasuya, K. 2007. Fungal degradation of poly(ethylene succinate). *Polym. Degradation Stability.* 92:44–52.

Jarrett, P., Benedict, C.V., Bell, J.P., Cameron, J.A., Huang, S.J. 1985. Mechanism of the biodegradation of polycaprolactone, In: *Polymers as Biomaterials,* S.W. Shalaby, A.S. Hoffman, B.D. Ratner, and T.A. Horbett (Eds.). New York: Plenum Press, pp. 181–192.

Jeong, KE., Park, T.H., and Shin, P.K. 1998. Measurement and acceleration of biodegradation in soil. *San'oeb misaengmul haghoeji.* 26:465–469.

Joo, J.E., Shin, P.K., and Bae, H.K. 1999. Effects of temperature and compost conditions on the biodegradation of degradable polymers. *J. Microbiol. Biotechnol.* 9:464–468.

Karim, A.A., Norziah, M.H., and Seow, C.C. 2000. Methods for the study of starch retrogradation. *Food Chem.* 71:9–36.

Kiatkamjornwong, S., Thakeow, P., and Sonsuk, M. 2001. Chemical modification of cassava starch for degradable polyethylene sheets. *Polym. Degradation Stability.* 73:363–375.

Kitano, W. and Yakabe, Y. 1993. Strategy for biodegradability testing in OECD. In: *Biodegradable Plastics and Polymers*, Y. Doi and K. Fukuda (Eds.), Elsevier Science B.V., Amsterdam, pp. 217–227.

Knowles, J.C. and Hastings, G.W. 1991. In vitro degradation of a PHB/PHV copolymer and a new technique for monitoring early surface changes. *Biomaterials.* 12:210–214.

Koenig, M.F. and Huang, S.J. 1995. Biodegradable blends and composites of polycaprolactone and starch derivatives. *Polymer.* 36:1877–1882.

Krzan, A., Hemjinda, S., Miertus, S., Corti, A., and Chiellini, E. 2006. Standardization and certification in the area of environmentally degradable polymers. *Polym. Degradation Stability.* 91:2819–2833.

Lei, L., Ding, T., Shi, R., Liu, Q., Zhang, L., Chen, D., and Tian, W. 2007. Synthesis, characterization and in vitro degradation of a novel degradable poly(1,2-propanediol-sebacate-citrate) bioelastomer. *Polym. Degradation Stability.* 92:389–396.

Lotto, N.T., Calil, M.R., Guedes, C.G.F., and Rosa, D.S. 2004. The effect of temperature on the biodegradation test. *Mater. Sci. Eng. C* 24:659–662.

Marqués-Calvo, M.S., Cerdà-Cuéllar, M., Kint, D.P.R., Bou, J.J., and Muñoz-Guerra, S. 2006. Enzymatic and microbial biodegradability of poly(ethylene terephthalate) copolymers containing nitrated units. *Polym. Degradadation Stability* 91:663–671.

Mehrabzadeh, M. and Farahmand, F. 2001. Recycling of commingled plastics waste containing polypropylene, polyethylene, and paper. *J. Appl. Polym. Sci.* 80:2573–2577.

Meier, M.M., Kanis, L.A., Lima, J.C., Pires, A.T.N., and Soldi, V. 2004. Poly(caprolactone triol) as plasticizer agent for cellulose acetate films: Influence of the preparation procedure and plasticizer content on the physico-chemical properties. *Polym. Adv. Technol.* 15:593–600.

Morancho, J.M., Ramis, X., Fernández, X., Cadenato, A., Salla, J.M., Vallés, A., Contat, L., and Ribes, A. 2006. Calorimetric and thermogravimetric studies of UV-irradiated polypropylene/starch-based materials aged in soil. *Polym. Degradation Stability.* 91:44–51.

Myllarinen, P., Buleon, A., Lahtinen, R., and Forssell, P. 2002. The crystallinity of amylose and amylopectin films. *Carbohydrate Polym.* 48:41.

Osawa, Z. 1992. Deterioration and stabilization of polymer. *Musashino.* 201–209.

Pandey, J.K., Reddy, K.R., Kumar, A.P., and Singh, R.P. 2005. An overview on the degradability of polymer nanocomposites. *Polym. Degradation Stability.* 88:234.

Pedroso, A.G. and Rosa, D.S. 2005. Mechanical, thermal and morphological characterization of recycled LDPE/corn starch blends. *Carbohydrate Polym.* 59:1–9.

Pitt, C.G., Chasalow, F.I., Hibionada, Y.M., Klimas, D., and Schindler, A. 1981. Aliphatic polyesters. I. The degradation of polycaprolactone in vivo. *J. Appl. Polym. Sci.* 26:3779–3787.

Preechawong, D., Peesan, M., Supaphol, P., and Rujiravanit, T. 2004. Characterization of starch/poly(ε-caprolactone) hybrid foams. *Polym. Test.* 23:651–657.

Ramis, X., Cadenato, A., Salla, J.M., Morancho, J.M., Valle, A., Contat, L., and Ribes, A. 2004. Thermal degradation of polypropylene/starch-based materials with enhanced biodegradability. *Polym. Degradation Stability.* 86:483–491.

Reed, A.M., and Gilding, D.K. 1981. Biodegradable polymers for use in surgery: poly(glycolic)/poly(lactic acid) homo and copolymers. 2. *In vitro* degradation. *Polymer.* 22:494–498.

Rosa, D.S. and Pantano Filho, R. 2003. *Biodegradação: Um Ensaio com Polímeros.* Itatiba: Moara; Bragança Paulista: Univ. São Francisco Editora.

Rosa, D.S., Bardi, M.A.G., Guedes, C.G.F., and Angelis, D.A. 2008. Role of poly-ethylene-graft-glycidyl methacrylate compatibilizer on the biodegradation of poly(ε-caprolactone)/cellulose acetate blends. *Polym. Adv. Technol.* DOI 10.1002/pat.1302.

Rosa, D.S., Calil, M.R.. Guedes, C.G.F., and Santos, C.E.O. 2002. The effect of UV-B irradiation on the biodegradability of poly-b-hydroxybutyrate (PHB) and poly-ε-caprolactone (PCL). *J. Polym. Environ.* 3:111–115.

Rosa, D.S., Guedes, C.G.F., and Carvalho, C.L. 2007a. Processing and thermal, mechanical and morphological characterization of post-consumer poly-olefin/thermoplastic starch blends. *J. Mater. Sci.* 42:551–557.

Rosa, D.S., Guedes, C.G.F., and Pedroso, A.G. 2004a. Gelatinized and nongelatinized corn starch/poly(ε-caprolactone) blends: Characterization by rheological, mechanical and morphological properties. *Polímeros: Ciência e Tecnologia.* 14:181–186.

Rosa, D.S., Guedes, C.G.F., Gaboardi, F., and Fogaccia, J. 2007b. Biodegradação de blendas de PHB/PEBD submetidas aos envelhecimentos natural e térmico. In *9° Congresso Brasileiro de Polímeros*. São Carlos: Associação Brasileira de Polímeros, 1:1–10.

Rosa, D.S., Guedes, C.G.F., Pedroso, A.G., and Calil, M.R. 2004b. The influence of starch gelatinization on the rheological, thermal, and morphological properties of poly (ε-caprolactone) with corn starch blends. *J. Mater. Sci. Eng. C 2*, 663–670.

Rosa, D.S., Lopes, D.R., and Calil, M.R. 2007c. The influence of the structure of starch on the mechanical, morphological and thermal properties of poly (β-caprolactone) in starch blends. *J. Mater. Sci.* 42:2323–2328.

Rosa, D.S., Lotto, N.T., Lopes, D.R., and Guedes, C.G.F. 2004c. The use of roughness for evaluating the biodegradation of poly-β-(hydroxybutyrate) and poly-β-(hydroxybutyrate-co-β- valerate). *Polym. Test.* 23:3–8.

Rosa, D.S., Volponi, E.J., and Guedes, C.G.F. 2006. Biodegradation and the dynamic mechanical properties of starch gelatinization in poly(ε-caprolactone)/corn starch blends. *J. Appl. Polym. Sci.* 102:825–832.

Sanchez, E.M.S., Ferreira, M.M.C., and Felisberti, M.I. 1999. Avaliação da degradação térmica e fotooxidativa do ABS Automotivo. *Polímeros.* 9:116–122.

Scott, G. and Gilead, D. 1995. *Degradable Polymers: Principles and Applications*. London: Chapman & Hall.

Semenov, S.A., Gumargalieva, K.Z., and Zaikov, G.E. 2003. *New Concepts in Polymer Science: Biodegradation and Durability of Materials Under the Effect of Microorganisms*. Boston: VSP.

Singh, R.P., Pandey, J.K., Rutot, D., Degée, P., and Dubois, P. 2003. Biodegradation of poly(ε-caprolactone)/starch blends and composites in composting and culture environments: The effect of compatibilization on the inherent biodegradability of the host polymer. *Carbohydrate Res.* 338:1759–1769.

Stevens, M.P. 1999.*Polymer Chemistry: An Introduction*, 3rd ed., New York: Oxford University Press.

Umare, S.S., Chandure, A.S., and Pandey, R.A. 2007. Synthesis, characterization and biodegradable studies of 1,3-propanediol based polyesters. *Polym. Degradation Stability.* 92:464–479.

Vilpoux O. and Avérous L. 2003. Plásticos a base de amido. In : *Tecnologia, usos e potencialidades de tuberosas amiláceas Latino Americanas,* M.P. Cereda, and O. Vilpoux (Eds.). São Paulo: Fundação Cargill, 3:499–529..

Westling, A.R., Stading, M., Hermansson, A.M., and Gatenholm, P. 1998. Structure, mechanical and barrier properties of amylose and amylopectin films. *Carbohydrate Polym.*, 36:217.

Wu, C.-S. 2003. Physical properties and biodegradability of maleated-polycaprolactone/starch composite. *Polym. Degradation Stability.* 80:127–134.

Yam, W.Y., Ismail, J., Kammer, H.W., Lechner, M.D., and Kummerlöwe, C. 2000. Thermal properties of poly(styrene-*block*-ε-caprolactone) in blends with poly(vinyl methyl ether). *Polymer.* 41:9073–9080.

Zee, M.V.D. 2005. Biodegradability of polymers: Mechanisms and evaluation methods. In: *Handbook of Biodegradable Polymers*, C. Bastiolli (Ed.). UK: Rapra Techology Limited.

Index

Printed and bound by CPI Group (UK) Ltd, Croydon, CR0 4YY

18/10/2024

01776243-0007